*Life Time*

# Life Time

### The New Science of the Body Clock, and How It Can Revolutionize Your Sleep and Health

## RUSSELL FOSTER

PENGUIN LIFE

AN IMPRINT OF

PENGUIN BOOKS

PENGUIN LIFE

UK | USA | Canada | Ireland | Australia
India | New Zealand | South Africa

Penguin Life is part of the Penguin Random House group of companies
whose addresses can be found at global.penguinrandomhouse.com.

First published 2022
003

Copyright © Russell Foster, 2022

The moral right of the author has been asserted

Set in 12/14.75pt Dante MT Std
Typeset by Jouve (UK), Milton Keynes
Printed and bound in Great Britain by Clays Ltd, Elcograf S.p.A.

The authorized representative in the EEA is Penguin Random House Ireland,
Morrison Chambers, 32 Nassau Street, Dublin D02 YH68

A CIP catalogue record for this book is available from the British Library

ISBN: 978-0-241-52930-0

www.greenpenguin.co.uk

This book is dedicated with love to Elizabeth, Charlotte, William and Victoria and to the memory of Doreen Amy Foster (17 August 1933–28 November 2020)

# Contents

# Contents

# List of Figures

# List of Tables

# Abbreviations

| | |
|---|---|
| Aβ | amyloid (plaques) |
| ADHD | attention deficit hyperactivity disorder |
| ADP | adenosine diphosphate |
| ANP | atrial natriuretic peptide |
| ASA | acetylsalicylic acid |
| ASD | autism spectrum disorder |
| ASPD | advanced sleep phase disorder |
| ATP | adenosine triphosphate |
| AVP | arginine vasopressin |
| BMI | body mass index |
| BPH | benign prostatic hyperplasia |
| BSB | Banking Standards Board |
| BST | British Summer Time |
| CBD | cannabidiol |
| CBTi | cognitive behavioural therapy for insomnia |
| COPD | chronic obstructive pulmonary disease |
| COVID-19 | coronavirus disease caused by the SARS-CoV-2 virus |
| CPAP | continuous positive airway pressure |
| CSA | central sleep apnoea |
| DDAVP | desmopressin |
| DSD | driver safety device |
| DSPD | delayed sleep phase disorder |
| DST | Daylight Saving Time |
| EEG | electroencephalogram |
| EMA | European Medicines Agency |
| EMF | electromagnetic field |
| FDA | US Food and Drug Administration |
| FSH | follicle-stimulating hormone |

| | |
|---|---|
| GABA | gamma-aminobutyric acid |
| GMT | Greenwich Mean Time |
| GnRH | gonadotrophin-releasing hormone |
| HbA1c | glycated haemoglobin |
| hCG | human chorionic gonadotrophin |
| HDL | high-density lipoprotein |
| HRT | hormone replacement therapy |
| IVF | *in vitro* fertilization |
| LDL | low-density lipoprotein |
| LE | light emitting |
| LH | luteinizing hormone |
| NAFLD | non-alcoholic fatty liver disease |
| NASH | non-alcoholic steatohepatitis |
| NDD | neurodevelopmental disorder |
| NREM | non-rapid eye movement sleep |
| nVNS | non-invasive vagal nerve stimulation |
| OCD | obsessive-compulsive disorder |
| OHS | obesity hypoventilation syndrome |
| OPN4 | opsin gene encoding melanopsin |
| OSA | obstructive sleep apnoea |
| PD | Parkinson's disease |
| PKC | protein kinase C |
| PMDD | premenstrual dysphoric disorder |
| PMS | premenstrual syndrome |
| PPI | proton-pump inhibitor |
| pRGC | photosensitive retinal ganglion cell |
| PRR | pattern recognition receptor |
| PTSD | post-traumatic stress disorder |
| RBD | REM sleep behaviour disorder |
| REM | rapid eye movement sleep |
| RLS | restless-legs syndrome |
| SBD | sleep-related breathing disorder |
| SCN | suprachiasmatic nuclei |
| SCRD | sleep and circadian rhythm disruption |

| | |
|---|---|
| SDB | sleep-disordered breathing |
| SRED | sleep-related eating disorder |
| SRMD | sleep-related movement disorder |
| SSRI | selective serotonin reuptake inhibitor |
| SWS | slow-wave sleep |
| THC | tetrahydrocannabinol |
| TIA | transient ischemic attack |
| TNF | tumour necrosis factor |
| TST | total sleep time |

# Introduction

Nothing in life is to be feared, it is only to be understood.
Now is the time to understand more, so that we may fear less.

Marie Skłodowska-Curie

Forty years ago, as an undergraduate studying zoology at the University of Bristol, I knew I wanted to be a scientist, but I had little real idea of what that meant or involved. The 'body clock' was just a fuzzy concept in my young, unfocused free-wheeling brain. However, during the final year of my undergraduate degree, I was a volunteer helper at an international meeting on biological rhythms. The job was not demanding, and I swanned about listening to the lectures and met the then leaders of the field. With the confidence – perhaps arrogance – of youth, I assumed that these scientific titans would want to speak to me just as much as I wanted to speak to them. Most were incredibly generous with their time, although I did learn not to approach one very senior professor over breakfast (it's amazing how much meaning can be conveyed in a stony silence and a fixed stare at a greasy sausage . . . ). It was a formative experience at many levels, and I soaked up the science like a sponge. Without my knowing it, this symposium defined my life-long interests and sparked an ambition to join this extraordinary group of international academics who were working on the fast-emerging science of biological time. My career as a scientist, from my undergraduate days to my current position as Professor of Circadian Neuroscience and Director of the Sir Jules Thorn Sleep and Circadian Neuroscience Institute at Oxford has allowed me to gain insights from, and occasionally share new knowledge with, colleagues from all over the globe. In a sense, this book represents the distillate of what I have learnt in studying the nature of biological time over the course of four decades. My hope is that I can convey to you some of the excitement, wonder and undiluted pleasure I have experienced over the years.

In the past few decades, there has been an explosion of thrilling new discoveries in and around the science of the body clock and the 24-hour biological cycles that dominate our lives. The most obvious of these cycles is the daily pattern of sleep and wake. Surprisingly, most books discuss the body clock and sleep separately. However, new research tells us that such a disconnected approach tells only part of the story. You cannot properly understand sleep without understanding the body clock, and sleep in turn regulates the clock. In the pages that follow, the body clock and sleep will be considered together as two intimately linked and intertwined areas of biology that define and dominate our health. In so many cases, your ability to succeed or fail, from driving home safely after work or dieting to achieve weight loss, will depend upon whether you are working with or against these 24-hour cycles. So much has happened in this area of science and medicine that it is often difficult to disentangle fact from fiction. In terms of health, sensible advice is frequently transformed into strident orders that resemble the commands shouted by a Regimental Sergeant Major on the parade ground: you *'must'* get eight hours of sleep; you *'must'* continue to share a bed with a partner who snores; you *'must not'* use a light-emitting eReader (LE-eBook) before bed. So rather than being recognized as a loyal friend, biological rhythms and sleep are frequently portrayed as the enemy that needs to be wrestled, subdued and defeated. Instead, we need to understand and embrace these rhythms.

In this book, I have attempted to unpack the science of body clocks and sleep, making some of the amazing and exciting discoveries accessible, and hopefully in a format that is fun and easy to read. I have been able to draw from my own personal experiences over the past forty years as a scientist in this field, and have benefited immensely from discussions with friends and colleagues, who have contributed directly to our current understanding of biological time. I provide the evidence behind our current knowledge of the science, and how this evidence can be used by each of us to make more informed decisions about improving our lives. From getting better sleep, to organizing our daily activities and even why we may benefit from taking medications or being vaccinated at a particular time of day. The information in this book will also give you a greater understanding of the behaviour of others, including why teenagers and the elderly might struggle to get restorative sleep, why your mood and decision-making skills may change from morning to afternoon, and why the risk of divorce is higher in those doing night shift work. I have emphasized throughout this book that we are all very different, and that while it *is* possible to make generalizations, taking an 'average value' can be misleading. Although the average length of the menstrual cycle in women is 28 days, only 15 per cent of women actually have a 28-day cycle. Your body clock and sleep biology can be likened to your shoe size: one size does not fit all, and making us all wear the same shoe size would be not only foolish but potentially harmful. The failure to recognize this variation is why some general advice in the media can be either overly simplistic or deeply unhelpful.

Sleep and daily rhythms emerge from our genetics, physiology, behaviour and the environment, and like most of our behaviours they are not fixed. These rhythms are modified by our actions, how we interact with the environment, and how we progress from birth to old age. From infancy to advanced adulthood, our body clock and sleep patterns change profoundly, but these age-related changes are not necessarily bad. We should stop worrying about our sleep and accept that 'different' is not necessarily worse. Some of the advice we are given can be just wrong, emerging from the murky world of 'received wisdom'. Such 'wisdom' can be ancient and extend back to the beginnings of recorded history. However, as we shall see in the following chapters, the repetition of an idea does not necessarily guarantee legitimacy. For example, *flipping babies helps sort out their sleep*. According to this old tale, flipping the baby forward, head over heels, will reset the child's internal clock so that it will sleep at night and be awake during the day. There is absolutely no evidence for this. Indeed, as a tale it may well have its origins in parental desperation. Chronic sleep deprivation, not least in parents, can badly affect judgement and our ability to act rationally! Another often-repeated myth is that the pineal gland hormone melatonin is a 'sleep hormone'. It is not, and in the following chapters I shall explain why.

My message throughout this book is that all of us, as individuals and as members of society, should make some effort to understand and act upon the new scientific knowledge of biological time. But why bother? For me, it makes sense in a complex and demanding world to achieve the best physical and mental health we can. Such knowledge will help us deal with the many and varied insults that life flings at us. However, there is more. If you want to embrace life, be creative, make sensible decisions, enjoy the company of others, and view the world and all that it has to offer with a positive outlook, then embracing biological time will help you do this. Why not make the most of the time we have, and maybe even extend that time?

## The Tick-Tock of the Biological Clock

The entrenched arrogance which goes with being human means that most of us assume that we are above the grubby world of biology, and that we can do what we want, at whatever time we choose. This assumption underpins the modern 24/7 society, and an economy that is dependent upon night shift workers to stock our supermarkets, clean our offices, run our global financial services, protect us from crime, repair our rail and road infrastructure, and, of course, care for the sick and injured when they are most vulnerable. All of this happens while most of us sleep, or at least try to sleep. Although night shift work is the most obvious disruptor of our body clock and sleep, many of us experience shortened sleep as we try to squeeze more and more work and leisure activities into a daily schedule that is already over-packed and bursting at the seams. So we push these 'additional' activities into the night. Our full-scale occupation of the night has been possible because of the widespread commercialization of electric light across the world since the 1950s. This extraordinary and wonderful resource has also allowed us to declare war upon the night, and, without really appreciating what we have done, we have thrown away an essential part of our biology.

We are, of course, *not* able to do what we want at whatever time we choose. Our biology is governed by a 24-hour biological clock that advises us when it is the best time to sleep, eat, think, and undertake a myriad of other essential tasks. This daily internal adjustment allows us to function optimally in a dynamic world, 'fine-tuning' our biology to the profound demands imposed by the day/night cycle generated by the 24-hour rotation of the Earth upon its axis. For our bodies to function properly we need the correct materials in the right place, in the right amount, at the right time of day. Thousands of genes must be switched on and off in a specific order. Proteins, enzymes, fats, carbohydrates, hormones and other compounds have to be absorbed, broken down, metabolized and produced at a precise time for growth, reproduction, metabolism, movement, memory formation, defence and tissue repair. All this requires a biology and behaviour that are prepared and ready at the right time of day. Without this precise regulation by an internal clock, our entire biology would be in chaos.

For a relatively new branch of biology, and an emerging branch of medicine, the science of body clocks has its roots much earlier than might be expected, going back to the late 1720s and the study of a plant with the Latin name of *Mimosa pudica* – meaning 'shy, bashful or shrinking' – also called the 'sensitive plant'. This member of the pea family, familiar to many gardeners, has delicate leaves which fold inward and droop when touched or shaken, and then reopen a few minutes later. In addition to responding to touch, the leaves fold up at night and open during the day. Jean-Jacques d'Ortous de Mairan, a French scientist, studied these plants.

De Mairan's seminal observation, for our story, was that *Mimosa* leaves continue to show this rhythmic folding and unfolding leaf movement for several days in complete darkness. He was amazed; it was clearly not the change in light and dark that was driving this cycle, so what could it be – could it be temperature? Daily changes in temperature were investigated in 1759 by another French scientist, Henri-Louis Duhamel du Monceau, who took *Mimosa* plants into a salt mine, where there were conditions of constant temperature and darkness, and found the rhythms continued. More than 100 years after this, in 1832, a Swiss scientist, Alphonse de Candolle, studied *Mimosa* plants under constant conditions and showed that these drifting or 'freerunning' rhythms in leaf opening and closing were not exactly 24 hours but around 22–23 hours.

Over the next 150 years, daily rhythms that continued under constant conditions with a rhythm close to, but not exactly, 24 hours were observed in many plants and animals. Such rhythms were later called *circadian rhythms* (*circa* means 'about', and *dia* 'a day'). However, it was fairly late in the game that circadian rhythms were studied in humans. Hints that they exist in us came from observations in the late 1930s by Nathaniel Kleitman. From 4 June to 6 July 1938, Kleitman and his student Bruce Richardson remained deep in Mammoth Cave, Kentucky. There was no natural light and the temperature was a constant and cool 12.2°C.

Light was provided by lanterns, so conditions were not completely constant. And they had to share the cave with a large population of curious rats and cockroaches. To stop them crawling into their bedding, the four legs of their bunkbed were placed into large tins containing disinfectant. They recorded sleep and wake times and measured their daily rhythm of body temperature. These observations showed that they continued to show roughly 24-hour cycles in body temperature and sleep/wake timing.

The true significance of these findings was not realized until the 1960s. One of the pioneers in the field, Jürgen Aschoff had an underground 'bunker' built in Andechs, a town in Bavaria with a Benedictine monastery that has brewed beer since 1455. University undergraduates, when not in the bierkeller, were housed underground in the bunker under constant dim light, and isolated from any external environmental time cues, but they did have access to a bedside lamp. So, again, they were not really under constant lighting conditions. Student sleep/wake cycles, body temperature, urine production and other 'outputs' were measured over many days and were shown to have a rhythmic daily pattern of about 24 hours under these semi-constant conditions. From such experiments, the human body clock was estimated to run at around 25 hours. More recent studies from Charles Czeisler's group at Harvard University suggest the average human clock ticks with a rhythm closer to 24 hours and 11 minutes. This difference in period was always a point of friction between Aschoff and the Harvard team. And the consensus today is that the difference was caused by the use of bedside lamps in the bunker experiments. Aschoff was a remarkable man and I learnt much from him – both scientifically and socially. About twenty-five years ago at a summer school party in Bavaria I opened a bottle of wine. Several minutes later there was a roar from Aschoff: 'Who has left the cork on the corkscrew?' I admitted that it was me and he said for all to hear: 'You *never* leave the cork on a corkscrew, it is the height of bad manners.' I have never done it since.

By the 1960s, circadian rhythms, which persist (freerun) under constant conditions and have a period close to but not exactly 24 hours, had been identified in many different plants and animals, including us. And everyone (well, nearly everyone) accepted that these rhythms were generated biologically – they were 'endogenous'. As in all branches of science, unless you live under a dictatorship, there is never complete agreement about anything. But dissent is good because it makes scientists refine their experiments to build an even stronger 'evidence base' for the hypothesis being tested. The most prominent dissenter was Professor Frank Brown at Northwestern University in Chicago. He believed that biological rhythms were driven by some natural geophysical cycle such as electromagnetism, cosmic radiation or some other, as yet unknown, force. Brown's central, and not unreasonable, argument was that no biological mechanism could be independent of temperature. When you increase temperature, biological reactions speed up, while cooling slows them down. But for a clock to keep time accurately it has to always run at the same speed. More observations were needed, and studies in plants and 'cold-blooded' insects showed that biological clocks would indeed keep good time – despite huge changes in environmental temperature. Brown was wrong, but his challenge led to experiments which showed conclusively that biological clocks were indeed 'temperature compensated'. Endogenous 24-hour biological clocks had to exist!

An internal clock allows you to not only know the time, but also to predict time, or at least predict regular events within the environment. As I mentioned, our bodies need the correct materials in the right place, in the right amount, at the right time of day, and a clock can anticipate these different needs. By anticipating the approaching day our bodies are prepared in advance so that the 'new' environment can be exploited immediately. Blood pressure and metabolic rate, along with many biological processes, rise before the new dawn. If we merely responded to the light of dawn to switch us from sleep into activity, then valuable time would be wasted getting our energy usage, senses, immune system, muscles and nervous system tuned up for action. It takes several hours to switch from sleep to activity, and a poorly adapted biology would be a major disadvantage in the battle for survival.

Two of the three essential features for an internal circadian clock have been touched upon so far – the ability to keep ticking with a period of about 24 hours under constant conditions, and to maintain a near 24-hour period even when environmental temperatures vary dramatically, showing temperature compensation. The third feature is called 'entrainment'; this capability is incredibly important and will be discussed in detail in chapter 3. I am, perhaps, a little biased about the importance of entrainment because this is what I have worked on for most of my career. As mentioned, circadian clocks do not run at exactly 24 hours, but tick a little faster or a little slower. In this way, circadian rhythms resemble an old mechanical grandfather clock which needs a slight daily adjustment to make sure the clock is set to the 'real' astronomical day. Without this daily resetting, the clock will soon drift and be out of alignment (freerun) with the environmental day/night cycle. A biological clock is of no use unless it is set to local time. For most plants and animals, including us, the most important 'entrainment' signal that aligns the internal day to the external world is light, especially the changes in light around sunrise and sunset. In us, and other mammals, the eye detects dawn and dusk to entrain our circadian rhythms, and eye loss prevents this resetting. People who have lost their eyes as a result of genetic disease or in combat, or because of a tragic accident, drift through time, experiencing periods when they get up and go to bed for a few days at the correct time, before they drift off again and want to sleep, eat and be active at the wrong time of day. A body clock with a period of 24 hours 15 minutes would take around 96 days to travel from 12 noon back around to 12 noon again, getting later by 15 minutes each day. Blind individuals experience something similar to constant jet lag. They become 'time blind', a state that I will discuss in detail in later chapters.

## The Big Sleep

Although the sleep/wake cycle is the most obvious of our 24-hour rhythms, hardly anybody talked about sleep at the meetings I attended in the early days. Sleep seemed to me, and to so many others at the time, too murky and nebulous a subject to get clear answers about. Sleep was also associated with abstract philosophical notions such as the 'mind', 'consciousness' and 'dreams'. It was too impenetrable for most of us. This notable lack of interest in sleep by most circadian researchers, including myself back then, reflected the divergent origins of the fields of circadian and sleep research. The science of circadian rhythms was established by biologists working on every sort of plant and animal. By contrast, sleep research has its origins in medicine and recordings of the electrical activity from the human brain – 'brainwaves'. Sleep was, and still is, studied intensively using electroencephalography (EEG), and interests were focused on how the EEG changed during different stages of sleep and disease. Based upon the size and speed of brainwave activity recorded from the brain by EEG, as well as eye movements and muscle activity, sleep is defined as either rapid eye movement (REM) sleep, or one of three stages of non-rapid eye movement (NREM) sleep. When we are awake our EEG shows small and rapid oscillations in our brain's electrical activity, but as we descend into NREM sleep these oscillations become larger and slower until we reach our deepest sleep, often called slow-wave sleep (SWS). From this state of deep sleep, the EEG transitions into faster and smaller oscillations once again and we enter REM sleep, which has been called 'paradoxical sleep' because it resembles the EEG seen during wake. During REM sleep we also experience paralysis from the neck down, while our eyes move rapidly under our eyelids from side to side – hence the name. This NREM/REM cycle occurs every 70–90 minutes, and across a night of sleep we experience four or five NREM/REM cycles, waking naturally from REM sleep. Around 15 years after the experiment in Mammoth Cave, Nathaniel Kleitman and another student, Eugene Aserinsky, discovered and named REM sleep in 1953 and linked REM to the time when we experience our most vivid and complex dreams. If you have a dog, you may have noticed that while asleep the dog may whimper or growl and make running movements as if chasing a rabbit. Such behaviours have led some to suggest that dogs, and indeed many mammals, also experience dreams during REM sleep. If you don't have a dog, you can always watch your sleeping partner in REM. It's fascinating, but a bit disconcerting for them if they wake and become conscious of your scrutiny!

It is only in the past 20 years, and most notably during the last 10, that circadian and sleep researchers have begun to talk to each other seriously and attend the same meetings. Indeed, meetings nowadays are designed to attract both groups of scientists, and today I consider myself to be both a circadian *and* a sleep researcher. So what got me into sleep? In my case there was a clear and defining moment, arising from a short discussion that irritated me intensely. In my former job, I spent quite a bit of time in the same building with neurologists and psychiatrists, and back in 2001 I bumped into a psychiatrist in one of the unreliable lifts at Charing Cross Hospital in west London. 'You work on sleep, don't you,' he said to me. 'No,' I replied politely, 'I study circadian rhythms.' He continued, oblivious of this subtlety, and said, 'My patients with schizophrenia have terrible sleep, and in my view that's because they don't have a job – so they go to bed late and get up late, which means they miss my clinic, are socially isolated and so can't make friends.' This 'unemployment' explanation made no real sense to me, so I teamed up with another psychiatrist to study patterns of sleep in a group of 20 individuals with a diagnosis of schizophrenia. We compared sleep in this group to unemployed individuals of the same age. The results left me gobsmacked. Sleep/wake patterns in people with schizophrenia were not just bad, they were smashed, and utterly different from the unemployed individuals, who showed similar sleep patterns to employed individuals.

Individuals with schizophrenia also had very little or no SWS and abnormal REM sleep. I wanted to know why sleep had fallen apart in these individuals, and this provided the starting point to study sleep in mental illness, and then later in other health conditions. Interestingly, many of my circadian colleagues, for a whole variety of different reasons, have also 'moved into sleep' in the past decade. Perhaps age has given us wisdom, or maybe courage? Even more importantly, a new generation of neuroscientists armed with multiple and powerful techniques to examine the brain have chosen to study sleep, and are now delivering amazing new information.

Although a host of fundamental questions remain, sleep today is no longer regarded as the 'black box' that it was when I started research. Remarkable new work has greatly improved our fundamental understanding of how sleep is generated within the brain, and how sleep is regulated by our environment. We also now appreciate that during sleep we establish most of our memories, solve problems and process our emotions; we remove dangerous toxins that build up during activity; we rebuild metabolic pathways and re-equilibrate energy reserves. And if we fail to get sufficient sleep, brain function, emotions and physical health all fall apart rapidly. For example, abnormal sleep makes us more vulnerable to heart disease, Type 2 diabetes, infections and even cancer. In short, our sleep defines our ability to function while we are awake, and lack of sleep and the circadian disruption of sleep impact enormously on our overall wellbeing and health. While the evidence demonstrating the importance of sleep is clear, this massive chunk of our biology, around 36 per cent of our life, is still not fully appreciated by many sectors of society. In five years of training, most medical students will have only one or two lectures on the topic, and the information covered is usually about EEG activity during sleep, rather than the new science of circadian rhythms and sleep that I will discuss in this book. In the public domain there remains a lot of sloppy thinking about sleep. Employers assume that their night shift workers will adapt to the demands of working at night. This assumption is wrong, and as a result employees can become dangerously ill, overweight and mentally impaired, and are at a higher risk of divorce and road accidents. As our society becomes increasingly 24/7, and as we squeeze more and more into an overcrowded day, our sleep has become the hapless victim.

## *What I Hope to Achieve*

My central aim is to empower you, the reader, by providing concrete information and guidance, based upon the latest science. You will be able to use the information in the following chapters to get a better understanding of what makes your body clock 'tick' and, critically, use this knowledge to develop an optimal personal routine that works for you, irrespective of age or circumstances. I want to cut through some of the myths, and maybe burst a few bubbles, including the view that teenagers are lazy and that the business executive who gets up at 4 a.m. and starts work is a role model. As you will see, this book encompasses a huge span of human biology and will hopefully stimulate you to dig deeper into many of the topics covered.

Each chapter will consider a central topic, define the science of that topic, and then address issues that impact upon our health and wellbeing. Some of the science can get a bit complex, but it is fundamental for gaining an understanding of our biology and health. The book is also structured so that you can jump back easily to earlier chapters and re-check information for a reminder. Finally, each chapter will finish with a short 'Questions and Answers' section designed to answer some questions that people frequently put to me and my colleagues. This Q&A section will also provide additional and sometimes oblique information. I stress that it is not my intention to provide medical advice; you should always seek this from your medical practitioner. But I will try to explain how some of your actions may be important in achieving optimal health and avoiding potential harm. Such actions include: why to eat at a particular time, when to exercise or when to take different medications, and why you should not drive in the early hours of the morning. This will not be 'finger wagging'. The aim is to provide you with the most up-to-date information that you can either adopt or ignore, but with a clear understanding of the consequences of your actions.

You will also find an Appendix I which provides some guidance on how you may want to develop your own sleep diary to monitor your sleep/wake patterns. Appendix I also includes a questionnaire which will allow you to estimate your 'chronotype', and whether you are a 'morning', 'neutral' or 'evening' person. Appendix II provides a brief outline of the immune system, digging a bit deeper into the complexity of this important part of our biology, which is covered in chapter 11. And in terms of detail, this book has been fully referenced, and I have been guided by one of my scientific heroes, Thomas Henry Huxley, who said: 'If a little knowledge is dangerous, where is the man who has so much as to be out of danger?' To help you build upon the 'little knowledge' in this book, I have cited the relevant scientific papers that have informed the discussion. Many of these scientific publications are, or will be soon, available online as a result of 'Open Access', whereby published research can be accessed free of cost. Indeed, most scientific papers are freely accessible from scientific journal websites 12 months after publication.

My hope is that you enjoy this book and become inspired by the emerging science of biological rhythms, and, importantly, that you will want to apply this science to your own health, happiness and wellbeing. I also hope that, after a suitable period of reflection, you will agree with me that by embracing this knowledge you will be more creative, make better decisions, gain more from the company of others, and view the world and all that it has to offer with a greater sense of curiosity and wonder.

Oxford, January 2022

# I.

# *The Day Within*

## *What is a body clock?*

I know who I was when I got up this morning, but I think
I must have been changed several times since then.

Lewis Carroll

Syncopation is a musical term meaning a variety of different rhythms played together to make a piece of music. By analogy, our biology is syncopated and we are the product. Everything about us is rhythmic. The electrical impulses generated in our nervous system, the beating of our heart, the release of hormones from glands, the contractions of the muscles regulating our digestion, along with a myriad of other processes, are all driven by rhythmic, endogenous changes in our bodies. And some of these rhythms relate to where we live.

One of the oldest intellectual challenges, faced by all civilizations, has been to discover the nature of our home. Our solar system established its current distribution of planets orbiting around the Sun around 4.6 billion years ago, and like other planets the Earth formed as a result of gravity pulling swirling gas and dust into a distinct solar body, making the Earth the third planet that orbits around the Sun. The early Earth was molten due to frequent collisions with other masses; indeed, the proto-Earth is thought to have sustained a massive impact with a Mars-sized body named 'Theia'. The Moon probably formed from the 'ejecta' from this collision about 100 million years after the formation of the solar system. This impact is thought to have knocked the Earth off its 'daily' rotational axis so that the Earth now tilts about 23.4° away from its orbital axis around the Sun, although there is a slight 'wobble' of a few degrees. This 23.4° tilt, as we orbit around the Sun, causes our yearly cycle of the seasons. During part of the year, the northern hemisphere is tilted towards the Sun (summer) and the southern hemisphere is tilted away (winter). Six months later, the situation is reversed. Critically, the Moon's gravitational pull stabilizes the Earth's axial tilt, moderating the degree of wobble. This has produced a relatively stable climate on Earth for billions of years, and many believe that life on Earth would never have got started without this stabilization by the Moon. Paraphrasing the song by the Rolling Stones, we are *all* children of the Moon.

The bottom line is that today we sit on a relatively stable and rhythmic planet that is around 4.5 billion years old with a daily rotational axis of 24 hours, or 23 hours, 56 minutes and 4 seconds to be precise. Around 600 million years ago, when complex life was emerging, a day lasted only 21 hours, so the Earth is slowing down. But that's another story. Our Earth currently orbits around the Sun every 365.26 days, and the tilt of the Earth on its rotational axis generates the seasons. The Moon orbits around the Earth approximately every 29.53 days and its gravitational interaction with the Earth and Sun produces the tides. Collectively, these geophysical movements generate the day, night, seasons and tides. Many animals, indeed most life forms, have evolved body clocks of various kinds that anticipate at least one, and sometimes all, of these environmental cycles: daily, annual and lunar.

Rhythmicity is such a ubiquitous feature of life and our everyday experiences that we take it for granted. Perhaps this detachment is unsurprising. Most of the time we cannot sense our internal workings, and in the industrialized nations, at least, the natural day/night cycle is overridden by electric light and artificial heat. The Sun never really sets for *most* of us, and the seasons no longer dictate our diet or where we live. Food is constantly available. In the UK we can buy strawberries from Kenya or southern California all year round, but as recently as 25 years ago the home-grown strawberry season was only six weeks long. Warmth at home or at work is accessed by the flick of a switch. We are now insulated from the environmental cycles that dominated our evolution. A major aim of this book is to reacquaint us with one of these cycles – the 24-hour cycle of day and night.

The study of physiology aims to understand how living things work. It's a huge discipline that includes the molecular processes within cells, how the nervous system works, the regulation of hormones, how the various organs of the body function and the generation of behaviour in all its varied forms. Human physiology, like that of most other animals, is organized around a 24-hour cycle of activity and rest. In the active phase, when food and water are sought and then consumed, organs need to be prepared for the intake, processing, uptake and storage of nutrients. The activity of organs such as the stomach, liver, small intestine and pancreas, and the blood supply to these organs, needs to be appropriately adjusted across the day and night. During sleep, we stay alive by mobilizing our stored energy. This energy is then used to drive many essential activities including the repair of body tissues, the removal of harmful toxins, and the formation of memory and the generation of new ideas in the brain. Because physiology shows such a marked daily pattern, it should be no surprise that our performance, the severity of disease, and the action of prescription drugs change across the 24-hour day. A few examples of these rhythmic 24-hour circadian changes are shown in Figure 1. Such rhythms have been observed for centuries, and, of course, the long-asked question has been 'Where do they come from?'

For hundreds of years, one of the main goals in understanding the brain has been to identify what parts of the brain do what, and it is truly a daunting task. You will read in many textbooks that there are 100 billion neurones in the human brain. Nobody

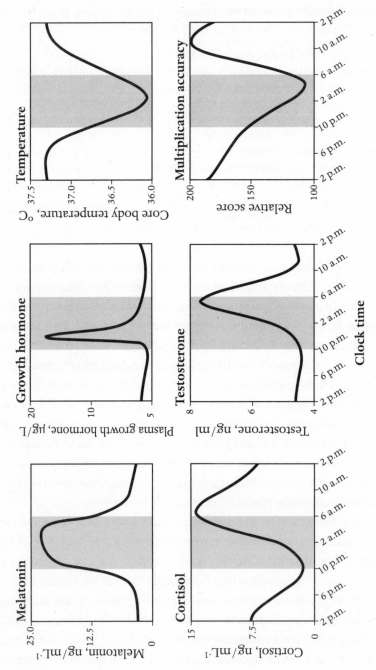

**Figure 1.**

**Figure 1. Examples of 24-hour daily changes in human physiology**. Shown here are representations of the daily changes seen in physiology: the hormone melatonin from the pineal gland (Figure 2)[1]; growth hormone released from the pituitary gland[2]; body temperature[1, 3]; the 'stress hormone' cortisol from the adrenal glands[1]; testosterone produced by the gonads (the testes in men and by the ovaries in women) and in small quantities from the adrenal glands[4]; and multiplication accuracy, representing one aspect of our cognitive abilities[3, 5]. Many hormones, such as cortisol, are released as 'pulses', and so shown here is the 'smoothed average' of hormone release. There are two important points to make about these rhythms. First, these are averages, and there will be differences between individuals relating to when these rhythms peak and their size or amplitude. The second point is that many of these rhythms were not recorded under constant conditions, and while they almost certainly have a circadian component, meaning they would persist for many cycles under constant conditions, they are more accurately called 'diurnal' changes. The significance of these changes will be discussed in later chapters.

seems to know quite where this figure came from, but anyhow it is wrong. The Brazilian researcher Suzan Herculano-Houzel undertook some careful studies to finally address this question, and her answer was that the average human brain contains around 86 billion neurones.[6] Now, I know this might sound a bit like the medieval debate on 'How many angels can dance on the head of a pin?', but a difference of 14 billion neurones is a lot of neurones. There are about 14 billion neurones in the *entire* baboon brain, and, for additional comparison, 75 million in the mouse brain, 250 million in the cat, and 257 billion in the elephant. So 86 billion is a lot of neurones, which is why the discovery that only 50,000 work together as a 'master biological clock',[7] coordinating our 24-hour circadian rhythms, is a truly remarkable achievement.

This 'master clock' in humans, and all mammals, is located in an area of the brain called the 'suprachiasmatic nuclei', or SCN (Figure 2). The discovery of this structure has a fascinating history. Researchers in the 1920s had noted that rats under constant conditions of darkness run in a running wheel (similar to the hamster wheels you can buy in a pet shop) with rest (sleep)/activity rhythms that are a little shorter than 24 hours. This observation was a bit of a shock because the prevailing view in the 1920s was that behaviour occurs as the result of a particular stimulus – a bit like a reflex. You provide a specific stimulus, and you get a specific kind of response. However, the rats showed a rhythmic pattern of daily activity without any obvious external stimulus. The activity pattern appeared to be generated from within the animal, and not driven by changes in light or other stimuli. So what was driving this rhythm?

Experiments in the 1950s and 1960s on rats removed different organs of the body to try to identify this 24-hour driver, but near 24-hour rest/activity rhythms persisted under constant conditions. The rat's brain was then examined. Small parts of the brain were surgically removed (lesioned), and rest/activity patterns examined. If you think it was bad for rats, remember that at this time lobotomy was a routine operation in humans, a procedure whereby most of the connections to and from the prefrontal cortex (Figure 2, p. 24) were cut in an attempt to 'cure' psychiatric conditions, and the chap that invented this technique got a Nobel Prize. The experiments on rats suggested that 'the clock' must reside somewhere deep in the brain, probably the hypothalamus (Figure 2), because destruction of this tiny area of the brain resulted in 'arrhythmicity', or the complete loss of any 24-hour patterns of activity and rest.[8] These studies were then followed up in the early 1970s, and the SCN (suprachiasmatic nuclei) emerged as the prime candidate.[9, 10] Almost 20 years later, the final and critical role of the SCN was confirmed in golden hamsters. In the late 1980s, Martin Ralph and Michael Menaker, who were close colleagues of mine at the University of Virginia, discovered a 'mutant' hamster, the '*Tau* mutant hamster', with an activity/rest pattern of 20 hours, compared to the non-mutant animals which had a pattern close to 24 hours. The SCN of a *Tau* mutant hamster (20 hours) was transplanted into the hypothalamus of a non-mutant hamster (24 hours) whose own SCN had been lesioned and showed complete arrhythmicity. Remarkably, the mutant SCN not only restored circadian rhythms in wheel-running behaviour, but, critically, the restored rhythms were 20 hours – not 24! Transplanting other parts of the hamster brain had no effect. These findings showed that the transplanted SCN must contain 'the clock'.[11] I remember these experiments vividly, and the incredible excitement we all felt as the data were collected on a daily basis and we observed that the restored rhythms were 20 and not 24 hours.

As mentioned, the SCN contains about 50,000 neurones,[7] and a remarkable finding was that each has its own clock. This was again first shown in rats, where the rat SCN was separated out into its individual cells and placed into cell culture. The electrical activity of individual SCN cells was then monitored and showed robust and independent circadian rhythms – all ticking away at a slightly different time to each other. What's more, these individual SCN neurones kept on ticking in a dish for weeks.[12] As SCN cells were shown to have a clock, the clock mechanism had to be located within the cell – there had to be a molecular clock! This was remarkable stuff indeed, and demanded an answer to the question: how is this rhythm generated?

In 2017, three researchers from the USA, Jeffrey C. Hall, Michael Rosbash and Michael W. Young, shared the Nobel Prize for discovering how the clock 'ticked'. They did this after almost 40 years of research, sometimes working together and at other times as rivals, with many young scientists all contributing a little piece of the jigsaw. I was working at the University of Virginia while some of the key discoveries were being made, and Hall, Rosbash or Young would visit and give a seminar about the latest progress. As scientists, they were equally brilliant, but as characters they were completely different, each with a very 'distinctive personality'. For example, Jeff Hall is also a notable scholar of the American Civil War, and on one memorable occasion visited the University of Virginia and gave a research seminar about his latest progress on the molecular clock dressed in a Northern Unionist Army tunic and cap. This choice of attire, which was possibly designed to provoke, was completely ignored by the Faculty in the heart of

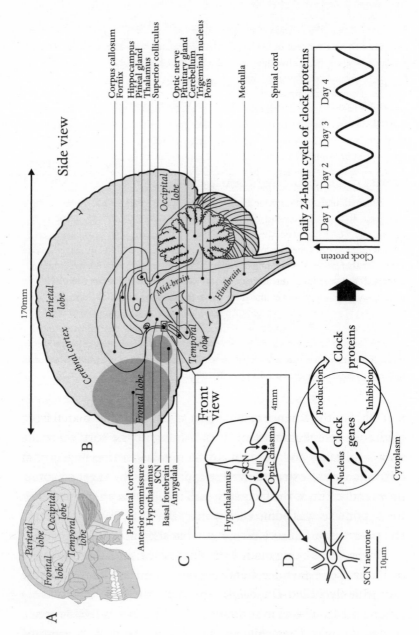

**Figure 2.**

**Figure 2. A.** Shows the location of the human brain within the skull and the most obvious brain lobes (parietal, frontal, occipital, temporal) that can be recognized from the outside of the brain. **B.** Provides a view of the brain along a mid-section from the side, and indicates the location of the key internal structures. The typical human brain comprises about 2 per cent of the body's total weight, but uses 20 per cent of our total energy intake. As little as five minutes without oxygen can cause brain cells to die, leading to severe brain damage. The brain is 73 per cent water, but it takes only 2 per cent dehydration for brain function such as attention, memory and other cognitive skills to be badly impaired. Our brains have usually finished developing by the age of 25. **C.** Shows an enlargement of the suprachiasmatic nuclei (abbreviated to SCN) from the front. The SCN represent the 'master biological clock'. The SCN sit either side of the third ventricle of the brain (III) and above the optic chiasma, which is where the optic nerves enter the brain and fuse. A small number of nerves within the optic nerve, called the retinohypothalamic tract, enter the SCN and provide light/dark information from the eye for entrainment (chapter 3). **D.** Shows a single SCN neurone, which is around 10μm (0.01mm) in diameter. The SCN has around 50,000 neurones and each can generate a circadian rhythm. Normally they are all connected to each other. Clock genes are located in the nucleus of each SCN neurone and create a message that guides the building of clock proteins. These proteins are made in the cytoplasm that surrounds the nucleus. The clock proteins then interact to form a protein complex which moves into the nucleus to inhibit or 'switch off' the production of further clock proteins. After a while, this protein complex is then broken down (degraded), which allows the clock genes to make clock protein once again. A roughly 24-hour cycle of protein production and protein breakdown is the result. This molecular feedback loop is converted into a signal (electrical or hormonal) that acts to coordinate the circadian clocks in the rest of the body.

the 'Old South'. Science is often portrayed as a linear march from ignorance to enlightenment. It is nothing of the sort; there are always mistakes and dead ends, and it is fascinating to recollect how often these extraordinary scientists got it wrong, and sometimes badly. But, as the facts accumulated, the lessons were learnt, the hypothesis was adjusted, the mistakes were quietly forgotten and progress once more resumed. That's science.

The progress being made by Hall, Rosbash and Young was not on humans or even mice, but on a very distant animal relative, the tiny fruit fly called *Drosophila*, the very same fly that swarms around the fruit bowl in summer and is often squashed without a second thought. *Drosophila* remains one of the most commonly used 'model species' to understand how genes give rise to physiology and behaviour, and has been studied for more than 100 years.[13] These flies are relatively cheap to care for, they breed rapidly and have exquisitely well-understood genetics, all of which has made them indispensable for basic research, including research into the circadian clock. So what did Hall, Rosbash and Young discover in *Drosophila*? At its core, the pathways in the cell that generate the 'molecular clockwork' consist of a 'negative-feedback loop', which consists of the following steps (Figure 2D): clock genes located in the nucleus of the cell create a message that provides the template to build clock proteins. These proteins are made in the cytoplasm (the matrix part of the cell that surrounds the nucleus). The clock proteins then interact to form a protein complex which moves into the nucleus to inhibit or 'switch off' the production of further clock proteins. After a while, this protein complex is then broken down, which allows the clock genes to do their stuff once more and make more clock proteins. A 24-hour cycle of protein production and protein breakdown is the result. And this is the molecular clock, well . . . more or less! The rates of clock gene activation, protein production, protein complex assembly, protein complex entry into the nucleus, inhibition of clock genes, protein complex breakdown and then reactivation of the clock genes all combine to produce a 24-hour rhythm, and changes (genetic mutations) in any one of these steps can either speed up, slow down or break the clock.[14] It was just such a mutation in the '*Tau* mutant hamster' that gave it a period of 20 rather than 24 hours.[15] The molecular clockwork of all animals, including you and me, is built in a very similar way. This is all the more remarkable when you think we shared a common ancestor with *Drosophila* more than 570 million years ago, when the Earth had a 22- to 23-hour day, suggesting that our biological clocks have had to slow down by a few hours over the past hundreds of millions of years.

The 24-hour rhythm of clock protein production and

degradation acts as a signal to switch on and off countless genes and the circadian production of their proteins that, in turn, regulate rhythmic physiology and behaviour (Figure 1). Our current understanding of the 'molecular clockwork' is the most complete example in any field of biology of how genes ultimately give rise to behaviour, and, for this first molecular characterization of a circadian rhythm in *Drosophila*, Hall, Rosbash and Young richly deserved the trip to Stockholm and their Nobel Prize, the presentation of which I was lucky enough to witness.

Interestingly, small changes in our clock genes (polymorphisms) have been linked to whether we are a 'morning', 'evening' or 'intermediate' body clock type. Morning types, or 'larks', like to sleep early and get up early, and it seems that they have faster body clocks, due to changes in one or more of their clock genes.[16] By contrast, evening types, or 'owls', have slower clocks and prefer to go to bed later and sleep in. So, by their contribution to our genes, our parents are still telling us when to get up and go to bed! Our body clock type is often referred to as our 'chronotype', and, as we shall discuss later, our chronotype is also influenced by our age and when we are exposed to light around dawn and dusk. You can explore your own chronotype with the information in Appendix I.

Although the SCN is the 'master clock' in mammals, it is not the only clock.[17] We now know that there are clocks within the cells of the liver, muscles, pancreas, adipose tissue, and probably in every organ and tissue of the body.[18] Remarkably, these 'cellular peripheral clocks' seem to use the same negative-feedback molecular clockwork as the SCN clock cells. This came as a big shock. I remember when Ueli Schibler, working at the University of Geneva, first presented his findings at a meeting in Florida in 1998 showing that non-SCN cells had clocks.[19] There was an audible gasp from the audience. Clock genes had previously been identified in non-SCN cells, but for many years these genes were thought to be doing something else, and the idea that cells outside the SCN could contain clocks was not seriously considered. The reason for this was that destruction of the SCN abolished the 24-hour rhythms of activity and hormone release of the sort illustrated in Figure 1. The conclusion from SCN lesion studies was that the SCN 'drives' 24-hour rhythms throughout the body. But we now appreciate that this idea was an oversimplification. Loss of observable rhythms after SCN lesions is due to two key factors: the first is that many individual peripheral clock cells 'dampen out' and lose their rhythmicity after several cycles – they run out of steam without a gentle nudge from the SCN. The second and more important cause is that without a signal from the SCN, individual clock cells in tissues and organs become un-coupled from each other. The cells continue to tick individually, but at slightly different times, so that a coordinated 24-hour rhythm throughout the entire tissue or organ is lost.[20] It's rather like visiting a stately home where all the antique clocks start to chime at slightly different times. This discovery led to the appreciation that the SCN acts as a pacemaker to coordinate, but not drive, the cir-cadian activity of billions of individual circadian clocks throughout the tissues and organs of the body. The SCN is rather like the con-ductor of an orchestra: it provides a time signal that coordinates the rest of the orchestra/body, and without the conductor/SCN everything drifts apart, so instead of a symphony you have a biological cacophony, and a failure to do the right thing at the right time.

The signalling pathways used by the SCN to synchronize, or entrain, these 'peripheral clocks' are still uncertain, but we know that the SCN does not send out countless unique signals around the body that target different tissues and organs. Rather there seems to be a limited number of signals which include the auto-nomic nervous system (that part of the nervous system responsible for control of the bodily functions not consciously directed) and several chemical signals. The SCN also receives feedback signals from other parts of the body, including the sleep/wake cycle, that adjust its activity, allowing the whole body to function in synchrony with the changing demands of the 24-hour day.[20-21] The result is a complex circadian network that coordinates rhythmic physiology and behaviour. The loss of synchronous activity between different circadian clocks, either within an organ or between organs such as the stomach and liver, is called 'internal desynchrony' and can cause serious health problems, which I will discuss in later chapters.

The circadian system fine-tunes our bodies to the varied demands of the 24-hour day/night cycle. But unless this internal timing system is set to the external world it is of no practical use, and it is this alignment of the 'internal' and 'external' day that I want to consider in chapter 3. But before that I want to take a look at sleep, that most obvious of our 24-hour patterns of behaviour, in the next chapter.

## Questions and Answers

**1. How many clock genes are needed to build a molecular clock?**
Long gone are the days when we talked about 'the' clock gene. It is difficult to put an exact number on this as it depends on what you mean by a clock gene. A workable definition would be that clock genes are like the cogwheels of a mechanical clock. They interact in a specific way to generate a 24-hour rhythm, and if you take one of these 'cogs' away, or damage a cog, the clock will be significantly altered or even stopped. Using this definition, there are around 20 different clock genes in us and other mammals like mice that drive the molecular clockwork.[22] However, this is a bit misleading as there are very many more genes that contribute to the regulation of the clock, the stability of the clock, and then how the clock drives circadian physiology. If we include these genes there may be hundreds. Also, it is worth being aware that all of these 'clock' genes have other roles, regulating biological key processes such as cell division and the regulation of metabolic processes.

## 2. Are human circadian rhythms influenced by electromagnetic fields (EMFs)?

Currently there is no strong evidence that EMFs can alter human circadian rhythms.[23] But absence of evidence is not evidence of absence. I think it fair to say that if there is an effect it is small.

## 3. Do humans have annual clocks?

We do show a variety of annual rhythms, including peaks in birth, hormone release, suicide, cancer and death. For example, in the northern hemisphere, and perhaps counterintuitively, suicide in spring is significantly higher than in the winter around December, when suicide rates are at their lowest.[24] Some argue that we are like sheep, deer and many other mammals that have an annual clock. But this is difficult to demonstrate experimentally as volunteers would need to be housed under conditions of constant light and temperature for at least three years; and the ethics of such experiments, let alone the practicality of finding volunteers, are unacceptable. Others argue that we do not have an annual clock, like a circadian clock, but are merely responding directly to annual changes in the environment such as day length or temperature.[25]

## 4. Do all animals have a suprachiasmatic nuclei (SCN)?

All mammals, including the marsupials (e.g. kangaroos) and egg-laying monotremes (e.g. duck-billed platypus), do have a structure in the brain that resembles an SCN. When experiments have been undertaken, the SCN seems to act as the 'master' clock, coordinating the circadian rhythms of the peripheral clocks. But this is not the case in birds, reptiles, amphibians and fish. In these animals there are several organs that can act as a 'master clock'. These are located in SCN-like structures within the hypothalamus, the pineal organ and even the eyes. A big puzzle is that in closely related species the importance of and interaction between the SCN, pineal organ and eyes vary considerably. For example, in the house sparrow, the pineal organ seems to be the dominant clock, while in the quail the eyes play this role. In pigeons, all three organs interact![26] This topic fascinated one of the pioneers of circadian research, Michael Menaker, who became a close friend and colleague when I was based at the University of Virginia.

**5. Do the genes and proteins of the molecular clockwork regulate non-clock behaviours too?**
Yes – and, as discussed in chapter 10, mutations in clock genes have been linked to cancer, and other conditions such as mental illness (chapter 9). Remarkably, an increased desire to consume alcohol has also been linked to changes in some of the 'clock genes'.[27] When a single gene is involved in more than one activity this is called a 'pleiotropic gene'. And this is the rule rather than the exception.

**6. Have humans evolved weekly or monthly biological rhythms?**
This has been much debated. While it is clear that life on Earth has evolved clocks to predict geophysical cycles such as the 24-hour rotation of the Earth, the seasons and the Moon-driven tides, evidence for internal clocks that predict man-made cycles such as the week or month is far less clear. Some have argued strongly in favour,[28] but most circadian biologists argue against the existence of biological clocks with a 7- or 31-day period on the basis that there is no robust evidence.

# 2

# A Heritage from Our Cave Days

## What is sleep, and why do we need it?

There is no scientific study more vital to us than the study of our own brain. Our entire view of the universe depends on it.

Francis Crick

In Greek mythology, Hypnos is the god of sleep. He is the son of Nyx (Night) and Erebus (Darkness), and his twin brother is Thanatos (Death). Hypnos and Thanatos live in the underworld (Hades). So even from ancient times sleep has been linked to darkness, death and hell. By association, sleep was hardly given a ringing endorsement by the ancients. And if we jump forward over 2,000 years to the twentieth century things don't get much better. The great entrepreneur Thomas Edison is reported to have said: 'Sleep is a criminal waste of time and a heritage from our cave days.' These may not have been his exact words, but Edison certainly would have agreed with another American, Edgar Allan Poe, who said: 'Sleep, those little slices of death – how I loathe them.'

Sleep, from the earliest times, has not been embraced. Indeed, during more recent centuries it came to be despised, partly because hard work was considered intrinsically virtuous and worthy of reward. Sleep stops us from working and so, by definition, sleep must be sinful. Of course, not all agreed with this view, and as you might expect, Oscar Wilde adopted a somewhat different attitude, explaining: 'Life is a nightmare that prevents one from sleeping.'

Sadly, the views relating to sleep held by Edison, Poe and many other like-minded individuals were adopted by decision makers throughout the nineteenth and twentieth centuries. Although attitudes have improved more recently, sleep today is still regarded as a type of 'sickness' in need of a cure. Something we have to tolerate, but would rather not. And without all the facts we have waged war on this essential part of our biology. Such an ill-conceived action has resulted in appalling consequences for our individual health and wellbeing, along with a major economic insult for the state.

Within our brain, the generation of the daily sleep/wake cycle involves a highly complex set of interactions between the hind-brain, mid-brain, hypothalamus, thalamus, cerebral cortex (Figure 2, p. 24), and all the brain neurotransmitter systems (e.g. histamine, dopamine, noradrenaline, serotonin, acetylcholine, glutamate, orexin, GABA (gamma-aminobutyric acid)) and a few hormones, none of which are unique to the generation of sleep. These systems combine to broadly flip the sleep/wake states, rather like a seesaw, to either sleep or consciousness. But sleep is not an 'off' state, but rather a condition of intricacy and change.

## The REM and NREM Sleep Cycle

For centuries, it was presumed that during sleep the brain was shut down and nothing much happened. Part of the reason for this assumption was that there were no real tools available to look at the sleeping brain until the 1950s. Since the 1950s sleep has been studied routinely in the laboratory by placing electrodes onto the skin of the scalp, stuck there by using an electrically conducting jelly, and measuring the pattern of electrical activity, called the electroencephalogram (EEG). I mentioned this in the introduction, but just to remind you: when awake, and during the early stages of sleep, the EEG pattern is fast (high frequency) and small (low amplitude). Think of the pattern you get when you rapidly oscillate a skipping rope that has been pulled taut between two people. But at the beginning of sleep, and the progressive descent into deeper slow-wave sleep (SWS), the electrical oscillations become slower (low frequency) and bigger (high amplitude). In this case, the skipping rope is held loosely and is gently oscillated. Sleep moves progressively through several stages (1–3) into deep sleep or slow-wave sleep (SWS), which is also called 'delta sleep'. From deep delta sleep, the pattern then changes, moving quickly from stage 3 sleep to stage 2, and then 1. This 'reversal' in brain wave activity, from stage 3 to stage 1, is then followed by another state of sleep. In this new state, the EEG/brainwaves are very similar to those in the awake brain, with high-frequency and low-amplitude oscillations in electrical activity. The eyelids are shut, but the eyes move rapidly. Heart rate and blood pressure increase, and the body is effectively paralysed from the neck down. This is called, for obvious reasons, rapid eye movement sleep, or REM sleep. After some several minutes of REM sleep, there is a switch back into non-rapid eye movement, or NREM, sleep. We then pass down through stages 1–3 into SWS and then back again to REM sleep once more. This cycling of NREM and REM sleep lasts approximately 70–90 minutes (depending on age), and in an average night we may experience five of these NREM/REM sleep cycles. But they are not identical. During the first part of the night we experience more SWS (stage 3), while during the second part of the night we undergo more frequent and longer periods of REM sleep. We usually wake naturally from REM sleep.

## NREM Sleep, Memory and Anxiety

NREM sleep has been linked with our ability to form memories and solve problems. This has been shown in a variety of different ways. One approach was to stimulate the brain to produce more

SWS while individuals slept in a controlled laboratory environment. This can be done using certain sound frequencies. The extra SWS during sleep was linked with the ability to remember more facts and events from the previous day.[29] Other experiments have deprived people of SWS. This is done by monitoring the EEG of sleeping individuals, and then waking them up when they start SWS. This loss of SWS reduces the ability to form memories.[29] Bursts of electrical activity called 'sleep spindles' in NREM stage 2 also seem to be important for the formation of memories.[30] In this case, drugs were used to reduce or increase sleep spindles and this, in turn, decreased or increased the formation of memories.[31] Another feature of NREM stage 2 sleep is large electrical events called 'K-complexes'. These seem to be particularly useful in preventing us from waking up in response to external noise or other events in the environment.[32] But, recent data suggest that K-complexes may also be involved in helping us generate memories. Most SWS occurs during the first half of the night, and this may be the basis of the often quoted saying that 'an hour of sleep before midnight is worth two hours of sleep after midnight'. But, for what it is worth, I think this is just another sleep myth. Poor sleep is associated with increased anxiety, and some recent studies have suggested that SWS during NREM sleep might be important in organizing the brain networks in the prefrontal cortex (Figure 2, p. 24) to reduce anxiety.[33] It is interesting that in schizophrenia there is a marked decrease in SWS during sleep. Perhaps this may be a contributing factor to the increased anxiety frequently reported in individuals with schizophrenia, and perhaps other mental health conditions.

## REM Sleep, Dreams and Mood

We dream in both NREM and REM sleep but dreams in REM tend to be longer, and more intense, complex and bizarre. As we naturally wake from REM sleep we may remember for a short time the last dream we experienced. Dreams seem to occur during all of the REM period, and the old idea that dreams occur in a flash when we wake is now thought not to be true. Dream content is very variable but usually involves the dreamer and individuals who are familiar to the dreamer such as friends, family or sometimes famous personalities. For most of us, dreams are usually visual experiences, and we rarely have dreams involving taste or smell. However, in people who have been blind since birth, their dreams are dominated by sound, touch and emotional feelings.[34] Dreams are often deeply bizarre but usually draw from our experiences at some very basic level. Importantly, loss of REM sleep is associated with increased daytime anxiety, irritability, aggression and hallucinations, supporting the idea that dreams and REM may be important for the processing of emotions and developing emotional memories.[35] I will come back to the subject of dreams later in this book, but need to say at this point that dreams are immensely difficult to study. They can't be quantified, they are entirely subjective and by their very nature they are self-reported. We can't measure them properly! Sigmund Freud believed that dreams represent a fulfilment of a repressed wish, and that the study of dreams provides a route to understand the unconscious mind. At the time of Freud, dream analysis played a key role in psychoanalysis. Today the importance of dreams in psychoanalysis has been strongly downgraded. The central problem is that without objective and reliable measures, understanding dreams is just speculation. As a result, dreams have often been hijacked by the dark practitioners of pseudoscience.

## Some Curious Facts about REM Sleep

Very strangely, and paradoxically, REM sleep deprivation in some people with depression (chapter 9) can cause a short-term improvement in the severity of the condition. For example, REM sleep deprivation for one whole night improves depressive symptoms in 40–60 per cent of individuals. However, after recovery sleep the depression returns.[36] So for practical purposes this is not a treatment for depression, although it might be a useful tool to explore how sleep and depression are linked in the brain.

Another surprising feature of REM is that during this stage of sleep men experience erections of the penis and women experience clitoral engorgement. These 'events' have been studied more extensively in men, presumably because there is a more obvious biological assay. Erections seem to last for most of the REM episode, either during night-time sleep or if sleeping during the day. Erections have been recorded during REM in male babies and even in individuals on life support.[37] It has also been suggested that the cave paintings in Lascaux in southern France (which should be on everybody's 'bucket list') depict sleeping males with pronounced erections. Sexual intercourse prior to sleep does not alter the extent of the erection, and excessive alcohol, which inhibits erections when awake, has little effect on these REM erections. Other studies suggest that there is no correlation between the sexual content of a dream and the occurrence of an REM erection.[38] We don't know why REM-related erections occur; one suggestion is that erections contribute to the health of the penis by increasing oxygen to the tissues and muscles, keeping the organ 'fit'. Intriguingly, REM erections have been observed in all mammals studied, with the exception of the nine-banded armadillo of North, Central and South America. This observation must surely be telling us something about REM erections? Another curious feature of the nine-banded armadillo is that it is one of the very few animals known to carry the bacteria that cause leprosy, and can transmit the disease to humans.[39] So, if you hit one driving, and this is a common event in Texas, I am told, take care how you deal with the carcass.

I mentioned that during REM sleep we experience our most

complex and vivid dreams. It is during this time that projections from the mid-brain to the spinal cord cause paralysis (also called 'atonia'), from the neck down. This is thought to prevent us from acting out our dreams. Support for this idea comes from a condition known as REM sleep behaviour disorder (RBD), where there is no or little atonia during REM sleep. I will discuss this in more detail later, but RBD is an early sign of the future development of Parkinson's disease.[40] At one end of the severity scale of RBD, individuals just move their arms and legs about, but some individuals can experience sleep talking, shouting, screaming or even physical violence. Unfortunately, RBD is most often acted upon after harm has been caused to a sleeping partner.[41] A famous, and widely reported case in the UK media, involved Brian Thomas, a 'decent and devoted' husband who strangled and killed his wife while on holiday. In his dream he was attacking an intruder, but in reality, and very sadly, it was his wife. The Crown Prosecution Service accepted he had not been in control of his actions and the jury at Swansea crown court were ordered to acquit Thomas. The only thing that Mr Thomas remembered of his dream was the break-in by an intruder.

## Switching Between Consciousness and Sleep

The countless interactions associated with the switch between consciousness and sleep, and the REM/NREM cycle, are normally regulated by two key biological drivers. Firstly, the circadian system (circadian drive), entrained by sunrise and sunset (chapter 3), 'tells' the brain circuitry when it is the best time for sleep and wake. This circadian drive acts to 'time-stamp' the sleep/wake cycle. The second driver, and perhaps the most intuitive regulator of sleep, depends upon how much sleep we need and has been called 'sleep pressure' or the 'homeostatic driver' for sleep. Sleep pressure builds from the moment we wake, rises throughout the day and reaches its highest level in the evening prior to sleep. The build-up of sleep pressure during the day is 'opposed' by a circadian drive for wake. Ironically, the circadian system produces the highest drive for wake just before we fall asleep. We fall asleep naturally when the circadian drive for wake drops and the sleep pressure is high. During sleep, the sleep pressure declines, and the circadian clock 'instructs' the brain to remain in the sleeping state – it provides a circadian drive for sleep.[42] We wake naturally when the sleep pressure has diminished, and the circadian system instructs the brain that it is now time to wake. Sometimes we may feel tired in the middle of the afternoon. This is often because sleep pressure builds faster than the circadian drive for wake. The circadian drive for wake can't keep up. This can happen, for example, after a bad or short night of sleep. Under these circumstances, we wake with a significant level of sleep pressure. Our response is to want to take an afternoon nap. A short nap will push back the sleep pressure and make us feel more alert. The amount of slow wave sleep (SWS), or 'delta sleep', we experience is a direct measure of sleep pressure and is proportional to how long we have been awake.[43] Of course, the circadian and homeostatic drivers of sleep do not act alone to determine the timing and length of sleep. Additional factors, including the demands of work and leisure, our genetics, age, and the consequences of mental and physical illness, along with our emotional and stress responses, all combine to deliver the sleep we get.

## Why Coffee Keeps You Awake

The build-up of several chemicals in the brain has been proposed as driving sleep pressure, and the best evidence is for a molecule called 'adenosine'.[44] Animal studies have shown that adenosine increases in the brain during wake, and sleep can be triggered by exposing the brain to adenosine. Caffeine in tea and coffee is very effective at keeping us alert and awake, and caffeine works by blocking the mechanisms in the brain (adenosine receptors) that detect adenosine.[43] Amongst other things, caffeine is an adenosine receptor 'antagonist', preventing the brain from detecting how tired it is. The short-term use of caffeinated drinks can be useful in keeping us awake during long motorway journeys,[45] but care must be taken because as the effects of caffeine wear off we can experience a wave of profound and overwhelming tiredness that can cause us to fall asleep at the wheel in the form of a 'microsleep'.[46]

## What is the Role of Melatonin?

Melatonin is often confusingly termed the 'sleep hormone', and this is very misleading. Melatonin is made mostly in the pineal gland, a structure in the middle of our brain (Figure 2, p. 24) and considered by René Descartes (1596–1650) to be the anatomical location of the soul and the spiritual part of a human being. Further discussion of the soul is beyond the scope of this book, and I refer you to your chosen deity. The pineal gland is regulated by the suprachiasmatic nuclei (SCN) via the autonomic nervous system to produce a pattern of melatonin release, with levels rising at dusk, peaking in the blood around 2–4 a.m. and then declining around dawn (Figure 1, p. 20). Bright light, detected by the eyes, also acts to stop melatonin production. As a result, melatonin acts as a 'biological signal of the dark'. While melatonin production occurs at night during sleep in day-active animals such as ourselves, in nocturnal animals, like rats and badgers, melatonin is also produced *at night* when these animals are active.[47] So melatonin can't be a universal 'sleep hormone'. But what does melatonin do in us? Certainly, the tendency to sleep in humans is closely correlated with the melatonin profile, but this may be correlation and not causation.

There are individuals who do not produce melatonin, notably tetraplegic individuals. The release of melatonin is regulated by a neural pathway from the SCN to the pineal gland, which passes through the cervical spinal cord in the neck. Severing this neural pathway, as in tetraplegic individuals, blocks the release of melatonin from the pineal, and these individuals have been reported to have poor sleep compared to controls. However, the poor sleep in tetraplegic individuals was very similar to the poor sleep seen in paraplegic individuals (paralysis of the legs and lower body) who had melatonin levels in the normal range.[48-9] These data suggest it was not the lack of melatonin but some other aspect of tetraplegia that caused the sleep problems. This conclusion is broadly supported by a small study that gave melatonin to tetraplegic individuals. Sleep in some individuals showed a small improvement with a reduction in the time to fall asleep (sleep onset latency) and fewer awakenings during the night; although, paradoxically, daytime sleepiness increased. The authors of this study stated that a randomized, placebo-controlled trial with a larger sample size would be required to confirm the findings.[50]

Beta-blockers, which are used to treat a variety of heart and blood pressure conditions, also show an 80 per cent reduction in melatonin production. They not only reduce blood pressure but also block signalling to the pineal gland, resulting in much lower night-time levels of melatonin. Individuals on beta-blockers are reported to have poorer sleep.[51] One study gave melatonin supplementation to individuals treated with beta-blockers. After three weeks, and compared to a placebo, total sleep time was increased by 36 minutes and the time it took to get to sleep was reduced by 14 minutes. So there was a small but significant effect.[52] Additional studies also suggest that taking melatonin may shorten the time taken to fall asleep and increase the total time spent asleep. However, these effects using synthetic melatonin[53], or drugs that mimic the effects of melatonin, are modest[54] (see chapter 14). In addition to the action of melatonin on sleep, it is also possible that the night-time rise in melatonin is detected by the SCN and provides an additional modulatory signal to entrain the 'master clock', reinforcing light entrainment signals from the eye and stabilizing the sleep/wake cycle.[55-6] So, in summary, the consensus, based on the data, is that melatonin seems to have a small direct action to promote sleep and/or it may provide an additional signal to tell the brain that it is night-time, and this is used to augment entrainment by light (chapter 3).

I have concentrated on sleep in humans, but would not like to create the impression that you must have a big and complex brain to show states of sleep and consciousness. Remarkably, sleep-like states have been observed in all vertebrates and invertebrates, including insects and even nematode worms. A recent and remarkable study in the octopus, a mollusc and related to snails, showed that these amazing animals have two different sleep states, resembling the NREM and REM sleep states of the vertebrates.[57] But what about animals with no brain at all and just a 'nerve net', like corals, hydras and jellyfish, collectively grouped in the phylum Cnidaria, also called Coelenterata. The first question is how would you recognize sleep in such animals? As it turns out, there are some well-established criteria for doing so. For example, if you prevent them being inactive/sleeping (theoretically increasing the sleep pressure), when given the opportunity, do they show more inactivity/sleep? When inactive/asleep do they show a reduced response to environmental stimuli such as touch or light? Is there evidence for regulation by a circadian clock or sleep pressure? And finally, do sleep-inducing drugs which act on adenosine or histamine receptors alter the patterns of activity and inactivity? The Cnidaria studied so far, such as box jellyfish (famous for its very nasty sting) and hydra (the animal many of us studied in school), fulfil all these criteria and so, by definition, sleep. The point I want to make is that you don't even need a brain to sleep.[58] Which leads me to the next question:

## So Why Do We and Other Animals Sleep?

I have already mentioned some important aspects of sleep, and I will discuss these in more detail in later chapters. The question I want to discuss here is why sleep evolved in the first place. In humans, for example, approximately 36 per cent of our entire lives is spent asleep, and, to state the obvious, while asleep we do not eat, drink or knowingly pass on our genes. This suggests that sleep provides us with something of profound value. When we are deprived of sleep, the sleep pressure becomes so powerful that it can only be satisfied by sleep. As a result, many researchers have assumed that there must be a single overarching role for sleep embedded deep within our biology. Others suggest that sleep has no intrinsic value at all but represents a by-product of some other truly adaptive trait yet to be discovered. I want to tackle this question with my own personal view on the subject, and I will start by asking two questions.

### Why has almost all life evolved a 24-hour circadian pattern of activity and rest?

Almost all life shows a 24-hour pattern of activity and rest, even bacteria.[59] It seems very likely that this rhythm evolved as a result of living on a planet that rotates once every 24 hours and that the resultant changes in light, temperature and food availability forced an adaptive evolutionary response.[60] Diurnal and nocturnal species have evolved numerous specializations that have allowed them to perform optimally under the different conditions of light or dark, but, critically, not both. Life seems to have made an evolutionary 'decision' to be active at a specific part of the day/night cycle, and, as a result, those species that are specialized to be active during the day will not be particularly effective at night. In the same way, nocturnal animals that are exquisitely adapted to move around and hunt under dim or no light fail miserably during the day. The struggle for existence has forced species to become specialists and not generalists, and no species can operate with the same effectiveness across the 24-hour light/dark environment.

*What are the important processes that occur during sleep?*

Given that there is a 24-hour rhythm of activity and rest, we need to address what happens during the physically inactive state of sleep. Overall, sleep may be the suspension of most physical activity, but critical and essential physiology occurs during this time at every level of our biology. For example, many diverse cellular processes associated with the restoration and rebuilding of metabolic pathways are known to be up-regulated during sleep;[61] toxins that build up as the by-product of activity are processed and made safe and ready for removal during sleep;[62] in humans and other animals with the ability to learn, information received during the day is processed during sleep and new memories and even new ideas are established. Indeed, 'sleeping on a problem' really can help the human brain find new solutions to difficult problems.[63] In short, during sleep the body performs a broad range of essential biological functions without which performance and health fall apart rapidly. Such critical activities are demanded for survival and need to occur at some point during the day/night cycle, and in my view evolution has allocated these key biological activities to the most appropriate slot of the sleep/wake cycle. So, if you have a complex brain, or even a simple nervous system, memory consolidation occurs *after* activity, during sleep, when the brain is not being swamped with new sensory information and has the capacity and available energy to perform the task optimally. In the same way, toxin clearance and the rebuilding of metabolic pathways need to occur after the toxins have built up and energy has been used. The compartmentalization and ordering of these events in time promote incredible efficiency. It's a bit like the production line in a factory, where manufactured objects pass through a set sequence of mechanical or manual operations in the correct order and at the correct time.

With these two questions in mind, I edge closer to my definition of sleep. It is not known why humans sleep on average eight hours a day, or why some animals are reported to sleep for 19 hours and others for only two hours. But surely this must relate to a complex set of competing interactions. To survive and thrive, individuals need to balance the essential requirements of gaining sufficient food and water, and the production and nurture of offspring, with the problems of physical survival such as encountering predators or pathogens. Once a stable rest/activity pattern has evolved for any species, then essential biological processes will be incorporated into this time structure at an appropriate point. In short, sleep has evolved as a species and developmentally specific response to a 24-hour world in which light, temperature and food availability change dramatically. So my answer to the question of why we sleep:

> *Sleep is a period of physical inactivity during which time an individual avoids movement within an environment to which it is poorly adapted, while using this time to undertake a range of essential biological activities that allow optimum performance during activity.*[64]

As a colleague commented to me recently after discussing this definition: 'So, sleep is a bit like the weekend: there is no single function. It is a time that is used for many different activities.' I agree, making it a highly flexible part of our biology that precludes a simple, one-dimensional definition. It's a bit like asking: 'Why are we awake?'

## Questions and Answers

### 1. What is 'local sleep'?

This is an important question. I have said that sleep and conscious-ness occur as a global brain 'flip-flop' between these different states. But this is not completely accurate. Fairly recently, a state called 'local sleep' has been described whereby a small area of the brain during the wake state shows an electrical activity that resembles the sleep state. My colleague in Oxford Vladyslav Vyazovskiy was one of the first to demonstrate this phenomenon. Rats were kept awake, and the electrical activity of brain cells in the cerebral cor-tex was monitored and shown to go briefly 'offline', with a local EEG resembling slow-wave sleep. Remarkably, while one group of neurones were 'asleep', an adjacent area was 'awake'. These periods of 'local sleep' increased the longer the rats were kept awake. So local populations of neurones in the cerebral cortex can fall asleep.[65] The reason for local sleep is unclear, but it probably allows some local restorative process as a result of extended sleep deprivation.

### 2. What is CBD and does it help you sleep?

CBD stands for cannabidiol. It is an active ingredient of cannabis (marijuana). Unlike tetrahydrocannabinol (THC), it won't get you 'high'. Early results are promising for reducing anxiety and for improving sleep, but large-scale trials are needed.[66] Some CBD manufacturers have come under government scrutiny for wild, indefensible claims, such as that CBD can cure cancer, for which there is no evidence. Because CBD is mostly available as an unregulated supplement, sometimes it is difficult to know exactly what you are getting, so if you decide to try CBD, talk to your medical practitioner, not least to make sure it will not affect other medications you are taking. Lifestyle changes to improve sleep are nearly always preferable to drugs or supplements, sub-ject to your medical practitioner's advice (chapter 6).

## 3. What is diphenhydramine and should you take it long term for sleep problems?

Diphenhydramine is an antihistamine used most commonly to relieve symptoms of allergy and hay fever, but it can also be used as a sleep aid. In the context of sleep, histamine acts as an excitatory neurotransmitter in the brain to promote wakefulness. Diphenhydramine, also known as 'Nytol', 'Benadryl' or 'Sleepeaze', acts both as an antihistamine blocking the action of histamine (promoting sleep) and as an anticholinergic, blocking the action of the neurotransmitter acetylcholine, again promoting sleep. Diphenhydramine is used for the treatment of allergic reactions, but because it blocks / reduces the action of histamine and acetylcholine it has sedative properties, and so is widely used as a non-prescription sleep aid. As a sleep aid it should only be used short term, and, as with other sedatives, lifestyle changes to improve sleep are always preferable. One concern is that because it is an anticholinergic drug, it impairs muscle action, alertness, and learning and memory. In a study on men and women aged 65 or older, people who used diphenhydramine-based drugs were more likely to develop dementia, and the dementia risk increased with extended use. Incredibly, taking diphenhydramine for three years or more was associated with a 54 per cent higher dementia risk than taking the same dose for three months or less.[67]

## 4. Does sleep vary much between mammals?

The short answer is yes! All mammals, including the egg-laying platypus and echidna, show alternating periods of REM / NREM sleep but the patterns vary greatly. For example, horses and giraffes can sleep while standing, but must lie down for short periods of REM sleep, which induces muscular paralysis (atonia), or otherwise they would collapse. The length of sleep varies strikingly between mammals, but there are some general trends. Overall, sleep times decrease with an increase in body size. Also, predatory species (e.g. lions) tend to show more sleep than prey species (e.g. zebra), and mammals that occupy relatively safe places during sleep (e.g. burrows or caves) tend to sleep longer. Large mammals such as giraffes and elephants in captivity spend approximately five hours per day sleeping, but it is unclear if such patterns occur in the wild, where these animals migrate great distances over long periods of time. Sleep studies in captivity versus in the wild on the brown-throated three-toed sloth (*Bradypus variegates*) show that in captivity sloths spend approximately 70 per cent of their time asleep, while in the wild sleep drops to 40 per cent. Similarly, activity/sleep patterns in mice alter dramatically in the laboratory compared to the wild, where they have to find their food and they experience dramatic changes in light and temperature. So we need more field observations and measurements of REM/NREM sleep in the wild to get a better understanding of the importance of sleep across the mammals and other animals.

**5. Is there any truth to the saying 'You need your beauty sleep'?**
Legend has it that if you want to be more attractive, you need to get a good night's sleep, and there may be some truth to this one. Studies have shown that when people are overtired they appear less attractive to other people.[68] This may be because tired people produce more of the stress hormone cortisol. Cortisol rises in response to inflammation in your body, and cortisol can lead to a breakdown in collagen, the connective tissue of your skin, and water retention, so your skin looks swollen and less attractive. In addition, studies suggest that good sleepers recover more efficiently from sunburn than poor sleepers.

**6. How do you sleep when you need to keep moving, like dolphins?**
Special forms of sleep have been described in marine mammals. In fur seals the EEG (brainwave activity) on land is similar to that of most other land-living mammals: both eyes are closed and there are cycles of REM/NREM sleep. In water, however, sleep

is often just in one half of the brain. This is called unihemispheric sleep, in which one side of the brain shows a sleep EEG, one eye is closed and one flipper is largely inactive. So one half of the body appears to be asleep while the other is awake. Fully marine mammals such as the whales and dolphins also show unihemispheric sleep, which seems to allow marine mammals to swim continuously. Recently, porpoises have been shown to undertake a particular dive that may be a potential period of sleep. During these dives, known as parabolic dives, porpoises emit few echolocation clicks which they normally use during hunting. Also, these dives are often to a shallower depth and seem to be intentionally slower.[69] Many birds fly non-stop for days or longer. For example, the great frigatebird can fly over the ocean without stopping for 10 days. And like dolphins they also show a pattern of unihemispheric sleep.[70] There is even a suggestion that crocodiles may show a form of unihemispheric sleep!

## 7. Has COVID-19 had a big effect upon sleep?

At the time of writing this (January 2022) it is still early days to know precisely what is going on, but a survey in 2020 from researchers at King's College London entitled 'How the UK is sleeping under lockdown' suggests that our sleep has certainly changed during lockdown, and the results are mixed. The King's College researchers conclude: 1. Half the population say their sleep has been more disturbed than usual; 2. Two in five people say they've slept fewer hours a night on average; 3. Two in five report having more vivid dreams than usual; 4. Three in 10 say they've slept for longer but felt less rested than they normally would; 5. A quarter say they've slept for longer and felt more rested; 6. People who say they're certain or very likely to face significant financial difficulties due to the disruption caused by coronavirus are more likely to have slept badly; 7. Those who find the coronavirus pandemic stressful are much more likely to experience poorer sleep; 8. Younger people are much more likely than older people to report experiencing changes to their sleep; 9. Men are sleeping slightly better than women. In addition, there are reports in the media suggesting that individuals with sleep problems before the pandemic have experienced worse sleep, and those who were good sleepers have started to experience poorer sleep. Indeed, the terms 'COVID-somnia' and 'coronasomnia' have been used to describe COVID-19-related sleep disorders. So the data emerging suggest that most, but not all, have had worse sleep. But until sufficient data are collected and analysed it is difficult to establish precisely what is going on with our sleep.

### 8. What have we learnt about the brain by studying the electro-encephalogram (EEG)?

This is a matter of fierce debate. Which means it can get a bit nasty. And I think the best answer to this question was relayed to me recently by a colleague as follows: 'Trying to understand the brain by recording the EEG is like trying to understand what happens in a building based on the lights going on and off and counting toilet flushes.' Harsh, but probably true.

### 9. If you die in a dream, does your heart actually stop, do you die for a short time in real life?

I was asked this question recently by a smart eight-year-old called Jacob. It's a fascinating question. I don't know the answer for sure, but I strongly suspect the answer is no. But the idea certainly made me think!

## 3.

# The Power of the Eye

### Entrainment and the cycle of dawn and dusk

Extraordinary claims require extraordinary evidence.

Carl Sagan

In the fourth century BC, Plato argued that we are able to see because light is emitted from the eye and that this light seizes objects with its rays. This was the 'extramission' theory of vision, and, as bizarre as it seems to us today, until the 1500s this was the widely held view in Europe of how the eye worked. To his credit Aristotle (384–322 BC) was one of the first to reject the extramission theory of vision, arguing in favour of the 'intromission' theory, whereby the eye *receives* light rays rather than projecting light into the world. Sadly, this eminently sensible theory from the ancient world was not embraced. Even Leonardo da Vinci in the 1480s first supported the extramission theory, but after dissecting the eye in the 1490s, he switched to the intromission theory. Early observations by Islamic physicians, notably Hasan Ibn al-Haytham, who lived from 965 to 1040 AD and is known in the West as Alhazen, documented that the pupil dilates and contracts in response to different levels of light and that the eye is damaged by strong light. He used these observations to argue correctly that light enters the eye and that light is not emitted from the eye.

The extramission theory was not considered seriously in scientific circles beyond the 1500s, but the idea would not die. A study published in 2002 identified a profound misconception in US college students that the process of vision involves emanations from the eyes, a view consistent with Plato's extramission theory of vision.[71] How can this be? It seems that in the absence of acquired knowledge all of us address a new problem based upon our personal experiences and fragmented stash of facts. This helps explain why around 10 per cent of us still believe that the Earth is larger than the Sun: to someone lacking education – our experience would suggest this – the Sun looks smaller. The point I am trying to make is that when we first try to solve a new problem we draw upon an assortment of personal experiences. Our view is inherently biased. What makes a 'good' scientist is the speed at which preconceptions are abandoned in the face of new knowledge. This chapter is all about how the eyes regulate our circadian rhythms, and, as we shall see, to gain this understanding key, and very long-held, preconceptions and dogma about how the eye works had to be abandoned.

As discussed, one most striking feature of circadian rhythms, across all life forms, is that they are not exactly 24 hours. The 'period' of the circadian rhythm is either shorter or longer than 24 hours. This is easily demonstrated in animals like mice. If a mouse is kept in constant darkness and given a running wheel it will show a daily circadian rhythm in running wheel activity. Mouse rhythms have a period that is a little shorter than 24 hours, and so mice start and end their circadian rhythms of activity in constant darkness a few minutes earlier each day – relative to the 24-hour outside world. As a quick reminder, this drifting pattern of the 'biological day' is called 'freerunning' (Figure 4, p. 103). The demonstration of a freerunning rhythm provides important evidence that circadian rhythms actually exist – they persist under constant conditions with a period that is a bit longer or shorter than 24 hours. If the rhythm were driven by some unrecognized geophysical signal arising from the Earth's 24-hour rotation, the biological rhythm would be exactly 24 hours, or, to be precise, 23 hours, 56 minutes and 4 seconds.

In the 1960s and 1970s, humans were kept under semi-constant conditions and their freerunning rhythms recorded. These experiments suggested that the human clock is around 25 hours. More recent studies, which corrected for some methodological issues, showed that for most of us the circadian clock is a little longer than 24 hours, by about 10 minutes or so.[72] For reasons we do not fully understand, the freerunning rhythms in constant darkness of day-active (diurnal) animals, like us, tend to have body clocks longer than 24 hours, while nocturnal animals, like mice, have freerunning rhythms shorter than 24 hours. This phenomenon even has a name: 'Aschoff's Third Rule'. The key point is that without daily resetting we would get up and go to bed about 10 minutes later each day, and the internal day would soon drift and be out of synch with the environmental day/night cycle. You experience a severe form of this mismatch with jet lag. After travelling across multiple time zones, the body clock and local time do not coincide, and we want to sleep and eat at the wrong local clock time. Given long enough, we catch up with the new time zone, but precisely how?

## The Role of the Eye

For most plants and animals, including us, the most important signal that aligns or 'entrains' our circadian rhythms to the day/night cycle is light, and especially light at sunrise and sunset. We know this because in us and other mammals eye loss prevents this resetting, and people who have lost their eyes because of some terrible accident or a genetic condition, or as a casualty of combat, drift through time, experiencing periods when they get up and go to bed for a few days at more or less the correct time, before they drift off again and want to sleep, eat and be active at the wrong time of day. Such individuals experience something similar to jet lag, but from which they can never recover. I will return to this topic in chapter 14, but to give you some sense of how devastating this can be, here is a quote from a participant in one of our studies who lost his eyes in combat: 'I'm at my wits' end, suffering from variable bedtimes and wake times and I am very often sleepy during the day and awake all night. I'm slowly becoming isolated from my family and friends.'

Let me stress, eye loss prevents light regulating the clock. However, a report in one of the top scientific journals, *Science*, in 1998 by researchers at Cornell University Medical College, New York, suggested that bright light applied to the skin behind the knee can shift circadian rhythms of body temperature and melatonin.[73] A media frenzy followed, and this paper was named by *Science* as one of the year's top studies, and two patented treatments for sleep disorders soon followed. Many of us were deeply sceptical about this study and challenged the findings, not least because eye loss in humans had been shown to block the effects of light on the clock.[74] But maybe we had missed something, perhaps our preconceptions were distorting our objectivity? The scientific method then kicked in, and other groups around the world attempted to replicate the findings using various approaches,[75-7] with one study in 2002 duplicating the methods precisely.[78] After hundreds of thousands of dollars, and a squandering of scientific talent over several years, all the efforts to replicate the findings failed – light applied behind the knee would not shift the circadian system. Errors in the methodological approaches, including the fact that subjects were exposed to light and not in complete darkness during knee illumination, are now thought to have been the basis for the original claim. For reasons I do not understand, this study has never been officially retracted – the formal process whereby a scientific publication is withdrawn due to errors or fraud that have come to light after publication. Many people remember the original study, but are unaware of the later work that failed to replicate the findings. The media tend not to report negative findings. Sadly, over 20 years later, I am still frequently asked: 'Well – what about those light sensors behind the knee?'

Although the eyes detect light for the regulation of the clock, it was not clear *how* this light is detected. This became a bit of an obsession for me. Until relatively recently, and before the research we undertook, the eye was considered thoroughly investigated and viewed as one of the best understood parts of the body. Years of painstaking research had explained how we see: light is detected and processed by a layered structure in the eye called the retina. The first layer of the retina consists of the visual cells, called photoreceptors, of which there are two types: the rods and cones. These cells detect light reflected from objects in the environment and pass their signals onto the cells of the inner retina, which assemble these signals into a crude image. The last layer of the retina is made up of retinal ganglion cells which act to integrate all the light information from the retina. The retina has been likened to a carpet, with the tuft representing the rods and cones, and the weave, or backing of the carpet, the inner retina and ganglion cells. The projections from the ganglion cells form the optic nerve, and send light information from the eye to the brain. The brain then constructs an image of the world in the visual cortex at the back of our brain in the occipital lobe (Figure 2, p. 24). If you get hit on the head, the cells in your occipital lobe get shaken and send out random electrical impulses, which your brain interprets as flashes of light – hence 'seeing stars' after a blow to the head. There are about 100 million rods and cones and one million retinal ganglion cells in the human eye, which indicates how much light information is processed and 'funnelled' as it passes from the rods and cones to the ganglion cells. Crucially, because 'how we see' could be broadly explained by the known understanding of how the eye works, it was naturally assumed that the rods and cones communicate light information to the clock. But this assumption was wrong.

Astonishingly, the visual cells of the retina – the rods and cones – are *not* needed for the detection of the light/dark cycle. There exists a third class of light sensitive cell (photoreceptor) within the eye. This discovery was driven forward by a very simple question in my mind – I simply could not understand how the rods and cones, so exquisitely evolved to generate an image of the world, could also be used to extract time-of-day information. For vision to work, the retina has to grab light, and then a fraction of a second later forget this experience and be ready for the next visual image. If this very rapid 'grab and forget' did not occur, then our world would *not* be a series of sharp images, but a constant blur of shades of light, dark and colour. Importantly, a sharp image is not needed by the clock. Instead, the circadian system requires an impression of the overall amount of light around dawn and dusk. Such broad changes in the light environment occur over minutes and hours. So my simple question was: 'How can the eyes be used for both the sensory tasks of vision and the regulation of biological time?'

## Discovering a New Light Detector in the Eye

With this question in mind, my team undertook a series of studies throughout the 1990s on mice that had genetic mutations which prevented the rod and cone photoreceptors developing or working properly. The mice were visually blind.[79-80] Such mice were being studied by other researchers at the time to try to understand the genetic basis of human eye disease. We used these animals to see if circadian entrainment (photoentrainment) to the light/dark cycle was affected by profound visual blindness caused by the loss of rods and cones. Over almost a decade of careful study, and using mice with different types of genetic disease, we found to our excitement that visually blind mice lacking rods and cones could still regulate their circadian rhythms to light perfectly normally. Not only could these mice entrain, but they did so with the same sensitivity to light. But when the eyes were covered, photoentrainment was abolished.[81-2] So, to be absolutely clear, mice lacking rods and cones were visually blind but not 'clock blind'. On the basis of these data we suggested that there was another photoreceptor in the eye, quite separate from the rods and cones, that detects light for the regulation of biological time. To my genuine surprise, this suggestion was first met with undiluted contempt by many in the vision community. On one occasion I gave a scientific talk, and a member of the audience shouted 'Bullshit!' before walking out, and another time a very angry individual shouted 'Are you seriously trying to tell us that after 150 years of research on the eye we have all missed an entire class of photoreceptor?' My early grant proposals were rejected because our results were just not believed. One particularly painful reason for rejection was: 'Why is Foster looking for novel photoreceptors within the eye, when we know the light sensors are located behind the knee?' It was not a happy time, and I bought many lottery tickets in an attempt to fund my research! I had no luck with that strategy either. However, this experience highlighted for me both the strength and the weakness of the scientific approach – the strength is that scientific progress demands overwhelming proof with verification by other scientists, which is why science can never be a solitary pursuit. But the weakness of the scientific approach is that there is usually embedded resistance to change, which can slow progress and stifle innovation. The challenge is to get the balance right between new scientific progress and entrenched dogma. The solution to the dogma we encountered was to do an ever-increasing number of experiments that provided more and more data. It was this that ultimately changed attitudes. The journal *Science* published two of our key studies,[81-2] and rapid progress soon followed.

Because I was originally trained as a zoologist, I have always been keen on the 'comparative method', which basically means comparing the features of one species to learn about some aspect of a second species. When I was based at the University of Virginia we worked on fish, lizards and mice. On one occasion 'Dan the Snake Man' from Louisiana sent us some *Anolis* lizards, which we were studying at the time. But by the time they got to the post office in Charlottesville the box containing the *Anolis* had fallen apart, releasing the lizards. These small, beautiful and completely harmless animals terrified the postal workers, who thought they were snakes, and the US postal service in Charlottesville was closed for several hours before the lizards were rounded up. If you visit Florida, these are the lizards you will often see running across the path before dashing up a tree. Our work on fish, which started in Virginia, continued when I returned to the UK, and our findings are highly relevant to this story. We showed that the eyes of fish had a new type of light-sensitive molecule, and that this photopigment was not found in the rods and cones, but in other cells of the eye, including the ganglion cells (those cells whose projections form the optic nerve). We had discovered a previously unknown photoreceptor in the eye.[83] This was really important, because if fish had novel photoreceptors in the eye, it was not such an absurd idea that mammals might also have a similar system. The comparative method had provided us with 'proof of concept' for the existence of another type of light detector within the eye. But mammals are not fish, and the vision community was still deeply sceptical. However, our work was finally being taken seriously, and a few other researchers got interested, joining the hunt for these new photoreceptors in mammals. Finally, studies on rats[84] and our studies on mice[85] showed that a small number of retinal ganglion cells respond to light directly. We have called these new photoreceptors 'photosensitive retinal ganglion cells', or pRGCs. A flurry of activity followed, by our group and others around the world, showing that about one in a hundred ganglion cells were pRGCs and that they form a 'photosensitive net' within the eye, capturing light from all directions in space to provide an overall measure of environmental brightness. Yet more work showed that the light sensitivity of the pRGCs is based upon a newly discovered blue-light-sensitive molecule (photopigment) called 'melanopsin', or OPN4.[86-7] This photopigment molecule was originally isolated by Ignacio ('Iggy') Provencio, from the light-sensitive pigment cells or 'melanophores' found in the skin of frogs and toads,[88] hence the name 'melanopsin', and the name stuck. But, sadly, it is often confused with 'melatonin', the main hormone produced by the pineal gland! Iggy was the first PhD student I supervised; indeed, it was Iggy who introduced me to 'Dan the Snake Man'. Iggy was, and still is, one of the nicest people on the planet and is now a full professor at the University of Virginia, back where we originally worked together.

In several key experiments our group showed that OPN4 and the pRGCs are maximally sensitive to blue light,[86-7] which also turns out to be true for all animals, including us. But why? The likely answer seems to be related to their role as dawn/dusk detectors. During daytime the sky is dominated by sunlight formed from light across the entire visible spectrum (violet to red), but as the Sun sets and the disc of the Sun disappears below the horizon, different colours (wavelengths) of light are scattered by particles in the atmosphere. The result is that light immediately at the horizon becomes enriched in yellow and red light, while the entire dome of the sky becomes enriched in blue light. Basically, the particles in the atmosphere at dusk act like a prism to split up the different wavelengths of light. The same happens at dawn. As the Sun rises above the horizon, blue light is scattered to fill the dome of the sky while red and orange dominate the horizon. We think that the pRGCs are most sensitive to blue light because this is the dominant colour of light at dawn and dusk. Which makes them ideal as dawn/dusk detectors.[89] The facts fit, but we can't know for sure.

## How Much Light Do We Need and When Do We Need to Get It?

Until 1987, the assumption was that human circadian rhythms are entrained by 'social cues', such as when we eat and interact with others. Part of the reason for this belief was that the early studies on humans failed to show any effects of light on human circadian entrainment. But we now know that the human circadian system is *very* insensitive to light compared to that in other animals like mice. Levels of light that will regulate the mouse circadian system can be as low as 0.1 lux (Figure 3, p. 61), but such intensities are completely ineffective in humans. The first clear demonstration of photoentrainment in humans used light levels of 5,000 lux for many hours.[90] The diagram in Figure 3 provides a rough approximation of the light levels in the environment and the sensitivities of different responses driven by rods, cones and the pRGCs in us.

Our notable insensitivity to light compared to mice and other rodents may result from the difference between being a day-active (diurnal) and a night-active (nocturnal) mammal. Diurnal mammals are exposed to light across the day, while nocturnal mammals, emerging from their burrows at twilight, experience low levels of light for a relatively short period of time before the Sun disappears. Because of this, increased sensitivity to light at dawn and dusk would be an advantage for a burrow-living nocturnal animal with a short window of opportunity to detect light. In humans there is also the additional problem of artificial light. Estimates vary but the controlled use of fire by our ancestor *Homo erectus* is thought to have begun some 600,000 years ago. If our ancestors were as sensitive to light as rodents, firelight at night would have been a major disruptor of circadian rhythms. So maybe our ancestors had to evolve decreased circadian light sensitivity because of the use of fire?

In 1800, most of human society in Europe, the USA and the rest of the world worked outside and was exposed to the natural cycle of light and dark. Today, in the UK, the proportion of workers in agriculture and fishing is approximately 1 per cent of the working population.[91] This illustrates how the vast majority of the workforce in the developed and developing economies have become profoundly detached from environmental light, and

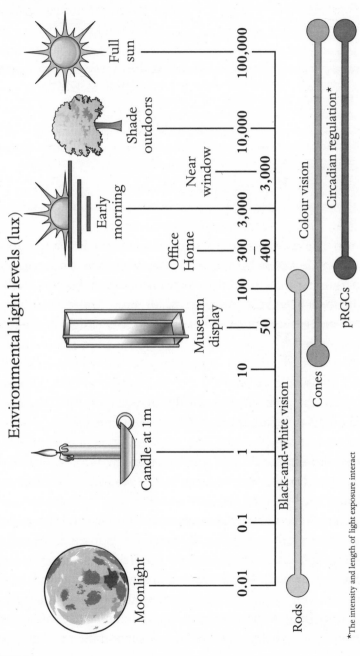

Figure 3.

**Figure 3. Levels of light encountered in the environment and the approximate sensitivities of rod, cone and pRGC photoreceptors in humans.** Levels of light are indicated in lux, which is a standard measure of light intensity. Under dim light our rods provide us with our 'black-and-white vision'. Colour vision needs brighter light and is provided by cone photoreceptors. Between 10 and 100 lux, both the rods and cones operate, but as light levels increase the rods become saturated, and cones give us our sense of high-contrast colour vision. By around 100 lux the pRGCs begin to operate, but require long-duration light exposure of many minutes and even hours. By contrast, the rods and cones achieve their sensitivity with light exposures in the millisecond (one thousandth of a second) range. So although there is plenty of light for colour vision in our domestic and work spaces inside (50 to 400 lux), there is often insufficient light for robust photoentrainment. Also see 'Questions and Answers', question 8, on page 69. You may want to buy a lux meter for around £20, or download a free lux app onto your mobile phone, and convince yourself how profound the differences in light intensity are between natural light outside and indoor artificial light.

experience only an attenuated light/dark signal for photoentrainment. Most of the time we simply don't get sufficient bright light for long enough to robustly entrain the clock to the solar day (Figure 3). Especially in winter. But it is not just the intensity of light that is important. The time of light exposure is absolutely critical too.

## The Timing of Light Exposure – the Importance of Dawn and Dusk

When we encounter light around dusk and in the early evening this will delay the clock in the SCN, and the effect is to make us go to bed later and get up later the next day. By contrast, early-morning light will have the opposite effect, advancing the clock, making us go to bed and get up earlier. Light in the middle of the day has little, and some would say no, effect in shifting the clock. This highlights the point that it is dawn and dusk light detection that is most important for entrainment. When we were all agricultural workers, we lived by the cycle of dawn and dusk, and the circadian system was 'pushed' backwards and then forwards on a daily basis and aligned appropriately to sunrise and sunset. But in an urban environment dawn and dusk exposure can be very variable. Our recent study on university students from around the world showed that the late chronotypes ('owls') were exposed to light in the evening (delaying light) but encountered little light in the morning (advancing light). The net effect was to shift the body clock to a later time.[92]

A more detailed, and now famous, study from the USA examined sleep/wake timing in individuals after one week of maintaining daily routines of work, school, social activities and exposure to electrical lighting compared to one week of outdoor camping (in tents) and exposed to natural light in the Rocky Mountains. After the week of exposure to natural lighting in the mountains, and especially morning light exposure, circadian timing and sleep/wake cycles had advanced by two hours – people were actually getting up and going to bed two hours earlier after just one week of natural-light exposure.[93] Before I finish this section, I should say that while light is the dominant signal to entrain the clock, exercise and eating at specific times can also influence entrainment, and I will discuss this in later chapters.

## What is the Effect of Light-Emitting Screens?

Just to remind you, exposure to light around dusk and in the evening will delay the circadian system (make you go to bed later and get up later the next day); while light in the morning will advance the circadian system (make you go to bed earlier and get up earlier the next day). This fact has been used to support the argument that computer or smartphone use prior to bedtime will disrupt sleep/wake timing, and so delay sleep. Additional support for this view comes from the fact that these devices emit light relatively enriched in blue light which will maximally stimulate the pRGCs.[94-5] In the popular press, fuelled by some scientists, the use of electronic devices before bedtime is 'known' to alter our circadian rhythms. It is remarkable to me that many researchers hold the mutually exclusive view that the human circadian system is relatively insensitive to light, yet at the same time is likely to be sensitive to the dim light emitted by electronic devices. And the data do not support the claims. The most detailed study to date compared reading a light-emitting (LE) eBook in dim room light for about four hours (6–10 p.m.) before bedtime for five consecutive evenings to reading a printed book under the same conditions. The light emitted from the LE eBook was around 31 lux, while the light reflected from the printed book was around 1 lux. The results showed that using the LE-eBook delayed the beginning of sleep by less than 10 minutes after five days, compared to reading the print book. Although the results were statistically significant (just), a delay of 10 minutes is almost meaningless.[96] Nevertheless, this paper is cited as evidence that LE eBook use before bed will have a major impact upon our circadian rhythms. Recently, computer programs have been developed that adjust a computer screen to make it more 'circadian friendly'. The program reduces both the intensity and levels of blue light from the screen in the evening compared to the rest of the day. The argument is that this will improve sleep and reduce circadian rhythm disruption. Although such programs have been widely and positively reviewed by technology journalists, bloggers and users, the program itself has not, as yet, been tested rigorously; though one recent study suggested that the effects were, at best, small.[97] Perhaps these programs have been welcomed so enthusiastically because they work well for visual comfort. In the evening our eyes are biologically adapted to lower light levels, so a dimmer computer or smartphone screen may be easier on the eye.

Although the likely impact of light exposure from screen use on photoentrainment is small, if there is any, studies *have* shown that late-night use of technology-related activities such as computer games, writing emails and using social media does increase brain alertness and delays sleep, leading to daytime sleepiness and poor performance.[98] This is a particular problem for teenagers,[99] many of whom actually fall asleep in class because of using social media or playing games into the early hours of the morning on a school night. The bottom line is that devices should not be used for at least 30 minutes before bedtime – not because of the light they produce, but because of the alerting activity they induce within the brain.

## A New Understanding of Blindness

The finding that we also have pRGCs that regulate our circadian system[100] is having an important impact for ophthalmologists in eye hospitals across the world. You can be visually blind, but not clock blind. There are many genetic diseases of the eye that result in the loss of the rods and cones and cause visual blindness, but these diseases often spare the pRGCs – as in the mice we originally used to discover these photoreceptors. Under these circumstances, individuals who have their eyes but are visually blind should be advised where possible to expose their eyes to sufficient light to regulate their circadian system. It is clear that every effort must be made to save damaged eyes with functioning pRGCs, even if the individual is visually blind. The growing appreciation that the eye provides us with both our sense of space (vision) and our sense of time (via clock regulation) is now redefining our definition and treatment of human blindness. I hope the person who shouted 'Bullshit' to me all those years ago now feels able to apologize!

## Questions and Answers

**1. Can other environment signals, other than light, entrain circadian rhythms?**

Yes. Although light is the most powerful signal to entrain the master clock in the SCN, exercise can also have an effect (see chapter 13). In addition, the 'peripheral clocks' can be entrained by eating food at a particular time (see chapter 13).

### 2. Can the rods and cones contribute to the regulation of the clock?

Yes. Originally the visual photoreceptors were not thought to play any major role in circadian entrainment, because their loss did not affect the ability of mice to entrain to light. There was no loss of sensitivity. So entrainment was all thought to be mediated by the pRGCs. But we now know that under certain circumstances the rods and cones can send light signals to the pRGCs and 'support' entrainment. The contribution of the rods and cones is still being worked out, but the current hypothesis is that there is an integration of light signals within the pRGCs so that the rods can be used for dim-light detection, cones can be used for the detection of higher light intensities and for the integration of intermittent (flickering) light exposure, and the pRGCs provide information regarding bright light over longer periods of time.[89]

### 3. Do you have to be exposed to 'blue light' to entrain the circadian system?

No. If the light is bright enough, as encountered in the natural world (see Figure 3), broad-spectrum 'white light' will be sufficient to bring about entrainment. The pRGCs are maximally sensitive to blue light, but their response is like a 'bell-shaped curve', and while less sensitive to green, orange and red light the pRGCs can still detect these wavelengths if the light is bright enough. Most light, either natural or artificial, is 'white', composed of multiple colours (all wavelengths). Blue light is only important when light levels become low, as at twilight (see Figure 3).

## 4. Can light change your alertness?

Yes, bright light, probably detected by the pRGCs, can change your levels of alertness and mood. In nocturnal animals such as mice and rats, light stimulates them to seek shelter, reduce activity and even sleep, while in diurnal species such as ourselves light promotes alertness and vigilance.[101] So circadian patterns of activity, and hence our sleep/wake cycles, are not only entrained by dawn and dusk but also driven directly by light itself. This direct effect of light on activity has been called 'masking', and, with the circadian system, restricts activity to that period of the light/dark cycle which is optimal for survival.[102] I will discuss the alerting effects of light in more detail in chapter 6.

## 5. In neonatal intensive-care units, the lights are nearly always on – is there any evidence that a light/dark cycle might be important for babies in intensive care?

There is increasing evidence that providing a light/dark cycle for preterm babies in the neonatal intensive-care unit has significant benefits, including increased weight gain and a reduced stay in hospital compared to infants who experienced constant light or constant near darkness. The SCN and the projections from the eye to the SCN appear to be formed by 24 weeks, so it is possible that light/dark information could be regulating the circadian system in preterm babies and influencing their development. It would seem sensible that where possible babies in the neonatal intensive-care unit should be exposed to a stable 24-hour light/dark cycle.[103] The same is, of course, true for adults in an intensive-care unit.

## 6. When we die our pupils dilate: has this got anything to do with the photosensitive retinal ganglion cells (pRGCs)?

The pRGCs regulate more than just our circadian rhythms. We have shown that the pRGCs are indeed important for regulating pupil size.[86, 100] They also detect light to regulate alertness, mood and a variety of other functions.[89] But I don't think they are involved in pupil dilation at death. One of the first things to happen when we die is that all of the muscles in the body relax, a state called 'primary flaccidity'. Pupils dilate, eyelids relax and often open, jaw muscles relax and the jaw may fall open, and other joints become flexible. From two hours after death, the muscles begin to stiffen and rigor mortis starts, peaking after twelve hours.

## 7. What happens during space flight and could our circadian clocks adapt to living on other planets, like Mars?

This is a serious issue, and one that NASA is studying in detail. The International Space Station travels at 27,520 kilometres per hour, and orbits the Earth every 90 minutes. Which means sunrise occurs every 90 minutes and astronauts experience 16 dawns and dusks every 24 hours. Sleep/wake cycles are highly disrupted and sleeping tablets are reported to be the most frequently used medication on the space station. I received funding from NASA in the 1990s and remember talking to engineers at a meeting in Huston who wanted to remove the windows from the space station and in any spacecraft travelling to Mars, to be replaced by artificial lights set to Earth time. Understandably, the astronauts flatly refused to have their windows taken away! Living on other planets with a different 'day' to Earth also represents a major problem. Mars rotates a bit more slowly than Earth, with a solar day ('sol') of 24 hours, 39 minutes. This may be within the 'range of entrainment' for many individuals with a longer chronotype, but those with a short chronotype might struggle to entrain to the Martian sol. When a new Mars Rover arrives on Mars, the scientists and engineers try to work on Mars time, but in the early days the conflict between trying to stay on Mars time while also experiencing Earth time caused major problems, including sleepiness, irritability, and decreased concentration and energy. More recently, scientists were given the opportunity to work exclusively on an artificial Martian sol, and were isolated from the Earth's light/dark cycle. Of those who took part, 87 per cent adapted to the Martian sol. So the circadian rhythms of most of us could probably adapt to Mars. The sol on Mercury is 1,408 hours and Venus is 5,832 hours. Putting aside the hideous nature of the physical environment on these planets, another reason not to go would be that our circadian rhythms could never adapt.

**8. So, it looks like the light from LE ebooks does not have much of an effect upon our circadian entrainment – but what about room light in the evening? Could this act to delay our circadian rhythms?**

This is quite a difficult question to answer, and my colleague in Oxford Stuart Peirson and I discuss it endlessly. Not least because the experiments that could answer this question have yet to be done! The first point to make is that because room light in the evening can be very variable (50 to 300 lux; see Figure 3) it is difficult to generalize. And in addition to room brightness, any effects on the clock will depend upon how long you are exposed to the light; the direction of the light; the colour (wavelength) of the light; your age; the time of day you detect the light – early evening versus late evening; how much light you have been exposed to during the day; the season – summer versus winter; and the intensity of light you encounter the following morning. In laboratory studies on young individuals, around 50 to 100 lux for more than two hours in the evening will have a small but significant effect upon the clock. More than 1,000 lux for two hours or more will have a big effect. Whether these findings translate to the 'real world', and how they apply to different age groups, remains unknown. The consensus at the moment is that domestic light in the evening is likely to have some impact upon our circadian entrainment, acting to delay the clock. This action of light will make us get up a bit later and go to bed a bit later the next day. As a result, it makes sense to keep light levels fairly low (less than 100 lux) in the few hours before bedtime. Keeping the lights low at this time has the added advantage of reducing our levels of alertness. Reduced alertness will prepare us for sleep.

# 4.

# Out of Time

## The nightmares of stress, shift work and jet lag

Science and everyday life cannot and should not be separated.

Rosalind Franklin

In 1956, John Foster Dulles, the American Secretary of State, made a long-haul flight to Cairo to discuss whether the USA would finance the Aswan Dam. Upon arrival in Cairo he was unable to concentrate during the negotiations. Dulles then flew directly back to Washington. When he arrived, he learned that the Egyptians had just bought a large amount of Soviet armaments. Without due consideration, Dulles cancelled his agreement with the Egyptian President to build the Dam. As a result, the Aswan Dam was built by the Soviet Union and this gave the USSR its first foothold in Africa during the Cold War. Dulles said his failure to keep an ally and deliver a lasting deal with Egypt was due to 'travel fatigue'. The term 'jet lag' had not yet been invented in 1956. In view of this early and famous example of jet lag, it is entirely fitting that Washington Dulles International Airport is named after John Foster Dulles.

Henry Kissinger was President Richard Nixon's national security assistant and Secretary of State between 1969 and 1977, and he travelled a great deal. But unlike Dulles he was more aware of the dangers of jet lag, and recalling his negotiations with the North Vietnamese he said: 'When I went directly from a transatlantic flight into talks, I found I was on the verge of losing my temper at North Vietnamese insolence – nearly falling into their trap by playing the role they had assigned to me. From then on I never began a negotiation immediately after a long flight.' The individuals who organized President George Bush's Asian tour in 1992 seem to have ignored, or been unaware of, the consequences of jet lag. The poor man arrived in Japan towards the end of a 12-day, 26,000-mile tour of four Asian countries. At a state dinner in his honour, the President suddenly fell ill and vomited on himself and his host, the Japanese Prime Minister. All this was captured on film and broadcast around the world. It is claimed that this memorable incident helped Bush's rival, Bill Clinton, win the presidential election later that year. Clearly, we ignore our circadian rhythms at our peril.

Jet lag is a familiar experience for very many decision makers in the political and business sectors, and is a classic example of sleep and circadian rhythm disruption, or SCRD. As I mentioned in the Introduction, it is often difficult to disentangle the impact of circadian rhythm and sleep disruption – the two are inextricably linked. This led my team to develop the term SCRD as a shorthand for these phenomena. SCRD will be used throughout the following chapters to indicate the collective impact of disrupting these biological systems, and, as we shall see, the impacts of SCRD are much more than the inconvenience of feeling tired at an undesired time. Short- and long-term SCRD can cause major problems across all the key areas of health.

As a brief reminder before we dig deeper, circadian rhythms fine-tune our physiology and behaviour to the varied demands of the 24-hour day. Sleep is an extension of this process, re-equilibrating and refining our biology, to allow an optimal performance during consciousness. These two interlinked and balanced systems define much of our ability to function, and if we work against this 'deep biology' we compromise every aspect of our health. I now want to explore the main consequences of SCRD in detail, and especially how SCRD and stress are interlinked.

## *Stress, Cortisol and Adrenaline*

Cortisol is released from the adrenal glands (adrenal cortex), which are located on top of each kidney. Cortisol plays a key role in regulating our metabolism by managing how we utilize carbohydrates, fats and proteins (chapter 12). It also acts to reduce inflammatory responses, increase blood pressure and increase levels of alertness. Cortisol is often called the 'stress hormone', which is a bit misleading as it is not just released during stress. The circadian system regulates cortisol release (Figure 1, p. 20), which rises before we wake, anticipating increased activity in the morning, and then declines towards sleep and overnight. In short, rhythms in cortisol helps our bodies anticipate the varied metabolic demands of activity and sleep.[104] If, however, we become 'stressed', then cortisol release overrides this daily pattern. Short-term stress, such as being chased by a hippopotamus (around 3,000 people are killed each year by hippos) or threatened by a mugger, will drive up cortisol, override the circadian control, and get the body ready for an emergency response of either fight or flight (run away!). Longer-term stress, or 'harmful stress', is defined as 'a physical, cognitive or emotional stimulus that results in impaired health or performance'. The problem with long-term stress is that the emergency response is switched to the 'on state' for long periods and this is unsustainable. Such stress is a bit like putting a car engine into first gear – it gives you an immediate and helpful acceleration, but if you leave the engine in first gear for too long you destroy the engine: an experience that will be familiar to many who have taught a family member to drive! Also very stressful. Let's explore the hormones associated with the stress response in more detail.

## Cortisol and SCRD

SCRD, of the sort experienced as a result of night shift work, jet lag or being chronically tired is a very potent and harmful stressor, and sustained SCRD, leading to increased continual cortisol release, contributes to:

### BLOOD SUGAR IMBALANCES AND DIABETES

Increased cortisol over long periods increases blood sugar levels.[105] Cortisol causes this by opposing the action of insulin produced in the pancreas. Insulin normally acts to remove glucose from the circulation (see chapter 12 and Figure 9, p. 275), so by opposing the action of insulin, there is increased production of glucose (gluconeogenesis) by the liver and other organs, leading to higher levels of glucose in the bloodstream. In an emergency, this would be immensely useful as it can fuel the muscles for 'fight or flight'. But if the glucose is not being 'burnt' as fuel for activity, it is converted into stored fat in the adipose tissues. With high levels of glucose in the blood, the pancreas attempts to produce more insulin to remove the glucose, and this can lead in turn to more cortisol production. Eventually the pancreas can't cope and this can result in major metabolic abnormalities such as Type 2 diabetes and obesity.

### WEIGHT GAIN AND OBESITY

High levels of cortisol lead to weight gain, particularly in the form of fat storage around the gut (chapter 12 and Figure 9). This is because cortisol mobilizes the liver to produce more glucose, but if the glucose is not metabolized it gets converted into stored fat in fat cells (adipose tissue) under the muscles and around the abdomen. The second way cortisol is involved in obesity is because cortisol directly alters appetite sensitivity and increases the cravings for high-calorie sugar-rich foods.[106] Cortisol also seems to influence two gut hormones, leptin and ghrelin.[107] Leptin is produced by fat cells and is a signal for *not* being hungry (satiation). Ghrelin is produced by the stomach and signals hunger, particularly for sugars (chapter 12 and Figure 9). Together, these hormones regulate hunger and appetite. SCRD via elevated cortisol causes leptin levels to fall and ghrelin levels to rise, driving an increased appetite for fatty and sugary foods.[108]

### IMMUNE SUPPRESSION

Long-term exposure to cortisol suppresses the immune system. The result can be an increased risk of infections, colds and even cancer (chapter 10). There is also an increased risk of developing food allergies, digestive problems and autoimmune diseases. This is linked to the fact that a healthy gut depends upon a healthy immune system.[109]

### GASTROINTESTINAL (GUT) PROBLEMS

Cortisol suppresses gut activity and digestion. Indigestion then develops, and the lining of the gut becomes irritated and inflamed. Ulcers can develop. Irritable bowel syndrome and colitis are more common in individuals experiencing SCRD.[110] Some of these problems come from altered gut bacteria, which I discuss in chapter 13.

### CARDIOVASCULAR DISEASE

Cortisol also seems to play a central role in the regulation of blood pressure, acting mainly on organs such as the kidney and the colon to increase the amount of salt (sodium) reabsorbed into the bloodstream and to increase the amount of potassium excreted in the urine. This causes water to be reabsorbed into the blood, which increases blood volume and hence blood pressure. Cortisol also constricts blood vessels which increases blood pressure even further. For 'fight or flight' this would be useful, increasing the delivery of oxygenated blood and nutrients to the muscles and the brain. Long-term blood vessel constriction and high blood pressure lead to blood vessel damage and the build-up of plaques, which are fatty and waxy deposits that form on the artery walls (chapter 10). These deposits narrow the artery and reduce blood flow, and this is called atherosclerosis, or 'hardening of the arteries'. Atherosclerosis is the most common cause of heart attack and stroke, and is frequently associated with SCRD.[111]

### MEMORY RECALL

Although cortisol under normal circumstances can increase our ability to acquire memories, high levels of cortisol prevent us from retrieving those memories.[112] This is something that many of us have experienced during an examination or interview. In addition, high levels of cortisol in the middle years, driven by SCRD, may even contribute to the development of dementia in later life.[113-14]

SCRD increases our cortisol-related health risks. But there is huge variation between individuals,[115] some of which can be linked to age. For example, cortisol levels increase as we age, and are higher in older women than in older men. These increased levels of cortisol in the aged have been linked to increased levels of stress in social situations and poorer cognitive performance. Indeed, age-related increases in cortisol may even contribute to atrophy of the structures in the brain that help us store memories – such as the hippocampus (Figure 2, p. 24).[116] In view of the problems associated with elevated cortisol, and particularly as we age, this may be a contributing factor to why many people find it more difficult to undertake shift work later in life – they find it more stressful and more damaging to health.[117]

### Adrenaline and SCRD

There is more to harmful stress than just cortisol. SCRD also activates the sympathetic branch of the autonomic nervous system (that part of the nervous system responsible for control of

unconscious bodily functions, such as breathing, heart rate, digestion, etc.), which sends connections to the inner part of the
adrenal gland, the medulla, and stimulates the release of adrenaline, also known as epinephrine. Like cortisol, adrenaline release
is regulated by the circadian system and is high during the day
and declines towards sleep,[118] preparing the body for the different
demands of activity and then sleep. SCRD overrides this timing
and raises adrenaline. Raised adrenaline and cortisol work
together to drive the 'fight or flight' responses. Sustained release
of adrenaline resulting from harmful SCRD-induced stress aug-
ments the action of cortisol and the problems listed above. In
addition, adrenaline causes the air passages in the lungs to dilate,
allowing more oxygen into the lungs to oxygenate the blood,
which can sustain muscle activity. Adrenaline also causes the
blood vessels to constrict, like cortisol, which increases blood
pressure, and adrenaline directs blood to the muscles of the legs,
arms, heart and lungs. All this leads to a marked increase in
strength and performance. Pain sensations are decreased, allowing you to keep running or fighting even after injury. Adrenaline
also causes a heightened awareness and edginess.[119]

The stress responses seen in people with SCRD and outlined
above[120] can be simulated in the laboratory. For example, in one
study, healthy young men were only allowed to sleep four hours
over six consecutive nights. After only six days, cortisol levels
increased very markedly, particularly in the afternoon and early
evening, a time when cortisol levels normally drop (Figure 1).[121-2]
It would be interesting to repeat these experiments, carefully, in
older people, who naturally have higher levels of cortisol, to see
if the effects of shortened sleep are even worse. The prediction
would be that cortisol responses would be greater in older than in
younger individuals.

Elevated levels of cortisol and adrenaline, as a result of sustained SCRD, can also drive psychosocial stress.[123] This is the

| Acute impact on emotional responses | Acute impact on cognitive responses | Chronic impact on physiology and health |
| --- | --- | --- |
| *Increased* | *Impairments in* | *Increased risk of* |
| Fluctuations in mood | Cognitive performance | Daytime sleepiness |
| Irritability | Ability to multi-task | Micro-sleeps |
| Anxiety | Memory consolidation | Cardiovascular disease |
| Loss of empathy | Attention | Altered stress response |
| Frustration | Concentration | Altered sensory thresholds |
| Risk-taking and impulsivity | Communication | Infection, lowered immunity |
| Negative salience | Decision making | Cancer |
| Stimulant use (caffeine) | Creativity and productivity | Metabolic abnormalities |
| Sedative use (alcohol) | Motor performance | Type 2 diabetes |
| Illegal drug use | Social connectivity | Depression and psychosis |

Table 1. **The impact of SCRD upon human biology.** The impact of SCRD on our acute emotional responses and acute cognitive responses and the chronic impact on physiology and health are shown. These effects are often due to the activation of the stress axis and the increased release of cortisol and adrenaline. Note that even short-term sleep loss (a few days) can have a big impact upon emotional performance and brain function. Longer-term sleep loss (months and years) as experienced by some shift workers has been shown to increase the risk of several key diseases, including cancer and cardiovascular disease. Such associations have long been a concern for shift workers, who suffer from extreme forms of SCRD. *References:* fluctuations in mood[125-8]; anxiety, irritability, loss of empathy, frustration[129-31]; risk-taking and impulsivity[132-5]; negative salience[63]; stimulant, sedative and alcohol abuse[136-40]; illegal drug use[141]; impaired cognitive performance and ability to multi-task[142-4]; memory, attention and concentration[145-8]; communication and decision making[138,149-52]; creativity and productivity[153-6]; motor performance[144,157]; dissociation/detachment[158-9]; daytime sleepiness, micro-sleeps and unintended sleep can happen after short-term SCRD but are more frequent in chronic SCRD[160-63]; altered stress response[164-5]; altered sensory thresholds[166-8]; impaired immunity and infection[169-70]; cancer[171-3]; metabolic abnormalities and Type 2 diabetes[61,121,174-6]; cardiovascular disease[176-8]; depression and psychosis[179-82].

phenomenon whereby we feel unable to cope with the demands placed upon us. This perceived inability to cope with the demands of life acts as an additional stressor, promoting the further release of cortisol and adrenaline (another feedback loop), and this can lead directly to behavioural changes, including frustration and low self-esteem, increased worry, anxiety and depression.[124] The consequences of SCRD upon our emotional, cognitive and physiological health as a result of sustained activation of the stress axis are summarized in Table 1.

## *The Slide into SCRD*

SCRD is a common feature across many sectors of society, from teenagers[183] to the business and public sectors, from night shift workers[184-5] to the elderly[186]. Inadequate sleep is an important part of SCRD and inadequate sleep in adults is usually defined as less than seven hours each night. However, there is considerable individual variation in sleep need, and an honest self-assessment of your specific sleep needs is the best way to establish if you are developing SCRD. I stress an 'honest' assessment because the tired brain is very good at fooling itself that it is *not* tired and perfectly able to function. Many people are just really bad at judging their skills and can greatly overestimate their abilities.[187] This general phenomenon has come to be known as the 'Dunning–Kruger Effect'. As Shakespeare said, 'The fool doth think he is wise, but the wise man knows himself to be a fool.' The more SCRD we experience, the less wise and more foolish we become.

In Appendix I, I provide some suggestions about how you may want to monitor your own sleep. In addition, the drift into SCRD is suggested if you:

- are dependent upon an alarm clock, or another person, to get you out of bed

- oversleep extensively (get up late) on free days
- notice you sleep much more on holiday
- take a long time to wake up and feel alert
- feel sleepy and irritable during the day
- feel you need a mid-afternoon nap to function properly
- are unable to concentrate and exhibit overly impulsive behaviours
- crave caffeinated and sugar-rich drinks
- receive advice from family, friends, work colleagues that your behaviour has changed, specifically you are:
  – more irritable
  – lack empathy
  – are less reflective
  – are more impulsive and disinhibited
- detect that you experience increased worry, anxiety, mood swings and depression.

If you experience several of these symptoms you may be developing SCRD, and in chapter 6 I have set out some general strategies to address this problem.

At this point I think it makes sense to discuss two groups who are particularly vulnerable to SCRD in a bit more detail, namely night shift workers and those who repeatedly fly across multiple time zones and experience repeated jet lag.

## The problems of the night shift

'Could be Wednesday, could be Friday, could be yesterday – I'm not really sure any more.' This was told to me by a police officer following a week of night shifts. Extended night shift work is associated with the severest forms of SCRD because individuals have to work against their own circadian biology, combined with shortened and disrupted sleep (chapter 2). The problem is that shift workers work when they should be asleep and try to sleep when their biology is prepared for wake. No matter how many years a shift worker spends on permanent night shifts, nearly all (97 per cent) of night shift workers remain synchronized to daytime.[188] This failure to shift is directly related to light exposure. As discussed in chapter 3, light exposure at dawn and dusk is essential for setting the circadian clock to the 24-hour rotation of the Earth. Artificial light in the office or factory is dim compared to environmental light (Figure 3). Shortly after dawn, natural light outside is around 2,000–3,000 lux compared to the 100–400 lux experienced at home or in the workplace, and by noon natural light can be as much as 100,000 lux.[189] After leaving the night shift, an individual will usually experience bright natural light in the morning, and the circadian system will lock on to this brighter light signal as daytime (which it is) and align circadian rhythms to the daytime state. In one study, night shift workers were exposed to 2,000 lux in the workplace and then completely shielded from natural light during the day. Under these circumstances, they shifted and became nocturnal.[190] However, this is not a practical solution for most night shift workers. Another study by Josephine Arendt at the University of Surrey, and a pioneer in this area, examined night shift work in individuals working on an oil rig platform in the North Sea. Those working overnight from 6 p.m. to 6 a.m. all shifted their clocks. This was attributed to the very bright artificial lights on the rig platform at night, along with sleeping quarters with no natural light.[191] These examples highlight the importance of light in setting the circadian clock, and why our clocks do not adapt to the demands of working at night. The assumption, long held by many employers and managers, is that our clocks *do* naturally adapt to the night shift. Indeed, this was the firmly held view expressed to me by a former chairman of the Confederation of British Industry. Nice chap, with good intentions, but he was completely wrong. Which I politely pointed out.

The health problems associated with SCRD, summarized in Table 1, are routinely observed in night shift workers. Most recently a study has shown that night shift workers are more likely to be admitted to hospital if they contract COVID-19.[192] Nurses are one of the best-studied groups, and many years of shift work have been associated with a broad range of health problems, including Type 2 diabetes, gastrointestinal disorders, and even breast and colorectal cancers. Cancer risk increases with the number of years of shift work, the frequency of rotating work schedules, and the number of hours per week working at night.[193] It is also likely that elevated cortisol, leading to immune suppression, will contribute to such increased risks of cancer.[109] The correlations between night shift work and cancer are now considered so strong that shift work is officially classified as 'probably carcinogenic [Group 2A]' by the World Health Organization. Other studies of shift workers show increased heart and stroke problems, obesity and depression (Table 1). A study of over 3,000 people in southern France found that those who had worked some type of extended night shift work for 10 or more years had much lower overall cognitive and memory levels than those who had never worked on the night shift.[194] Just to twist the knife a bit more, and as discussed earlier in this chapter, SCRD impairs glucose regulation and metabolism, and increases hunger and the risk of Type 2 diabetes and obesity.[121] Night shift workers have elevated levels of cortisol, which has also been shown to suppress the action of insulin and raise blood glucose.[122] At another level, there is also a striking association between SCRD and smoking. Independent of social background and geographic region, smoking increases with higher levels of SCRD.[195] Along with smoking, the consumption of alcohol and caffeine also increases with SCRD.[195] Finally, the tendency towards depression increases when work times are not aligned with biological sleep times.[196]

SCRD can be a real problem for airline pilots. On 22 May 2010,

Air India Express flight 812 from Dubai to Mangalore crashed during landing. The aircraft overshot the runway, fell down a hillside and burst into flames. Of the 160 passengers and six crew members on board, only eight passengers survived. The Civil Aviation Ministry reported hearing snoring recorded on a cockpit voice recorder, and it looks like the pilot was chronically tired and had fallen asleep (had a microsleep) during the critical point of landing. Sleep loss has been linked to multiple notable industrial accidents, including the Chernobyl nuclear power plant explosion, the *Exxon Valdez* oil tanker spill, the explosion of the *Challenger* space shuttle and the Bhopal chemical plant disaster[197] (see chapter 9). But such accidents are not new. At 4.02 a.m. on the morning of 2 November 1892, near Thirsk railway station in the North Riding of Yorkshire, an express train crashed into a goods train. Ten people were killed and 39 injured. James Holmes was the signalman who ultimately triggered the disaster. The day before the crash, his baby daughter, Rosy, became ill and died. Before the shift, Holmes had been awake for 36 hours, caring for the child, trying to find a doctor and supporting his grief-stricken wife. James reported to the stationmaster during the afternoon, before his night shift, saying that he would be unable to work that night. The stationmaster was not sympathetic, and James was forced to work or lose his job. In the early hours of the morning a goods train came to a halt just outside Holmes's signal box. At about this time James fell asleep, then woke, but forgot that the goods train was on the line, and could not see it due to fog. He then accepted a passenger express train onto the line, which crashed into the back of the goods train at 60 miles an hour. Holmes was charged with manslaughter and found guilty but was given an absolute discharge. The railway company, however, was severely criticized because it ignored James's concerns about his lack of sleep. This shows a level of sympathy towards an employee that we do not normally associate with our Victorian ancestors.

## *The impact of jet lag*

Almost everyone who has flown across a few time zones has felt at least some of the symptoms of jet lag: tiredness, an inability to fall asleep in the new time zone, cognitive problems and memory loss, body aches, digestive problems and general disorientation, and possibly vomiting into the lap of your distinguished host like poor President Bush. In their advice to travellers, several airlines warn that the effects can be severe with decision-making ability downgraded by up to 50 per cent, communication skills by 30 per cent, memory by 20 per cent and attention down by 75 per cent. Somewhat disconcertingly, the cognitive problems found in night shift workers have also been shown in long-haul airline pilots and aircrew. One study found that airline crew who often travel across many time zones, and experience sustained SCRD, have high levels of cortisol, which was associated with cognitive defects, including impaired reaction times. Another study looked at female flight attendants who had worked for international airlines for five years and frequently crossed multiple time zones with rapid turnaround times and little chance to rest between flights. These individuals were compared with female flight attendants with a less demanding schedule and lower levels of SCRD. Brain scans were used to measure the sizes of the flight attendants' temporal lobes (Figure 2, p. 24), brain areas crucially important for language and memory. The temporal lobes of attendants undertaking multiple trips with short recovery times were significantly smaller. These individuals also had higher levels of cortisol in their saliva, and, interestingly, the higher the cortisol in the saliva, the smaller the temporal lobes. Along with smaller temporal lobes, these individuals showed impaired cognition and slower reaction times.[198-9] 'Become a flight attendant – see the world and shrink your temporal lobes' is, I suspect, a slogan that is unlikely to appear on recruitment posters.

The faster you cross time zones, the worse the effect on most people. Incidentally, there is some evidence that 'boat lag' affected transatlantic passengers in the days of the large ocean liners. Between 1905 and 1955 it took 4–5 days to cross the Atlantic. Today, in a jet, it is about 6–7 hours. Our circadian system simply cannot adapt quickly enough to such rapid changes. On average it takes about a day to adjust for each time zone crossed, so about five days if crossing five time zones. There are, however, notable differences between individuals, and it makes a big difference if you fly east or west. As discussed below, most of us, possessing a later chronotype, find it easier to travel west.[200]

### DOES MELATONIN WORK FOR JET LAG?

Melatonin has been used extensively as a potential treatment for jet lag. In most, but not all, studies, melatonin was found to decrease the symptoms of jet lag in people crossing five or more time zones when taken close to the local bedtime at the new destination.[201] However, the overall effects were shown to be very moderate, and it is worth pointing out that there are marked individual differences in both the sensitivity to melatonin and the effects of jet lag.[202] The point is, for some people melatonin is probably helpful for jet lag and for others not at all. For this reason, and because melatonin can cause sleepiness in some sensitive individuals, the general advice about taking melatonin is to not drive or operate heavy or dangerous machinery for 4–5 hours after taking it. In addition, long-haul pilots and crew (and others crossing multiple time zones repeatedly) are not advised to use melatonin due to potential difficulties with timing the dose. By taking melatonin at the wrong time, and if you are sensitive to melatonin, you may confuse the circadian clock even more, possibly blunting the effects of light. In addition, if you have, or if you have a family history of, psychiatric illness or migraine, you are not advised to use melatonin.[202-3] For a final discussion of the role of melatonin see chapter 14.

## SEEKING OUT (OR AVOIDING) LIGHT
### TO CURE JET LAG?

Do I take melatonin to help with jet lag? Personally, no – I use light instead. My experience after many trips travelling east to Australia from the UK was that melatonin made things worse, but natural light exposure did help. While there are no guarantees, an effective way of minimizing the problems caused by jet lag is to use light to 'shift' the clock. The basic 'rule of thumb' is that if you travel west from the UK, you seek out light in the new time zone; if you travel east from the UK for more than 6–8 time zones, then avoid morning light in the new time zone but seek out afternoon light. As we discussed in chapter 3, light shifts the clock differently at different times of the day. Light around sunset (dusk) delays the clock, so we go to bed later and get up later the next day. Light around sunrise (dawn) advances the clock so we go to bed earlier and get up earlier the next day. Because light at different times can either advance or delay the clock, it is essential when crossing several time zones that light exposure in the new time zone moves the clock in the correct direction. When travelling west, e.g. from the UK to New York (five hours behind the UK), you seek out light upon arrival in the new time zone. The light will hit your clock when it 'thinks' it is dusk in the UK and the clock will be delayed to New York time. So, take a stroll outside when you arrive in the Big Apple! When travelling east, e.g. from the UK to Sydney (around 11 hours ahead of the UK), you need to advance your clock, and this is a bit more complicated. To adjust quickly, avoid morning light and seek out late-afternoon light for the first few days. This is because dusk in the UK will be morning in Sydney, and light exposure at this time will delay your clock and not advance it to the new time zone. But by late afternoon in Sydney it will be morning in the UK (dawn), and light exposure at this time will advance the clock and help drag it forward to the new time zone. A pair of dark sunglasses is the simplest means of dealing with the problems of inappropriate light exposure.

So, to recap, to reduce jet lag, calculate when it is dawn (advancing light) and dusk (delaying light) in the new time zone, then aim to seek out or avoid light in the new time zone to either delay or advance your clock for the first few days after arrival. There are several apps available online for additional help if you search 'How to prevent jet lag using light' – but as with all apps some will be better than others, and you may want to check user recommendations online before that all-important purchase. Combined with appropriately timed light, eating at the same daily 'local' mealtime appears to help shift the peripheral circadian clocks to the new time zone. This is particularly important for the peripheral clocks in the liver and pancreas and their regulation of our metabolism[204] (see chapters 12 and 13). Exercise at different times can also help you adjust, as I discuss in chapter 13.

### BRITISH SUMMER TIME (BST), DAYLIGHT SAVING TIME (DST), CIRCADIAN RHYTHMS AND 'SOCIAL JET LAG'

The natural cycle of dawn and dusk is essential for setting our internal circadian rhythms. The Earth rotates eastwards, moving from west to east, in what is called a 'prograde' motion. As viewed looking down upon the Earth from the north pole, the Earth turns counter-clockwise which is why the Sun appears to rise in the east and set in the west. In addition to this solar day, our societies have constructed 'social time', which systematically hacks the rotation of the Earth into minutes and hours. An important reason for the implementation of a standardized clock time was the development of the railway system, and the need to have a standardized timetable based upon a mechanical clock and not local time. Local time defines noon when the Sun is at its highest point above the horizon, but local noon is later the further west you live in the same time zone. The Earth is divided into lines of longitude, and it takes the Sun four minutes to pass from one line of longitude to the next, with the zero line running through Greenwich in south-east London. All this was agreed, more or less, by consensus at the International Meridian Conference in 1884. The discussion got quite heated, and the French delegation abstained in the final vote, arguing that Paris should have the honour not Greenwich. Most European countries aligned their clocks to Greenwich within 10 years, although the French maintained Paris time until 1911.

It takes one hour for the Sun to move across 15 longitude lines, which is usually defined as one time zone. However, country and state borders rarely coincide with time zones, and so time zones may be artificially modified. China is the most marked example. China spans five geographical time zones, but there is a single official national standard time, called Beijing Time. As a result, in Kashgar, in western Xinjiang, solar noon can be as late as at 3.10 p.m. clock time. Crucially, our circadian clocks still follow solar time. Evidence for this comes from several studies showing that circadian cycles, such as the sleep/wake cycle, are at an earlier mechanical clock time at the eastern edge of a time zone and become progressively later the further people live west in the same time zone.[205] So people in Poland (east), on average, get up earlier than people in Spain (west) – relative to their alarm clocks.

Daylight saving time (DST) is the practice of advancing clocks forward by one hour (skip or 'lose' one hour) during the warmer months of spring so that a 6 a.m. clock time becomes 7 a.m., and 6 p.m. clock time becomes 7 p.m. (an hour is lost). 'Spring forward.' After this change in spring, your alarm clock will get you out of bed at 7 a.m., but your body clock 'thinks' it's 6 a.m. For clarity, in the UK the period when the clocks are advanced one hour forward of Greenwich Mean Time (GMT) is called British

Summer Time (BST). In the autumn, setting the clocks back by one hour ('gain' an additional hour) so that 7 a.m. becomes 6 a.m. and 7 p.m. becomes 6 p.m., returning to standard time ('Fall back'). The original logic was that people would use less electric light in the 'lighter' evenings during the spring and summer months and save energy. Germany led the way with DST (*Sommerzeit*) in 1916 during the First World War to alleviate massive wartime coal shortages. Britain, France and Belgium, along with other states, soon followed. Some recent studies suggest there is in fact no energy saving with DST in the economies of today.[206] While DST in the spring provides another hour of daylight after work or school for recreation and gardening, it comes at a cost. The circadian system remains locked on to the solar cycle (just like night shift workers) and does not adapt to the new social clock time.[207] So, individuals are being driven out of bed, and forced to be active, an hour earlier than their internal circadian time. This is 'social jet lag', a term developed by Till Roenneberg at Ludwig-Maximilian University (LMU) in Munich. Social jet lag is a mismatch between when our circadian system would like to wake us up and when we are forced out of bed due to social demands such as work, school or DST. This was originally thought not to be important. However, during the first days after the clocks change, DST has been associated with increased irritability, less sleep, daytime fatigue, mental health problems, and even decreased immune function and sleep.[206, 208-9] Even worse, an increased number of heart attacks[210], strokes[211], and accidents and injuries in the workplace are reported during the first weeks after a clock change[212]. A very recent study showed there is a 6 per cent increase in fatal car crashes the week following our 'spring forward'.[213] Although I could not find any detailed systematic studies, carers and organizations that look after the elderly frequently report that DST, and the associated changes in the hours of daylight, and the change in mealtimes and sleep schedules, all contribute to an increase in emotional, behavioural and cognitive difficulties. Such difficulties seem particularly bad in individuals with dementia and Alzheimer's, whose symptoms of sundowning are worse following circadian rhythm disruption.[214] For clarity, sundowning in individuals with dementia and Alzheimer's refers to a state of confusion, anxiety and aggression, with pacing or wandering about, which occurs in the evening and early night-time.

The consensus, based upon the emerging data, is that we should *not* fight against our circadian rhythms and the alignment of our body clocks to the solar cycle.[206] Rather than causing social jet lag, we should abandon DST, returning to Standard Time, which is when the solar day, our circadian rhythms and the social day can be most closely aligned. I appreciate that many will disagree – not least Scottish golfers, who send irate letters to me saying they want to play golf late in the day while it is still light. At the risk of alienating the members of The Royal and Ancient Golf Club of St Andrews, I'm not sympathetic.

So until DST is abandoned what can we do? Currently, in the spring an hour is lost, so your alarm clock will get you out of bed at 7 a.m., but your body clock 'thinks' it's 6 a.m. In the autumn an hour is gained because 7 a.m. becomes 6 a.m. You get an extra hour in bed. As we lose an hour in the spring, and this is associated with an increased rate of car accidents, be particularly careful during the morning commute. You may want to think about an extra-large cup of coffee in the morning during the first week after the shift to keep you more alert; also, during the days before the DST shift, try and adjust your bedtime to about 10–15 minutes earlier (in the spring) or later (in the autumn) each day, which helps your body to gently adjust to the new schedule and eases the shock to the system. Exposure to morning light in the spring will also advance your clock (chapter 3) – dragging the circadian system earlier in time and helping you get up earlier. As will be discussed in chapter 6, it is always important to maintain 'good sleep practice', and you will find additional tips there to help you deal with DST.

## Questions and Answers

**1. Why does shift work become more difficult as we age?**
Multiple studies have shown that young people, in general, cope better with night shift work compared with middle-aged and older individuals. The reasons are not clear, but one link may be with cortisol. Cortisol levels are higher in older individuals, and as a result older individuals may be more vulnerable to SCRD, showing increased stress, poorer cognitive performance, and even a reduction in the size of memory-related structures in the brain.[116] Another link is that young people tend to be later chronotypes (owls) and so can cope better on late shifts, compared to older people, who tend to be earlier chronotypes (larks).[215]

**2. Are there any food supplements that we can add to our diet to help deal with shift work?**
Multivitamins, vitamin D, vitamin B12, melatonin, magnesium and tryptophan (which is the precursor of serotonin and melatonin) are often recommended for night shift workers, but there is no really solid evidence that these supplements help. The consensus is that it is probably most useful to focus on an overall good diet rather than on individual supplements. However, a vitamin D supplement could be important. Shift workers, and indoor workers in general, are consistently reported as being those groups most likely to suffer from a vitamin D deficiency.[216] The greatest proportion of vitamin D, around 90 per cent, is synthesized by exposure of the skin to sunlight (UVB energy), and although vitamin D deficiency is commonly associated with issues such as bone health, low vitamin D is also linked to other illnesses, including immune problems, metabolic abnormalities, some cancers and even mental health conditions.[216] So a vitamin D supplement at an appropriate level may be a good precautionary

measure, and especially during pregnancy.[217] As with all supplements, you should discuss this with your medical practitioner. It is important not to take more than the recommended daily dose, as high levels of vitamin D can be toxic, causing nausea and vomiting and kidney problems, including the formation of calcium stones. Another food supplement that has been much discussed for shift workers is tryptophan. Tryptophan is an amino acid that is a building block for many proteins, and it is also a precursor for the neurotransmitter serotonin, the pineal hormone melatonin and vitamin B3. Opinions are still a bit mixed, but there is some evidence that taking tryptophan supplements or eating tryptophan-rich food can make a small improvement in sleep, reducing the time it takes to fall asleep and increasing the total time of sleep. The reasons are unclear – perhaps it increases brain serotonin, which decreases anxiety, or maybe because it increases levels of melatonin. We don't know.[218] As is almost always the case, more research is needed.

### 3. Why do I yawn?

I dread this question at the end of a talk as my angst interprets the enquiry as a thinly veiled criticism! I remember how I felt early in my career when, during one of my first lectures to undergraduates, I looked out onto a sea of student faces and one face disintegrated into a huge yawn! I was crushed. But, people yawn for a variety of reasons and we yawn when we are tired, bored, anxious, hungry, or about to start a new or difficult activity. The real purpose of a yawn remains a mystery. Until a few decades ago, yawning was explained as a mechanism to inhale a large amount of air to increase oxygen levels in the blood to compensate for oxygen deprivation. But the 'oxygenation hypothesis' now seems unlikely. One favoured idea is that the act of yawning 'cools the brain' to promote arousal and alertness. The act of yawning increases blood flow to the skull, which cools the brain and counteracts feelings of sleepiness, making us more alert, particularly when we feel sleepy. However, no physiological effect of yawning has been demonstrated so far, and it is possible that yawning doesn't really have a role – although I intuitively think this unlikely. But what is certain is that yawning is contagious. One person yawns and it spreads through the group. And it's not just us. Contagious yawning has been documented in chimpanzees, wolves, domestic dogs, sheep, pigs, elephants and recently lions. The argument currently is that yawning improves alertness and that contagious yawning evolved to boost group vigilance among animals that live cooperatively, increasing collective awareness for threat detection or coordinated social action.[219]

**4. I have read that living at the western edge of a time zone can cause health problems. Can this really be true?**

Crazy as it sounds, this is true. Social jet lag is a mismatch between when our circadian system would like to wake us up and when we are forced out of bed due to social demands such as work or school. Sunrise in the eastern part of a time zone occurs earlier than at the western edge of the same time zone. So the circadian system of those living in the east will be set earlier by dawn than it will for those in the west. Easterners will be better adapted to their time zone than those in the west, resulting in more social jet lag for westerners. The same will be true of individuals living on the western border of a time zone and others on the eastern border of the adjacent time zone. Remarkably, if you compare the health of individuals living in the western edge of a time zone you see evidence for greater levels of circadian rhythm disruption such as obesity, diabetes, cardiovascular diseases, depression and breast cancer.[220]

## 5.

# Biological Chaos

### Generating sleep and circadian rhythm disruption

That we are not much sicker and much madder than we are
is due exclusively to that most blessed and blessing of
all natural graces, sleep.

Aldous Huxley

The nature versus nurture debate tried to resolve the extent to
which human traits arise from either nurture – environmental
factors, either in the womb or during an individual's life – or
nature – from a person's biology and specifically their genes. Such
debates can become very polarized and very misleading. In the
1980s the notion of 'genetic determinism' (also known as 'biological determinism') became popular in some quarters, along
with the belief that human behaviour is controlled directly by an
individual's genes rather than being influenced by the environment. It all became rather tribal, with elements of the political
right embracing determinism, while the left utterly rejected the
notion. The discussion moved from biological questions relating
to the role of genetics in shaping our biology, to the issue of
whether it was politically or ethically permissible to admit that
genes contribute to our behaviour at all. It always puzzled me at
the time how bright and well-intentioned people could poison
debate by polarizing it in this way. In reality, of course, there is a
seamless and continuous interaction between our genes and our
environment throughout life. This interaction has been clearly

demonstrated by studying behaviours such as schizophrenia. The overall lifetime risk of developing schizophrenia in the general population is around 1 per cent (1 in 100). Estimates vary a little, but in identical twins that are 100 per cent genetically identical, if one twin develops schizophrenia, the other twin has a 1 in 2 (50 per cent) chance of developing the condition. In non-identical twins raised together, with 50 per cent shared genetics, if one twin develops schizophrenia, the other has a 1 in 8 (around 12 per cent) chance of developing the condition. So clearly there is a genetic contribution to schizophrenia, but the environment plays an important role too. There are likely to be multiple environmental factors, many difficult to define, but for schizophrenia some links have been identified, including: malnutrition or stress in the mother, abuse or head injury as a child or teenager, drug abuse or difficult social circumstances.[221]

Today, most scientists think that it is obvious that both genetics and the environment define our development, not least because of an increased understanding of the mechanisms of 'epigenetics'. Epigenetics means 'above' or 'on top of' genetics and refers to external modifications that regulate the activation of genes. These modifications do not change the DNA sequence (the genetic code) itself, but instead they alter how cells 'read' their genes. The environment can alter slightly the way that DNA is folded up, or the proteins (histones) that surround the DNA. The net result can be either an increase or a decrease in the activation of a specific gene and, as a result, the levels of the protein it encodes. Recent studies have shown that epigenetic modifications can be passed from parent to child. This means that the altered epigenetic pattern of turning a particular gene 'on' or 'off' in a parent could be inherited by their children and grandchildren. So each of us could be more vulnerable to a disease because of the environment of a parent or grandparent, even though we have never experienced that environment ourselves. Remarkably, new evidence is emerging that sleep and circadian rhythm disruption (SCRD) may cause epigenetic modifications that increase the chances of meta-bolic disease, obesity, heart disease, stroke and high blood pressure and may even affect our cognitive abilities.[222] The details are still being worked out but the implications are clearly enormous.

The point I want to stress is that sleep is an immensely complex behaviour, and, like all other behaviours, the sleep we get will be profoundly influenced by an intricate interaction between our gen-etics and our environment. In our own families, many of us will have noted that similar sleep patterns can be seen across the gen-erations. Indeed, there are now clear studies illustrating a familial contribution to the type of sleep we get.[223] But the relative role of genes versus the epigenetic modification of genes versus the direct impact of the environment on physiology will vary enormously, and the challenge in the coming years will be to disentangle this biology. However, before we can work out the mechanisms, we have to know what we are studying. In short, we need to be able to classify the diverse and varied patterns of SCRD experienced.

Some of the names and classifications used to describe SCRD can be confusing, not least because the 'labels' are associated with a certain degree of ambiguity and carry unhelpful historical bag-gage. Many people ask me about these various descriptions of SCRD, and so I have included a short overview here. However, I am keen to stress that while 'self-diagnosis' can often be a helpful guide, and a possible incentive for corrective action, it must not replace a formal clinical diagnosis. What follows relates to the clas-sifications that your medical practitioner or health care provider may use. You may also like to take a look at Appendix I to generate your own sleep diary, and perhaps get other family members to join in so you can compare similarities or differences.

There are currently, and officially, 83 different types of sleep and circadian rhythm 'disorder' which have been grouped into seven dis-tinct categories.[224] As with all classification schemes, it is not perfect, and is frequently revised as our knowledge and understanding grow. It is also important to stress that in most cases these categories do not define the underlying mechanism or problem causing the disorder. The conditions listed can arise from a variety of different and often overlapping causes, perhaps related to changes within the brain, the demands of work and/or leisure, the demands of family, our genetics and/or epigenetic changes, our age, the consequences of mental or physical illness, and our emotional and stress responses to all of these factors. The seven major classifications of SCRD currently used are summarized here.[224]

## Category 1: Insomnia

A diagnosis of insomnia is used to describe difficulties falling asleep or staying asleep for as long as you would like. Insomnia is also frequently the reason for daytime sleepiness (illustrated in Figure 4, p. 103). Apparently, Adolf Hitler was a notorious insomniac, and would finally fall asleep in the early hours of the morning. Indeed, he was fast asleep at the Berghof when the Allies invaded France on D-Day on 6 June 1944. His generals would not send reinforcements to Normandy without Hitler's permission, and no one dared wake him, so he slept until noon. The delay is thought to have saved many lives and been critical for the Allied invasion.

Insomnia is one of the most frequently used classifications of a sleep problem because it is very broad. It is also important to stress that insomnia has multiple causes that produce the same sort of problem, ranging from stress, the demands of work and/or leisure, poor sleep habits, side effects from medication, too much caffeine, nicotine and alcohol, pregnancy, age and mental illness. And, of course, the 24/7 society. In chapter 6, we will see that in many cases actions can be taken to help deal with insomnia. Also note that many of the classifications of SCRD listed below can cause insomnia and daytime sleepiness, emphasizing that insomnia is a descriptive term and does not provide any information about causation. Let's consider some issues around the topic of insomnia.

## *Sleepiness and fatigue*

Insomnia can be associated with both sleepiness and fatigue, and it is important to distinguish between the two. One of the consequences of insomnia is daytime sleepiness, but it is incredibly important to determine if this is 'sleepiness' rather than 'fatigue'. Fatigue describes an overall feeling of tiredness or lack of energy and is not the same as sleepiness. Sleepiness is resolved by restorative sleep. When you have severe fatigue, you are not motivated and have no energy and, critically, the feelings of tiredness are overwhelming and cannot be relieved by sleep.[225] Fatigue can be a symptom of serious underlying health problems. Indeed, a key symptom of COVID-19 and other viral infections is fatigue, which is not cured by sleep, even though you often sleep much more. Months after infection with COVID-19, some people are still battling the overwhelming fatigue associated with 'long COVID'.[226] Because fatigue can be an important symptom of some form of chronic health problem, you should seek medical advice if fatigue persists.

## *Waking in the middle of the night – biphasic and polyphasic sleep*

Another important issue associated with insomnia is waking several, or many, times in the middle of the night, as illustrated in Figure 4. As discussed in chapter 2, the basic idea, that there is an on/off switch that drives the sleep/wake cycle, arising from an interaction between the circadian clock and sleep pressure, is broadly correct. But it must be more complicated because sleep in humans and other animals is often not a single consolidated block of sleep (as we are so often told). Sleep can occur in two episodes (biphasic sleep) or even multiple episodes (polyphasic sleep), with sleep separated by short periods of wake.[227] How these sleep patterns are generated is not fully understood, but they raise a very important point about our expectations regarding getting a 'good night's sleep'. A single period of sleep without waking (monophasic sleep), is considered 'normal' sleep (Figure 4). But this may not be the 'normal' state after all. It has been suggested that because of the 24/7 society, and a reduction in the time available for sleep at night, sleep has become compressed and squeezed into a single episode. During the COVID-19 pandemic of 2020 and 2021, and the need to self-isolate, many have had the opportunity to sleep longer, and the result has been either biphasic or polyphasic sleep. Interestingly this is often self-diagnosed as worse sleep. In actuality, it's certainly a different pattern of sleep, but it may not be worse. It is also important to stress that biphasic or polyphasic sleep is the normal situation for most animals, and may have been for humans before the Industrial Revolution.[228]

There is not universal agreement about biphasic or polyphasic sleep being the ancestral state of human sleep[229], the original idea was partly developed on the basis of historical research, and diary records of biphasic sleep[230-31]. This has been investigated by Roger Ekirch and detailed in his book *At Day's Close*. Ekirch has documented many accounts of biphasic sleep, from medical texts, court records and diaries, and found evidence that in pre-industrial Europe biphasic sleeping was considered the normal practice. The whole household would prepare for sleep a few hours after dusk, then wake for a few hours in the middle of the night, and then have a second sleep until dawn. Ekirch found multiple references to a 'first' and 'second' sleep: he cites prayer manuals from the late fifteenth century which offered special prayers for the hours in between sleeps, and a doctor's manual from sixteenth-century France advising couples that the best time to conceive was not at the end of a long day's labour but 'after the first sleep', when 'they have more enjoyment' and 'do it better'.

Miguel Cervantes wrote in his novel *Don Quixote*, published in 1615: 'Don Quixote followed nature, and being satisfied with his first sleep, did not solicit more. As for Sancho, he never wanted a second, for the first lasted him from night to morning.' However, references to biphasic sleep became less common from the late seventeenth century and had disappeared altogether by the 1920s as artificial lighting and modern industrial practices became the norm. Such observations from history stimulated laboratory-based studies. Participants were given the opportunity to sleep longer by imposing a 12-hour light and 12-hour dark lighting schedule. The result was polyphasic and biphasic sleep.[227, 232] This is a nice example of how historical studies, and the social sciences, can inform contemporary science. The bottom line is that for many of us, when we have the opportunity to sleep longer, we revert to polyphasic sleep.

This raises the important point, that if the natural state of human sleep is polyphasic, then we need to rethink our interpretation of 'insomnia' and disrupted sleep at night. New research suggests that if we wake up at night, sleep is likely to return, *if* sleep is not sacrificed to social media and / or other alerting behaviours. The key point is that waking at night need not mean the end of sleep. If you wake in the middle of the night, it is important not to activate stress responses (chapter 4). Don't remain in bed getting increasingly frustrated by the failure to fall back to sleep. Some individuals find it useful to leave the bed, keep the lights low and engage in a relaxing activity such as reading, listening to music, then returning to the bed when sleepy again (chapter 6). And while I am on the subject . . .

### Self-imposed polyphasic sleep

Another area of confusion has been the mini-craze for self-imposed polyphasic or segmented sleep. This is not a good idea. Because biphasic and / or polyphasic sleep may be the norm for many individuals, and because REM / NREM cycles are on average every 70–90 minutes throughout the night[233] (chapter 2), an idea has emerged that we should impose a pattern of polyphasic sleep upon ourselves. In practice this means that the time spent sleeping is deliberately broken into multiple sleep periods across the 24-hour day, with a main period of sleep during the night and shorter sleep bouts during the day. The net result is a major reduction in sleep time over the 24-hour day/night cycle. Many different polyphasic sleep schedules have been proposed, but all these schedules are completely different from the daytime nap or siesta patterns seen in many countries around the world, which *do not* aim to reduce total sleep time. One schedule, the 'Überman' polyphasic sleep schedule, recommends six 20-minute sleep periods spread evenly across the 24-hour day, to achieve a total of 2 hours sleep in 24 hours. The 'Everyman' polyphasic sleep schedule suggests a three-hour sleep at night with three 20-minute sleep periods during the day, achieving a total of four hours of sleep across the 24-hour day. The advocates of such schedules claim that by adopting this pattern you can improve your memory and mood, allow better dream recall and live longer. Unsurprisingly, such claims are not supported by the evidence. Indeed, imposed polyphasic sleep schedules, with accompanying sleep loss, have been linked to poorer physical and mental health, and poorer daytime performance. In short, and despite reports in the media, the science does not support the hype and such schedules are not recommended by the National Sleep Foundation.[234]

## *Category 2: Circadian Rhythm Sleep/Wake Disorders*

These are an important group of diverse conditions which can give rise to insomnia, daytime sleepiness and of course the problems listed in Table 1 (p. 77). Such disorders arise because of the central role of the circadian clock in regulating sleep (chapter 2). Causes include abnormal light exposure, which can be corrected easily (chapter 6), genetic abnormalities, and some of the problems associated with blindness or neurodevelopmental disorders which currently can be difficult or impossible to correct (chapter 14).

The key circadian rhythm disorders are illustrated in Figure 4 and are discussed here.

## Advanced sleep phase disorder (ASPD)

ASPD is characterized by difficulty staying awake in the evening and difficulty staying asleep in the early morning. Typically, ASPD individuals go to bed and get up about three or more hours earlier than the social norm. ASPD has been associated with key changes in the genes that drive the molecular clockwork.[235-7] ASPD can also occur in some individuals during ageing.[238]

## Delayed sleep phase disorder (DSPD)

DSPD is the opposite of ASPD, and is characterized by a near three-hour delay (or more) in sleep onset and wake. Because of the demands of work, this condition often leads to greatly reduced sleep duration during the working week, along with considerable daytime sleepiness and then long periods of sleep on free days. ASPD and DSPD can be considered as extremes of having a morning (lark) or evening (owl) chronotype (Appendix I). The use of timed light exposure can help markedly with these conditions (chapter 6). DSPD has been associated with key changes in the genes that contribute to the molecular clockwork.[239-40] And gene/environmental interactions can cause forms of DSPD in adolescents[241], depression[242], mental illness[243] and neurodevelopmental disorders[244].

## Freerunning, or non-24-hour sleep/wake disorder

This describes a condition in which an individual's sleep/wake cycle occurs at a different social clock time each day – usually later and later each day. The circadian rhythm is not synchronized (entrained) to the 24-hour day. This is typically seen in individuals with severe eye damage or eye loss[245], or in other conditions such as schizophrenia[246], neurodegenerative disease or brain trauma[247]. A highly structured routine of sleep, exercise, mealtimes and, where appropriate, light can help synchronize the circadian system to the 24-hour day, but interventions can sometimes be difficult and only partially effective at best (chapter 14).

### Fragmented, or arrhythmic, sleep

This is a rare condition but is sometimes observed in individuals with mental illness or individuals with brain damage arising from trauma, stroke or a tumour.[247] As with non-24-hour sleep/wake disorder, a highly structured routine of sleep, exercise, mealtimes and morning light can help synchronize the clock to the 24-hour day, but, again, interventions can sometimes be difficult and only partially effective (chapter 14).

## Category 3: Sleep-related Breathing Disorders

This term refers to a broad range of breathing problems that can give rise to insomnia, the most familiar of which is chronic snoring and obstructive sleep apnoea. These conditions can be a real problem, and as Anthony Burgess is reported to have said: 'Laugh and the world laughs with you, snore and you sleep alone.'

### Obstructive sleep apnoea (OSA)

OSA is very common and is caused by a relaxation of the muscles in the back of the throat, which prevents normal breathing while asleep. These throat muscles support the back of the roof of the mouth (soft palate), comprising the uvula, which is a triangular piece of tissue hanging from the soft palate (during swallowing,

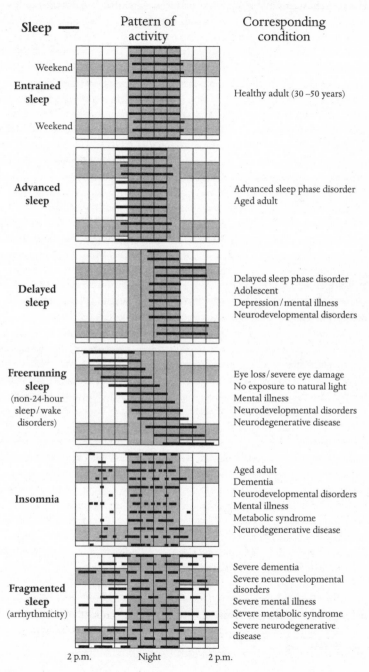

Figure 4.

**Figure 4. Illustration of sleep/wake patterns.** Abnormal patterns of sleep/wake arise from multiple causes and influences, both genetic and environmental. Filled horizontal bars represent periods of sleep on consecutive work days and at the weekend (horizontal bar). **'Normal' entrained sleep**, which may not be the normal state for many, shows a stable and single sleep episode that occurs for about eight hours, and at roughly the same time each day. For social reasons, there may be a slightly delayed sleep onset and offset at weekends. **Advanced sleep phase disorder (ASPD)** is characterized by difficulty staying awake in the evening and difficulty staying asleep in the early morning. Typically, individuals go to bed and rise about three or more hours earlier than the societal norm. ASPD is observed in the genetic condition of Advanced Sleep Phase Disorder[248] and in aged individuals[238]. **Delayed sleep phase disorder (DSPD)** is characterized by a three-hour delay or more in sleep onset and offset. This often leads to greatly reduced sleep duration during the working week and extended sleep on free days. ASPD and DSPD can be considered as extremes of morning (lark) or evening (owl) preferences. Both ASPD and DSPD are not merely shifted sleep/wake patterns, but conditions that cause distress or impairment because they conflict with the schedules demanded by societal pressures or personal preferences. DSPD is frequently observed in: the genetic condition Delayed Sleep Phase Disorder[240]; adolescents[241]; depression[242] and mental illness[243]; and neurodevelopmental disorders[244]. **Freerunning or non-24-hour sleep/wake disorder** describes a condition whereby an individual's sleep occurs systematically later or earlier on subsequent days. Because the body clock of most individuals is longer than 24 hours, the usual pattern is that sleep/wake occurs later and later each day. Rarely, the freerunning pattern may run in the opposite direction, with sleep occurring earlier and earlier each day. A freerunning pattern has been observed in individuals with eye loss[249]; weak exposure to natural light[250]; mental illness[246]; neurodevelopmental disorders[244]; and neurodegenerative disease[251]. **Insomnia** is a term used to describe a condition that leads to difficulty falling asleep or staying asleep, even when a person has the chance to do so. Insomnia is frequently associated with reduced sleep (hyposomnia) and can arise from multiple causes,[252] including circadian rhythm disruption. The former US president Bill Clinton is reported to have had insomnia, and he partly blames his heart attack on the fatigue associated with the condition. Insomnia is observed in: aged adults[253]; dementia[254]; depression[255] and mental illness[246]; neurodevelopmental disorders such as attention deficit hyperactivity disorder (ADHD)[244]; neurodegenerative disease[256]; and metabolic syndrome[257]. **Fragmented sleep**, or **arrhythmicity**, is typically observed in individuals who lack a functioning circadian clock, as has been seen in individuals with tumours located in the hypothalamus.[89] It is also a feature of severe dementia[258]; mental illness[246]; neurodevelopmental disorders[244]; neurodegenerative disease[259]; and metabolic syndrome[260]. Understanding the basis for these different circadian abnormalities provides the substrate for developing new drugs to correct these defects (chapter 14).

the uvula moves and helps prevent food and liquids from going up your nose cavity), the tonsils and the tongue. During an episode of OSA the muscles relax, and this causes the airway to narrow or close when you are breathing in. The result is that breathing may be poor or prevented for about 10 seconds or longer. This lowers the level of oxygen in the blood and causes a build-up of carbon dioxide. In adults, the most common cause of OSA is obesity. There is also evidence that we deposit more fat in our tongue as we age and this makes the tongue heavier and more inclined to flop back and cover the airway.[261] Alcohol can also relax the muscles at the back of the throat and this increases the chances of OSA. There is a 50 per cent greater chance of developing OSA if you are male. Medical conditions that are associated with obesity, such as hypothyroidism, and polycystic ovary syndrome in women, are linked to OSA. Common symptoms of OSA include snoring, stopping breathing during sleep, frequent awakenings during the night, wakening with a dry mouth or sore throat, morning headache and daytime sleepiness. During an episode, when breathing has stopped and carbon dioxide rises in the blood, the brain detects that it is not getting enough oxygen and triggers wake, and individuals wake and gasp for breath. OSA can be very dangerous, not least in making heart and high blood pressure problems worse. Some studies have suggested that eye conditions, such as glaucoma and optic-nerve damage are higher in individuals with OSA.[262] OSA is also a risk factor for stroke and heart conditions including: **bradycardia**, where the heart rate is too slow; **supraventricular tachycardia**, where your heart suddenly beats much faster than normal; **ventricular tachycardia**, where the lower chamber of the heart (ventricle) beats too fast to pump blood effectively and the body doesn't receive enough oxygenated blood; and **atrial fibrillation**, where the heart rate is irregular and often rapid. All of these conditions are risk factors for dementia.[254] An additional problem is that some benzodiazepine-based sleeping tablets (chapter 6) and general anaesthesia can relax the upper airway and may make OSA worse, increasing the risk of surgery using a general anaesthetic. The good news is that in most cases OSA is easily treated using devices that gently and continuously push air into the airways; this is called continuous positive airway pressure/power, or CPAP. Such devices can take a bit of getting used to for some people, but for most they solve the problems of OSA-induced insomnia. So if your snoring is loud enough to disturb your sleep or the sleep of others, if you wake up gasping or choking, if your sleeping partner notices intermittent pauses in your breathing during sleep and if you fall asleep while you're working, or even worse when driving, then you must visit your medical practitioner for a proper diagnosis and treatment. There is a simple online questionnaire which can help quantify daytime sleepiness, called the Epworth Sleepiness Scale that you may want to try. There is a strong argument that many serious motor vehicle crashes could be prevented by treating OSA.[263] In addition to OSA, there are a number of other sleep-related breathing disorders. For example:

## Central sleep apnoea (CSA)

This is very like OSA, but rather than a physical obstruction blocking the airway, in CSA the brain stops sending, or sends abnormal, signals to the muscles that regulate breathing. A stroke, or other conditions, such as brain trauma resulting from a road traffic accident, that affects the hindbrain (that part of the brain important for breathing) (Figure 2, p. 24) can cause CSA.[264]

## Sleep-related hypoventilation disorders

These conditions occur when the lungs are not sufficiently ventilated, causing carbon dioxide levels in the blood to increase. Obesity, genetic abnormalities, medications (e.g. opioids and benzodiazepines) and infection have all been linked to sleep-related hypoventilation.[265] Obesity hypoventilation syndrome (OHS) is also known as 'Pickwickian syndrome' and is named after the character Joe from the 1837 Charles Dickens novel *The Pickwick Papers*. Unsympathetic nicknames were a feature in medicine until relatively recently. For example, poor Joe, and you may want to look up the original illustrations from *The Pickwick Papers*, had many of the symptoms that were later described as OHS, including obesity and sleep apnoea, feeling sleepy or fatigued during the day, swelling or a bluish colour in the fingers, toes or legs (known as cyanosis), and morning headaches due to high levels of carbon dioxide in the blood. Individuals with Pickwickian syndrome have also been called 'blue bloaters', which described a person who is blue, is overweight, has shortness of breath and may have a chronic cough. 'Blue bloaters' are often contrasted with 'pink puffers', meaning people who are thin, breathe fast and are pink in colour. It's an old term for what would be described today as severe emphysema. The bottom line is that 'blue bloaters' and 'pink puffers' are examples of chronic obstructive pulmonary disease (COPD), and sleep in these individuals is invariably greatly disrupted.[266]

### Sleep-related hypoxemia disorder

This condition refers to below-normal blood oxygen levels during sleep, which may be a symptom of medical conditions such as pulmonary hypertension, which is increased blood pressure in the blood vessels that supply the lungs (pulmonary arteries), or neurodegenerative diseases, stroke and even epilepsy.[267]

## Category 4: Central Disorders of 'Hypersomnolence'

These are an interesting set of conditions that lead to severe daytime sleepiness even though the individual experiences an

apparently normal night of sleep and does not complain of difficulties falling asleep or staying asleep (insomnia). The most
common form is narcolepsy. **Narcolepsy** affects an estimated 1 in
every 2,000 people in the United States, so about 200,000 Americans and approximately 3 million worldwide. As narcolepsy can
vary in its severity, it is estimated that only 25 per cent of people
who have narcolepsy have been formally diagnosed; it is only
when it is severe that people seek help. Narcolepsy causes a person to suddenly fall asleep at inappropriate times because the
brain is unable to regulate the sleep / wake cycle normally. The
result is often excessive daytime sleepiness and feelings of drowsiness throughout the day, with difficulty concentrating or staying
awake. Some individuals may fall asleep suddenly and without
warning, and experience narcolepsy with **cataplexy**, which is the
temporary loss of muscle control leading to collapse. Such cataplexy can be triggered in response to emotions such as laughter and
anger. Yet other individuals with narcolepsy may also experience
**sleep paralysis**, which is a temporary inability to move or speak
when waking up or falling asleep. Narcolepsy is also associated
with excessive dreaming, particularly as individuals fall asleep
(**hypnogogic hallucinations**) or just before or during waking
(**hypnopompic hallucinations**).

The causes of narcolepsy are complicated.[268] In some cases,
and I stress some cases, it arises due to a lack of the brain neurotransmitter orexin (also called hypocretin), which normally acts
to drive the wake state.[269] It seems that the lack of orexin is caused
by an autoimmune response in which the immune system attacks
the cells that make orexin in the hypothalamus, or maybe even
the cells which respond to orexin. Support for an autoimmune
origin for some types of narcolepsy comes from the finding of
an increased risk of developing narcolepsy following vaccination
with Pandemrix, an H1N1 influenza vaccine that was produced
for the pandemic of 2009 and used in several European countries.[270] It seems that the adjuvant used for the vaccine (adjuvants

are substances added to a vaccine to increase the body's immune response to that vaccine) triggered an overly aggressive immune response in a small number of people, around 1 in 52,000.[271-2] This type of adjuvant is no longer used, and other H1N1 flu vaccines have not been associated with narcolepsy. Important lessons have been learnt from this unexpected and sad event for the development of new vaccines such as the COVID-19 vaccines. While there is currently no cure for narcolepsy, the condition can be managed using medications such as Modafinil. This medication can only be prescribed by a clinician, and is used to great effect to increase daytime alertness and reduce daytime sleepiness. In addition to narcolepsy, Modafinil is also frequently taken by individuals who want an alertness 'boost'. It has been called the 'world's first safe smart drug'. Currently we do not know what the short- or long-term side effects might be from taking Modafinil, and buying this drug online for enhancing alertness is strongly not advised.

## Category 5: Parasomnias

Parasomnias are a complex and varied group of sleep disorders that involve unwanted experiences while you fall asleep, when you are sleeping or as you are waking up. Parasomnias may include abnormal movements, behaviours, emotions, perceptions or dreams. In many cases you may remain asleep during the event and, when asked, will often have no memory that it occurred. Parasomnias come in a variety of different types,[273] including:

### Confusional arousal

This is where you act in a very strange and confused way as you wake up or just after waking. You may experience slow speech, confused thinking, poor memory and a foggy state of mind. They can occur when someone tries to wake you, particularly when woken from slow-wave sleep (SWS).

### Sleepwalking

Also called 'somnambulism', this is when you get up from bed and walk around even though you are still asleep. You may talk and your eyes are usually open and have a confused, 'glassy' look to them. Actions while sleepwalking can be crude or strange, such as urinating in your mother's shopping basket, as (apparently) I did several times as a child! You may move furniture or climb out of a window. Sleepwalking most often occurs during SWS. The best advice if you encounter a person sleepwalking is gently to guide them back to bed. Individuals may wake naturally from sleepwalking and then find themselves in a strange place, or they may return to bed and be entirely unaware of what has happened.

### Sleep terrors

Also called 'night terrors', these parasomnias typically involve sitting up in bed and screaming or shouting, possibly lashing out, and with a look of horror on your face. These episodes also typically occur during SWS. The best thing to do if your child or sleeping partner has an episode of night terrors is to stay calm – it can be awful to watch – and wait until they calm down. You should not interact with them, unless they are not safe, and you shouldn't attempt to wake the individual when experiencing an episode. They may not recognize you and become more agitated if you try to comfort them. Individuals will probably have no recollection of the event in the morning, but it may be useful to discuss the episode to find out if anything is worrying them, as anxiety and stress can trigger sleep terrors.

## *Sleep-related eating disorder (SRED)*

SRED involves repeated episodes of compulsive binge eating and drinking after waking up in the night, and may only last 10 minutes. Episodes occur almost every night, and can occur several times a night. There may be no recall of the binge eating in the morning, and trying to stop a person during an episode tends to provoke anger and resistance.

## *REM sleep behaviour disorder (RBD)*

RBD occurs when you act out vivid dreams while in REM sleep. Normally during REM sleep the body is paralysed (chapter 2), but in RBD this paralysis is lost so dreams are accompanied by lots of action, and they may even be violent. As previously mentioned, devoted husbands have attacked and even killed their wives, mistaking their wife for an intruder during the dream state (chapter 2), and a diagnosis of RBD has been the basis for acquittal if murder has taken place during an RBD episode.[40] RBD should not be confused with sleepwalking, when the sleeper is usually in SWS and does not act out their dreams. When woken from an episode of RBD, the individual can often recall clear details of the vivid dream. RBD is predictive of more serious conditions, with approximately 50 per cent of individuals with RBD developing Parkinson's disease or dementia within 10 years of developing RBD, and the likely mechanism is altered levels of the neurotransmitter dopamine in the brain due to the loss of dopamine-producing neurones.[274]

## *Sleep paralysis*

This is where you are awake but unable to move your body when either falling asleep or, and more commonly, from waking from sleep. In REM sleep the body is paralysed, and in sleep paralysis this paralysis persists after waking. You may be unable to speak, or unable to move your arms, legs or body. You are fully aware of what is happening, and an episode can last for seconds or minutes.[273] Sleep paralysis can be very alarming but it is harmless, and most people will only have an episode once or twice in their lifetime, but to reduce the chances of sleep paralysis make sure you are getting enough sleep, and practise good 'sleep hygiene', of the sort I will discuss in chapter 6.

### *Bruxism*

This is an involuntary grinding or clenching of the teeth while sleeping which is often related to stress or anxiety. Some people experience pain in the face or jaw and headaches. If severe and prolonged, teeth can be damaged, and one solution is to use a medically approved mouth guard which keeps the teeth separated.[275]

Other parasomnias include: simple **sleep talking**; **sleep-related hallucinations**, which are imagined events that seem very real; frequent **nightmares** that prevent you from getting a good night's sleep which often result in insomnia; **bedwetting** (**nocturnal enuresis**), which is common in children and occurs when a person urinates by accident during sleep.

The causes of this diverse array of parasomnia conditions are largely neurological (within the brain) and there are multiple risk factors that can increase their occurrence. Some people are more likely to sleepwalk or have other parasomnias when they are stressed. Sleepwalking or bedwetting often occur in childhood and are usually outgrown. There may be a genetic contribution if conditions such as sleepwalking or night terrors run in the family. Around 80 per cent of patients with **post-traumatic stress disorder** (**PTSD**) have distressing dreams/nightmares during the three months following the trauma in which the trauma is visualized. Nightmares are also a common side effect of some medications, such as antidepressants and some blood pressure medications, so check the information leaflet that accompanies your medications. Sleepwalking, night terrors and other parasomnias are also more likely in individuals who use alcohol excessively or take mind-altering drugs.

## Category 6: Sleep-related Movement Disorders (SRMDs)

SRMDs are relatively simple, often stereotyped, movements that may disturb your sleep and give rise to insomnia.[276] The most common is **restless-legs syndrome** (RLS), where there is an overwhelming urge to move your legs, and the sensation gets worse in the evening or at night. RLS is also associated with involuntary jerking of the legs and arms, known as **periodic limb movements in sleep** while asleep. Symptoms vary from moderate to severe, and can be daily or infrequent. In severe cases, RLS can be very distressing and disrupt the sleeper and the bed partner. In most cases, there is no obvious cause, although some neurologists have suggested that the neurotransmitter dopamine, which has also been linked to REM sleep behaviour disorder, may play a role. Periodic limb movements can be a common disorder during pregnancy, making sleep miserable, but disappearing shortly after giving birth. Several studies have shown that people with Parkinson's disease (a condition associated with a deficiency in dopamine) are more likely to have RLS.[277] In some cases, it is caused by an underlying health condition, such as iron deficiency, and taking iron supplements in various forms has been shown to help reduce the severity of the condition in some instances.[278] Kidney failure has also been linked to the syndrome.[279]

## Category 7: Other Sleep Disorders

For the sake of completeness, I include this final category. However, it is not very useful as it encompasses those sleep conditions which do not neatly 'fit' into any of the other six classifications: it is a category of last resort! For example, **environmental sleep disorders** would be included here and refer to a sleep disturbance caused by environmental issues such as aircraft or road traffic noise or even tobacco smoke. These environmental factors disrupt sleep and can lead to insomnia and excessive daytime sleepiness. These 'other' sleep anomalies may also arise from an undiagnosed neurological problem which is very rare and/or very poorly studied.[280]

## Questions and Answers

### 1. How do I know if I have some sort of sleep disorder?

The most common symptoms that you may recognize in yourself, or in a partner or family member, are: excessive drowsiness or tiredness throughout the day; difficulty waking in the morning; difficulty staying awake during the day, even when driving or working, and the need for a nap; mood changes or irritability; trouble falling asleep several times per week; difficulty staying asleep all night; loud snoring that may wake you or your sleeping partner; frequent headaches or a dry throat when you wake. You can use the information above, along with your own sleep diary (Appendix I), to help get a better understanding of the type of sleep or circadian rhythm disorder you may have. Use this information to generate an informed discussion with your medical practitioner.

### 2. What's the crust/sand in my eyes when I wake up?

The Sandman (Ole Lukøje) is a mythical character in Western and Northern European folklore who puts people to sleep and encourages and inspires beautiful dreams by sprinkling magical sand onto their eyes. There is certainly no Sandman, but when we wake in the morning, we often have 'sleep crust', or 'sand', in the corners of our eyes. The medical term for this is 'rheum'. Rheum is a mix of mucus, dead cells from the surface of the eye, oils, and tears produced by the eye during sleep. During the day, debris is washed away from the surface of the eye and the eye is lubricated by blinking and tears. We don't blink while we are asleep, so rheum collects in the corners of our eyes – it feels like sand in your eyes. There has also been speculation that the side-to-side movements of the eye during rapid eye movement sleep (REM) is to keep the eyes lubricated while we sleep and can't blink. If the rheum is yellow or green this might be a sign of infection, and if it persists you should seek medical advice.

### 3. I think I have a sleep disorder but the only help I get from my medical practitioner is sleeping pills. What can I do?

This is a common problem and occurs because of three key factors. (i) Your medical practitioner has probably had very little training in the areas of sleep and circadian medicine and may not be aware of how to advise you. (ii) In fairness to your medical practitioner, pharmacological treatment options are very limited. In many cases, a prescription of sleeping pills is currently the only option, which can be useful short term (chapter 6). (iii) Those treatment options that are available, other than sleeping pills, such as cognitive behavioural therapy for insomnia, or CBTi, require specialist practitioners and 1:1 sessions. Such specialists are rare and many health care systems do not have the resources needed for such treatments. There are, however, actions that we can undertake to improve our sleep. In the same way that diet and exercise can improve health, good sleep practices can be adopted to improve sleep, which is the topic of the next chapter (chapter 6).

**4. A lot of people are talking about 'kava kava' to help with sleep and relaxation. Is there any evidence to support this?**

The history of kava kava is that it comes from the Pacific Islands and is consumed throughout the Pacific Ocean cultures, including in Polynesia, Hawaii, Vanuatu and Melanesia. The root of the plant is used to produce a drink with sedative and anaesthetic properties, and active ingredients are called 'kavalactones'. Interestingly, these kavalactones seem to enhance GABA (gamma-aminobutyric acid) pathways in the brain, which are also the targets of both the benzodiazepine (e.g. Diazepam (Valium), Chlordiazepoxide (Librium), Alprazolam (Xanax) and Z drugs (e.g. zopiclone, zaleplon and zolpidem), which are discussed in the next chapter. Kava kava drinks have become popular in Australia, New Zealand, the USA and Europe to help with relaxation and sleep. A major, although fairly old, review of published papers available up until 2003 concluded: 'Compared with placebo, kava extract appears to be an effective symptomatic treatment option for anxiety. The data available from the reviewed studies suggest that kava is relatively safe for short-term treatment (1 to 24 weeks), although more information is required. Further rigorous investigations, particularly into the long-term safety profile of kava, are warranted.'[281] Studies on rats also suggest that kava extract can improve sleep in sleep-disturbed animals.[282] So it is still early days, but kavalactones may provide a useful aid for sleep in the not too distant future.

# 6.

# *Back in the Rhythm*

## *Solutions to sleep and circadian rhythm disruption*

Perhaps the most valuable result of all education is the ability
to make yourself do the thing you have to do, whether
you like it or not.

Thomas Henry Huxley

So many people feel that there is nothing they can do about their
poor sleep, and that sleep is 'what you get'. Until relatively
recently, the only response to a declaration of 'I'm not sleeping
well' was a prescription of sleeping tablets. Between the 1920s
and the mid-1950s, practically the only drugs used as sedatives and
hypnotics (sleeping pills) were barbiturates such as pentobarbital
(Nembutal). Barbiturates cause nervous system suppression by
stimulating the inhibitory neurotransmitter system in the brain
called the gamma-aminobutyric acid (GABA) system. These
drugs were used extensively in the 1960s to the mid-1970s despite
being highly addictive and dangerous. They were dangerous
because there is only a small difference between a dose of barbit-
urate that has the desired effect of sedation and an overdose that
can cause coma or death. The barbiturates were gradually
replaced by the benzodiazepines.

In 1955, Leo Sternbach (who fled from the Nazis to the USA in
1941), working for Hoffmann-La Roche as a research chemist,
discovered the first benzodiazepine, chlordiazepoxide, marketed
in 1960 as Librium. A modified form of chlordiazepoxide with

enhanced activity, called Valium (diazepam), followed in 1963. The benzodiazepines are less dangerous than barbiturates as they are less addictive and less likely to cause death by accidental overdose. Benzodiazepines, like barbiturates, work largely by enhancing the release of the inhibitory neurotransmitter GABA, to calm the individual. In the mid- to late 1970s, benzodiazepines were the most frequently prescribed of all medications. They can be helpful if taken short term and intermittently. But it took 15 years to appreciate that they *can* also be addictive if taken chronically, and they can cause memory loss and depression. So by the 1980s they began to fall out of favour.[283] Benzodiazepines that have been approved for insomnia include estazolam, flurazepam (Dalmane), temazepam (Restoril), quazepam (Doral) and triazolam (Halcion).

Most recently, the benzodiazepines have been replaced by another class of drugs, the 'non-benzodiazepine, or Z, drugs' such as zolpidem, zaleplon and zopiclone. These are not benzodiazepines, as they are based upon another chemical compound, but also enhance GABA release. The Z drugs were initially thought to be safer than benzodiazepines, but they are now known to have essentially the same long-term-use problems as benzodiazepines – addiction, depression and memory loss. There are also reports of sleepwalking and even sleep driving.[284]

The benzodiazepines and Z drugs can be used short term for the effective treatment of sleep problems such as insomnia, but should not be used long term. However, before such drugs are prescribed, alternative approaches to improve sleep should be attempted first. Today, there *are* evidence-based alternatives to benzodiazepines and Z drugs. Such alternative 'corrective actions' have been given the general name of 'cognitive behavioural therapy for insomnia', or CBTi. The aim of CBTi is to use methods for treating sleep and circadian rhythm disruption (SCRD) without the use of sleeping pills. CBTi is designed to change unhelpful sleep habits and to encourage individuals to adopt behaviours that promote falling asleep or staying asleep and that prevent daytime sleepiness. CBTi can be undertaken by individuals on their own, and/or with regular, often weekly, visits to a specialist medical practitioner, and/or by using some form of app-based digital CBTi such as Sleepstation or Sleepio. A helpful approach, if you do decide to undertake some form of CBTi, is to keep a sleep diary so that you can assess whether a change in your behaviour actually delivers better sleep. You can easily develop your own sleep diary, and an example is provided in Appendix I. It really is worth trying to keep a record of your sleep as many of us underestimate our sleep and think it is worse than it actually is.[285]

Some suggestions to alleviate or mitigate aspects of SCRD are listed below for your consideration, and I want to stress that there is no 'one perfect solution'. It's a bit like exercise: there are lots of different approaches you can take – you just need to get on with it! And again, like exercise, you have to stick at it – sadly there is often no 'quick fix'. These suggestions are based upon current thinking and address the questions I get asked most frequently. The questions have been grouped into four sections: What should I do during the day? What could I do before bed? How do I make my bedroom a haven for sleep? What should I do in bed? And the information has been summarized in Table 2 (p. 135). After considering these questions, the final part of the chapter considers what employers can do now to help mitigate some of the problems faced by their employees resulting from workplace-induced SCRD. First, let's start with the actions we can take.

## What Should I Do During the Day?

### Morning light

Most of us should get as much morning natural light as possible. As discussed in chapter 3, this has been shown to move the circadian clock to an earlier time (advance the clock).[93] This will make you feel sleepier earlier, and an earlier bedtime will help you get longer sleep. A small group of individuals (around 10 per cent of the population) who are very early chronotypes (larks), who go to bed and get up very early, may benefit from experiencing late-afternoon/evening light exposure (see Appendix I if you want to assess your chronotype). This acts to delay the clock (make you get up later and go to bed later) and will align these individuals more closely to the rest of us in the population. In the absence of natural light, morning light exposure using a light box has also been shown to be helpful for circadian timing problems.[286] There are many light boxes available on the market, and the main thing to look for is that they produce enough bright light, in the region of 2,000 lux or more (Figure 3, p. 61). The key point is that if you are an owl and want to become more of a 'morning person', seek out advancing morning light, but avoid delaying evening light.[92]

### Taking a nap or siesta

A siesta or short nap in the early afternoon is historically common throughout the Mediterranean countries, Southern Europe and Central China. The siesta has also been adopted in many countries where Spain had a historical influence like the Philippines and Hispanic America. A feature that unites all these places is a warm climate and a large midday meal. Accommodating the siesta into modern industrial schedules is proving difficult in the cities of Spain and other countries, and there have been repeated calls by governments and the business sector to abandon the practice altogether. However, if you live in places like rural Spain, and the siesta is still part of the daily schedule, then embrace it! Sadly, most of us cannot. North American, Northern European and the English-speaking nations, have largely adopted the view that a one- or two-hour siesta in the afternoon is not appropriate and cannot be accommodated into modern work schedules. I sense some whiff of puritan meanness here . . . So if you are not living in a warm rural environment, what do you do if you feel tired in the afternoon and crave a nap? The first point to make is that if you want a nap, you are probably not getting enough sleep at night, and that needs to be attended to first. However, if you enjoy a nap, then the occasional nap for no longer than 20 minutes is probably fine. Such short naps in sleepy individuals have been shown to improve the alertness and performance across the afternoon. Longer naps may be counterproductive as recovery from an extended nap can lead to feelings of grogginess and lowered alertness for some time after the nap. This is called 'sleep inertia'.[287] In addition, naps close to bedtime (within six hours or so) will act to reduce sleep pressure (chapter 2), and this will probably delay sleep at bedtime. This can be a big problem in some teenagers who go to sleep late, get up tired and struggle through the school day. A recent study in US adolescents suggested that SCRD affects around 24 per cent of students,[288] and we find similar numbers in the UK. After school in the afternoon, a significant number of students arrive home and then sleep for several hours. This reduces the sleep pressure and delays sleep that night, generating a vicious circle of later sleep times followed by longer naps in the late afternoon.[289] The bottom line is that for most of us the occasional nap is fine, but be careful you don't become dependent upon long daytime naps. This will delay sleep at night, leading to shortened night-time sleep, especially on workdays, when a lie-in is not possible. Not all of us have to follow a conventional Northern European work/sleep regime. Winston Churchill adopted the Spanish tradition of a siesta, starting around 4.30 p.m. after a good lunch. Throughout the Second World War he would take off his clothes and go to bed for up to two hours. At 6:30 p.m. he would wake and have the second bath of the day, and this set him up for a long dinner. Around 11 p.m. he would work for a few hours before finally going to bed. Churchill was an unapologetic night owl, and as the Prime Minister during that era, he could decide his own work schedule.

## Exercise

The relationship between exercise and sleep is complex, but, overall, exercise is good for our sleep.[290] For most of us, exercise in some form can help sleep/wake timing and reduce insomnia, particularly if it is outside under natural light in the morning. It may be that exercise and light combine to improve sleep and sleep/wake timing.[291] However, exercise close to bedtime (1–2 hours) may be a problem. The transition from wake to sleep involves, and may even require, a small drop in core body temperature.[292] Moderate to vigorous exercise may override this circadian-driven change in body temperature and delay sleep onset in some, but not all, individuals.[293-4] In addition, vigorous exercise can cause a 'runner's high'. And for those of us that don't run, it is described as 'pure happiness, elation, a feeling of unity with one's self and/or nature, endless peacefulness, inner harmony, boundless energy, and a reduction in pain'. I mentioned this to a friend and, rather missing the point, she said: 'That's exactly how I feel after eating chocolate.' Anyway, exercise-induced elation and boundless energy just before bed may not help with sleep, but peacefulness and inner harmony could be good for it. The basis for this 'high' was originally thought to be the release of endorphins from the pituitary gland. However, more recently, it has been suggested that another group of natural compounds called the 'endocannabinoids' are responsible. These are molecules made all over the body with a similar chemical structure to plant cannabinoids, and they increase in the blood in response to exercise.[295] Interestingly, dark chocolate may stimulate endocannabinoid activity. The bottom line is that exercise is overwhelmingly good for our health, but a 'rule of thumb' indicates that exercise very close to bedtime may delay sleep in some people. I will discuss exercise again in chapter 13, and the role of timed exercise in helping regulate our metabolism.

## When to eat

I will also discuss 'when to eat' in more detail in chapter 13. So, in brief, eating late in the day has been shown to increase the chances of weight gain[296] and increase the susceptibility to metabolic problems such as Type 2 diabetes.[297] As discussed in chapter 5, weight gain can lead to obstructive sleep apnoea (OSA) and all the problems associated with that condition.[298] In addition, digestive processes are reduced toward bedtime, so if the major meal of the day is prior to bed, this can lead to digestive health issues such as excessive stomach acid production and a greater risk of peptic ulcers.[299] Stomach pain can then act to disrupt sleep. And to complete the circle, SCRD increases the risk of peptic ulcers.[300]

## Coffee and tea

Caffeine in tea (black or green) and coffee can have a major alerting effect on the brain as it blocks the receptors in the brain that respond to adenosine, which, as we discussed in chapter 2, helps drive sleep pressure. In addition, caffeine increases adrenaline release, which promotes the 'fight or flight' response (chapter 4), increasing heart rate, breathing and alertness. There is considerable individual variability in responses to caffeine, depending on body weight, pregnancy status, medication, liver health and prior caffeine drinking history, but in healthy adults significant levels of caffeine remain in circulation for 5–6 hours after taking a drink. As a result, a strong coffee or tea in the afternoon could delay sleep at night.[45] A good overall strategy is to be aware of how much caffeine you are drinking. Start the day with the strongest caffeine-rich drinks and then taper off after lunch, switching to decaffeinated by mid-/late afternoon. It is surprising how much caffeine there is in the drinks we consume. An average cup of coffee (240ml) contains around 100mg or more of caffeine. A single espresso will have around 75mg caffeine; a cup of black tea (240ml) 40–50mg; green tea (240ml) 20–30mg; a 330ml can of Coca-Cola or Coca-Cola Zero sugar, 32mg of caffeine. So stick to decaffeinated drinks or herbal tea in the late afternoon and evening. My family are avid coffee and tea drinkers but switched a few years ago to decaffeinated drinks in the afternoon and evening and now we all get to sleep sooner.

### Stress

As discussed in chapter 4, don't let stressful experiences build throughout the day, and if you find that you have unmanageable levels of stress consider stress management or mindfulness techniques. Short-term emotional stress, spilling over from daytime activities, is a very powerful agent for sleep disruption.[301] It really is critical to keep stress under control during the day. But try to avoid sedatives (see below).

## What Could I Do Before Bed?

### Light levels and computer screens

As discussed in chapter 3, the circadian system needs relatively bright light, in the 100–1,000 lux range, and for an extended period of time of 30 minutes or more to robustly shift the clock (chapter 3, Figure 3). So under normal conditions, low levels of domestic lighting (100–200 lux), or the light emitted from most screens, which is less than 100 lux, will have only a minor or no effect on shifting our circadian rhythms. However, in addition to shifting the clock, light, and especially blue light, has a direct alerting effect upon the brain. The levels of light needed to alert the brain seem to be lower than the intensities needed to shift the clock.[302] Increased light-induced alertness before bed will delay sleep, but the actual levels of light, and the impact of different colours of light, on alertness are complex and not fully understood.[303] Although the precise data are lacking, I think it makes sense to lower light exposure around two hours before going to bed, both in terms of overall room light and in specific settings such as looking directly at a computer screen. It is noteworthy that for most of us, one of the last things we do before bed is spend time in the bathroom, which is often the most brightly lit room in the house, looking directly into an illuminated mirror while we clean our teeth! In addition to reducing alertness, reducing light levels before bed can form part of the bedtime routine of 'psychological sleep preparation' which, with other activities (see below), prepares us for sleep.

## Sleeping tablets and sedatives

As I mentioned at the start of this chapter, the use of prescription sedatives to aid sleep can be useful short term to readjust sleep patterns. However, long-term use, especially in night shift workers, can cause problems because of side effects. For example, the chronic use of benzodiazepines (e.g. Xanax, Valium, Ativan and Librium), which are anti-anxiety medications and increase drowsiness, are potentially addictive when used chronically and can lead to impaired memory formation and reduced attention and alertness during the day.[304] There is also a suggestion that long-term use, greater than three years, may increase the risk of developing dementia.[305] However, in another study, no such associations have been found between benzodiazepines and Z drug use and dementia.[306] Non-prescription sedatives such as alcohol and antihistamines (e.g. diphenhydramine and doxylamine) should be avoided, as the side effects, especially with alcohol, can have a poor impact upon health and our ability to function during the daytime.[307] So sleeping tablets should ideally not be used, but may help with a short-term correction of sleep. Long-term use should be avoided.

## Difficult discussions

I appreciate that this may be the only time in the day to talk to your partner about pressing matters, but it is really important to avoid any discussion or consideration of stressful topics immediately before bed. The acute elevation of cortisol and adrenaline will increase alertness and delay sleep (chapter 4). Avoid topics like personal finance or sad topics in the news. You could ask your partner about the best thing that happened to them during the day, or tell them something funny you have read or been told, or mention something that your partner did that you enjoyed or appreciated. Be nice! And I resurrect our old family motto: 'If you can't say anything nice, don't say anything at all!' As an aside, I always liked the corruption of this motto: 'If you haven't anything nice to say about anyone, come sit by me' – which I thought was by Dorothy Parker, but was in fact said by the Washington socialite Alice Roosevelt Longworth.

## Taking a bath

A relaxing behaviour, such as a bath or shower, or warming the hands and feet,[308] can be very useful before bed. Again, it can be part of your routine of 'sleep preparation', but, in addition, warming the skin will also help dilate the blood vessels in the skin (peripheral vasodilation). This increases blood flow from the core of the body to the skin, where heat is radiated to the environment. Why is this important? Some really interesting studies have shown that skin vasodilation, which causes heat loss, reduces the time it takes to fall asleep.[309] So keep your hands and feet warm in bed – especially if you have a condition like Raynaud's. A few years ago, a friend with Raynaud's complained of difficulty getting to sleep, but after I suggested she wear bed mittens and bed socks, her sleep improved greatly. I predict a whole new line of luxury sleepwear mittens and socks in the coming years. Thick enough to keep the hands and feet warm, but not too thick to prevent heat loss.

## How Do I Make My Bedroom a Haven for Sleep?

### The bedroom

Making the bedroom or sleeping space suitable for sleep is a much overlooked yet critical part of getting the sleep you want. If the bedroom is too warm, this will affect your ability to lower core body temperature, and this will delay sleep onset. Ideally, the bedroom should promote sleep by minimizing distractions and stimuli that alert the individual. The sleeping space should be quiet[310] and dark, and devices such as televisions, computers and smartphones should be removed. Smartphones are now used routinely as alarm clocks, and so removing them from the bedroom can be a problem. However, if the phone is a distraction, then it should be replaced by an alarm clock; but this is also not straightforward. Many of us 'clock watch' and get anxious about the amount of time left available for sleep and so constantly check and re-check the alarm clock, generating more anxiety.[311] Under these circumstances, the alarm setting can be used, but the face of the clock can be covered. Bedside lights should be bright enough to read, but kept as low as possible to reduce alertness.

### Sleep apps

Sleep apps can be useful in providing a record of roughly when you went to sleep, when you got up, total sleep time and how many times you got up during the night, and for this most are reasonably accurate. However, measures of REM versus non-REM or even 'deep sleep' are more difficult to assess from the currently available devices, and may be profoundly misleading. In theory, such monitoring systems could be useful in showing you that changes in your behaviour have indeed had an impact and improved your sleep. But because most of the commercial apps available fail to provide an accurate measure of overall sleep[312], individuals can become anxious if their device inaccurately reports 'insufficient restful sleep' or 'low levels of REM sleep' [313-14]. It is worth noting that no sleep apps are currently endorsed by the national sleep academies or sleep specialists.[315] As a result, it would be wise and prudent not to take sleep apps too seriously. So many anxious individuals have approached me after a seminar worried about what their sleep apps report. One individual told me that his app was stating that he had 'too little deep sleep'. His response was to set his alarm for 3 a.m. so that he could wake up and check his app to see how much 'deep sleep' he had experienced. I spent some time explaining why this was a bad idea.

## What Should I Do in Bed?

### Routine

Keeping a good bedtime routine of getting up and going to bed at the same time is very helpful. Especially if the routine is optimal for your own sleep needs in terms of sleep timing and duration.[316] Such a schedule reinforces the exposure to those environmental signals that entrain the circadian system, most especially light (chapter 3), but also food and exercise (chapter 13). Sleeping in for an extra few hours on a Saturday or Sunday morning may feel like a good idea to try and catch up on the sleep you missed during the working week, but sadly it usually does not help you make up the sleep debt you have accumulated. True, you may be a bit less drowsy or even stressed during that day, so it can help short term, but sleeping in will not wipe out the accumulated effects of sleep loss on your health. In addition, by sleeping in you will not get the necessary morning light exposure to set your circadian rhythms. There is a real problem for those individuals who are 'natural long sleepers', needing nine hours or more of sleep each night. Because of the commute to and from work, family demands and a multitude of other pressures, it may not be possible during the working week to get nine hours of sleep, and currently we do not know if for these individuals sleeping in at the weekend might be helpful or not. Some sleep experts predict catch-up sleep will be beneficial for long sleepers.

### Consensual sex

The question 'Does sex help you sleep?' is something so many people want to ask me, but are usually too afraid to enquire in public. Though this question is usually asked anonymously in the 'Chat Room' when I give online talks. Interestingly, this was an especially common online question during the COVID-19 lockdown. There's even a French expression that has been adopted by some of the English-speaking world for how quickly men fall asleep after orgasm: *la petite mort*. So, what is the evidence? The short answer seems to be a clear yes, sleep is good for sex, and sex is good for sleep. A fairly recent study examined women over a two-week period and showed that a one-hour increase in sleep length corresponded to a 14 per cent increase in the probability of engaging in partnered and consensual sexual activity. OK, good sleep seems to promote sex,[317] but does sex promote good sleep? At first sight there may seem to be a problem here: how can sex, which is arousing, at least for most people, promote sleep? A recent large survey examined the perceived relationship between sexual activity and subsequent sleep in the adult population: 778

participants (442 females, 336 males; around 35 years of age) completed an online anonymous survey, and the results indicated that orgasm with a partner was associated with the perception of good sleep. In addition, orgasm achieved through masturbation was also associated with better sleep and getting to sleep.[318] The authors of this study conclude: 'Promoting safe sexual activity before bed may offer a novel behavioural strategy for promoting sleep.' Interesting. Although I am not convinced that this represents an entirely 'novel' strategy – but the point they make is noted.

The basis for feeling sleepy after sex seems to be related to the release of a specific set of hormones that do similar things in both men and women. Sex increases the release of oxytocin from the posterior lobe of the pituitary gland.[319] Oxytocin does multiple things, and its action can depend upon what you are doing, but in the context of sex and sleep it makes you feel more connected to your partner and it also lowers cortisol, so reducing stress.[320] In addition, having an orgasm releases a hormone called prolactin which can remain elevated for at least an hour after orgasm,[321-2] and makes you feel relaxed and sleepy. The combined effects of oxytocin and prolactin, in men and women, mean that you are more inclined to cuddle up to your partner and then fall asleep.

## The mattress

A good mattress, pillows and bedding make intuitive sense for good sleep, but historically there are relatively few scientifically controlled studies regarding mattress type and sleep quality.[323-4] Research has suggested, however, that a mattress and bedding that can conduct heat away from the body, and so lower core body temperature, can reduce the time it takes to get to sleep and increase deep slow-wave sleep.[325-6] There is a huge variety of mattresses and bedding to choose from. And, similarly, there is huge variation in the type of mattress and bedding people find comfortable. The easy part is deciding if you need a new mattress, and if you answer yes to several of the following questions, you may want to consider a change in your mattress, pillows and bedding. Does your mattress feel saggy or unsupportive? Do you wake up with aches and pains in your back or limbs? Do you feel your partner moving around at night? Has it been more than seven years since you purchased your mattress? Do you become more allergic or develop asthma symptoms in bed? Is it uncomfortable when you have sex? Do you get too hot in bed but the bedroom itself is cool? Do you sleep badly? It is also good to discuss these questions with your sleeping partner. As we spend approximately a third of our lives in bed, it really is important to find the bedding arrangements that work for us and our sleeping partner. Ask friends about their mattress recommendations, and go to a variety of showrooms and try them out, obviously being alert to the sensitivities of other customers.

### Lavender and relaxing oils

Relaxing oils are often proposed to help improve sleep. However, the evidence that such oils really help sleep is not strong,[327] and there is the suspicion that any effects are probably placebo. Although there is some evidence that lavender does have a 'greater than placebo' effect for sleep improvement,[328] and, to be fair, 'absence of evidence is not evidence of absence'. More research is needed, but for some individuals relaxing oils do anecdotally improve sleep, perhaps because the association of a distinctive 'conditioning' smell, such as lavender, can be part of a bedtime routine that psychologically prepares you for sleep. When sleeping away from home, the smell of a partner's perfume or aftershave might also remind you of home and help you sleep. Marilyn Monroe, when asked what she wore in bed,

famously replied: 'I only wear Chanel N°5.' Sadly, chronic barbit-
urate use, and not Chanel N°5, appears to have been Ms Monroe's
primary sleep aid. With tragic consequences.

### Earplugs

Earplugs, including wax earplugs, can help if a sleeping partner
snores, or if there is external noise.[329] If a partner's snoring
becomes too disruptive then you may want to consider sleeping
in another room.[330] This does not reflect upon the strength of the
relationship, and may even improve the bonds of the partnership
by improving sleep, making you more empathetic, loving and
happy! As I have mentioned previously, ensure your partner does
not have obstructive sleep apnoea (OSA) (chapter 5) by asking
them to visit their medical practitioner.

### Lying in bed awake

I will mention several times in this book that waking up in the
middle of the night can occur for multiple reasons (see e.g. chapter 8), but this need not mean the end of sleep. Under such
circumstances, it is important not to activate stress responses by
remaining in bed and becoming increasingly frustrated. You may
want to leave the bed, keep the lights low and engage in a relaxing
activity such as reading or listening to music. For the reasons I
mentioned earlier, consensual sex might also provide an additional option.

### Worrying about dreams

From our earliest records, we humans have been fascinated by
dreams and what our dreams might mean. Originally, they were
thought to be linked to the spirit world, until Aristotle, Plato and
the later European psychoanalysts of the nineteenth and twentieth

centuries suggested that dreams are a way to act out unconscious desires in a safe setting which would otherwise be unacceptable.[331] But, frustratingly, today we are still very unsure why we dream (chapter 2). At some level dreams probably help us process information related to memory formation and/or try to work out our emotional status/problems. Part of this processing might involve cleaning-up operations and the removal of 'junk', perhaps weakening rather than strengthening memories, which then leads to bizarre associations.[332] We do know that increased anxiety can be linked to more vivid dream images.[333] For example, at times of heightened anxiety many people report having more frightening and realistic, vivid dreams. After the Twin Towers were destroyed by the terrorist attack in New York on 9/11, many New Yorkers reported dreams of being overwhelmed by a tidal wave or being attacked and robbed, but there was no increase in dream content involving aeroplanes or tall buildings. There was no 'exact replay' in dream form of the actual events of 9/11, despite the repeated showing of these events on the television.[334] The authors conclude from their findings that it is the underlying emotional state of the individual that generates the dream imagery.[334] Indeed, some dreams may simulate threatening events, which then allows us to play out different possible responses for more creative decisions.[335] Try not to worry about your dreams. Take comfort in the fact that your brain is doing what it should – trying to make sense of a deeply complicated world. And finally, despite the persistent belief held by many that dreams can predict the future, this has never been demonstrated scientifically.[336] Just to emphasize, what REM sleep and perhaps dreams can do is help us clarify and resolve some of the emotional and stressful issues we experience. It has been estimated that we have no recollection of more than 95 per cent of our dreams, and, with this in mind, if dreams were truly essential, and used to guide us during our wake state – then why don't we remember more?

## *Post-traumatic stress disorder (PTSD) and nightmares*

Nightmares are particularly vivid dream events thought to be the brain working through intense emotional experiences, and some nightmares can cause us to wake up. However, nightmares are different from PTSD. Nightmares are usually abstract in nature. But PTSD develops following a specific traumatic event and individuals with PTSD have severely disturbing recurrent and involuntary recall of the event itself, either during the day in the form of flashbacks or during sleep in the form of nightmares that visualize the event.[337] It is thought that less than 10 per cent of individuals who have experienced traumatic events develop PTSD. However, the term PTSD is being used increasingly to describe any emotional disturbance following a traumatic event. This acts to diminish the severe nature of PTSD in those individuals suffering from the condition. As sleep enhances the consolidation of memory (chapter 10), and especially as REM sleep has been associated with the consolidation of emotional memories,[338] there is currently debate regarding whether individuals should be encouraged to sleep or be kept awake following a traumatic event. There are some data to suggest that sleep deprivation following a traumatic event may be useful for the reduction of flashbacks, but more studies are needed to establish whether this should be introduced into clinical practice.[339-40] Individuals with PTSD usually recover over time, particularly if the advice for 'good sleep' outlined in this chapter is followed. However, if PTSD persists, clinical help should be sought.

The discussion above considered multiple ways to help improve your sleep. Nevertheless, it is so important to stress that there is huge variation in sleep duration, timing and structure, not only between individuals but also within each of us across our lifespan (chapter 8). This means that you have to identify what works best for you, and then defend those behaviours!

| During the day | Before bed | The bedroom | In bed |
|---|---|---|---|
| For most individuals, get as much natural morning light as possible. The morning use of light boxes can also help regulate sleep | Reduce light levels approx. two hours before bedtime | Not too warm (18–22°C) | Try to keep to a routine – go to bed and get up at the same time each day, including the weekend and free days |
| If you nap ensure it is not longer than 20 minutes, and not within six hours of bedtime | Stop using electronic devices approx. 30 minutes before bedtime | Keep it quiet, or use 'white noise' or a relaxing sound such as the sea | Ensure the bed is large enough with a good mattress and pillows |
| Exercise – but not too close to bedtime | Ideally avoid prescription sedatives/sleeping tablets | Keep it dark. Use black-out curtains if street light is a problem | Keep bedside lights low |
| Concentrate food intake to the first and middle parts of the day | Don't use alcohol, antihistamines or other people's sedatives | Remove TV, computers, tablets, smartphones | Consider using relaxing oils (e.g. lavender) |
| Avoid excessive consumption of caffeine-rich drinks especially in the afternoon and evening | Avoid the discussion or consideration of stressful topics immediately before bed | Don't 'clock watch' – consider removing an illuminated clock | Use earplugs or an alternative place to sleep if your partner snores. Ensure snoring is not due to sleep apnoea |
| Make time to step back from stressful situations – don't let stress accumulate. Consider relaxation techniques immediately after work | 'Wind down' before bed. Adopt behaviours that relax you: listening to music, reading, mindfulness or a relaxing bath can be useful | Don't take sleep apps that monitor 'REM' vs 'NREM' sleep too seriously. Currently none have been endorsed by the major sleep societies | If you wake stay calm: consider leaving the bed, keep the lights low and find a relaxing activity, then return when tired |

**Above all – define what works best for you and stick to your routine**

Table 2. Summary of what actions can be taken to alleviate or mitigate aspects of SCRD.

## SCRD in the Workplace – Some Easy Solutions

Employers are in a position to introduce simple measures in the workplace to help address SCRD-related problems. For night shift work alone, 1 in 8 UK workers now work nights, and the number of workers has risen by 250,000 in the past five years to reach more than 3 million individuals, and the trend is expected to increase. So what could be done now to develop 'best practice' to ease SCRD and improve employee safety, health and welfare? I discuss a few easy-to-implement suggestions here.

### Guarding against the loss of vigilance while driving home

Night shift work, with its associated problems of both circadian misalignment and sleep loss, represents a real problem for alertness and increased sleepiness. But it is not just night shift work. Many employees undertake extended working outside the hours of 9 a.m. to 5 p.m., let alone the demands of family life. Work and family life are frequently accompanied by tiredness, loss of vigilance and a high level of micro-sleeps (uncontrollable falling asleep) – dangerous, both in the working environment, but especially if driving is part of your job. Driver fatigue has long been recognized as a major cause of road accidents.[341] A recent report by the National Highway Traffic Safety Administration in the USA stated that every year about 100,000 police-reported crashes involve drowsy driving. These crashes result in more than 1,550 deaths and 71,000 injuries. However, the real number may be much higher, as it is difficult to establish whether the driver was drowsy at the time of a crash. Indeed, this is strongly suggested by another study, commissioned by the American Automobile Association Foundation for Traffic Safety. This organization estimated that 328,000 drowsy-driving crashes occur annually, which is more than three times the number reported by the police. Of these, 109,000 resulted in injury and around 6,400 in death. For these sorts of reasons, some car factories in Bavaria provide buses to get the night shift workers home safely. For many years, the rail industry has utilized some form of 'dead man's handle', or driver safety device (DSD), to alert the driver that they have lost vigilance or fallen asleep, but analogous preventative measures have not been widely adopted in either domestic or commercial motor vehicles until recently. Part of the problem has been the lack of availability of non-invasive driver drowsiness detection technology. However, a range of devices, including steering pattern monitoring, vehicle position in-lane monitoring, and driver eye / face monitoring to detect drowsiness have been developed and are available as units, and are increasingly incorporated into new cars.[342] Employers could provide or subsidize the purchase of such devices.

### Guarding against the loss of vigilance in the workplace

Increased tiredness and loss of vigilance in the workplace during the evening or night shift have been linked to an increased level of accidents. In one study, the average risk for injury was 36 per cent higher after four consecutive night shifts compared to four consecutive day shifts. In another study, work-related injuries increased more than 15 per cent on the second consecutive night shift and 30 per cent on the third, compared to the first shift. Injury risk also increases as the number of breaks decrease.[197] Again, some form of drowsiness detection technology, e.g. identifying a driver's head nodding, could be used to warn individuals that they are falling asleep. In addition, alertness has been shown to improve by illuminating the working environment with sufficiently bright light to promote alertness. Most recommendations suggest that a room lit with 300 lux is sufficient for visual needs. While this level of light will increase alertness compared to dim light, it is not enough to achieve peak alertness, which is around

1,000 lux or more. More studies are needed to define precisely when and how much light is needed in different settings,[343] but an employer should be aware that light levels affect alertness and ensure that light in the workplace is closer to 1,000 lux rather than 300 lux. So it is now clear that multiple consecutive night shifts, a night shift with too few breaks and a dimly lit working environment all contribute to poor performance and a higher risk of accidents.

## Guarding against illness

SCRD is associated with a range of physical and mental health problems (Table 1, p. 77), and detecting such problems early allows the use of interventions which help prevent chronic health conditions developing. As a result, those at higher risk of SCRD should have higher-frequency health checks. If cancer is detected early, the chances of effective treatment and survival increase dramatically. In the same way, the early detection and treatment of conditions such as Type 2 diabetes or depression can prevent these conditions spiralling out of control.

As we know that metabolic abnormalities and cardiovascular diseases are higher in people with SCRD, the appropriate nutrition should be made available to help reduce the promotion of these conditions in the workplace. Vending machines and canteens invariably supply sugar- and fat-rich foods, and, as we shall discuss in chapter 12, this is deeply unhelpful. Sugar-rich foods or sugary beverages at the beginning of a night shift will produce a 'sugar rush' followed by a crash as insulin rapidly reduces blood glucose (Figure 9, p. 275), all of which helps to increase tiredness. So don't have a blow-out in the canteen or fast-food outlet before you start work. Ideally the canteen should help you by providing snacks and easy to digest foods with a high protein content, such as: soups, nuts and seeds, peanut butter, hard-boiled eggs, chicken and fish. I know this is 'boring', but the choice is stark – some boring food or early death.

## Impact on relationships

Multiple studies have shown that the divorce rate is higher for night shift workers. One study by US researchers found that among men working night shifts, and who had children, separation or divorce was six times more likely in the first five years of marriage than if they were day workers. Before COVID, one US law firm claims that the proportion of divorces among night shift workers has risen by 35 per cent in the past three years, indicating the scale of the issue.[344] Part of the problem may be the failure of a partner to appreciate some of the damaging effects of SCRD, especially on behaviour. The provision of information and educational materials to support employees and, critically, the people they share their lives with could be immensely helpful.

## Chronotype and the best time to work

There is considerable variation across the population in terms of chronotype (chapter 1). Just to remind you, chronotype is the tendency for individuals to sleep at a particular time during a 24-hour day, and the population can be divided into early types or 'larks' (10 per cent of the population); late types or 'owls' (25 per cent of the population); and those in the middle called by some 'intermediates', or 'doves' (65 per cent of the population). Studies have shown that the greater the mismatch between your circadian-driven sleep/wake timing and the time when you are required to work, the greater the risk of developing health problems. To describe this mismatch, and as mentioned in chapter 4, my old chum Till Roenneberg coined the catchy phrase 'social jet lag'.[345] Put simply, employers could match an individual's chronotype to specific work schedules. Larks would be better suited to the morning shifts, and owls to the night shifts. This is not a complete solution to night shift work, but it could reduce the social jet lag along with some of the significant problems of working against internal time.

## Why Does Society Ignore SCRD?

I find it really worrying that although there has been a major increase in our awareness of the importance of sleep, and of the consequences of SCRD, little is being done to deal with these issues. At a societal level, we are not training our health care professionals in this critical area of medicine. A recent survey by the insurance company Aviva reported that as many as 31 per cent of the UK population say they have insomnia. Two thirds (67 per cent) of UK adults suffer from disrupted sleep and nearly a quarter (23 per cent) manage no more than five hours a night. There is clearly a problem, yet in a five-year basic training medical students will be lucky to get one or two lectures on sleep, and probably nothing on circadian rhythms. Many of our medical practitioners lack training in sleep and have no idea about the importance of circadian rhythms to our biology. There are very few clinically qualified individuals in the area of sleep and circadian medicine because training opportunities are very limited. Governments are aware of the problems at a superficial level, but have failed to address the broad issues of SCRD with legislation or clear evidence-based guidelines. Funding bodies seem unwilling to provide the serious money required for detailed and long-term studies to develop new knowledge to understand SCRD, or even to apply what we currently understand to improve health and wealth across all sectors of society.

Despite the current lack of interest in SCRD and the importance

of circadian rhythms in particular, I remain optimistic. By analogy, 30 years ago in the UK diet and exercise were considered the
preserve of 'health fanatics' and hardly ever discussed by medical
practitioners, or indeed government. But now these topics feature large in National Health Service advice and practice. I think
there will be a similar change in attitudes towards sleep and circadian rhythms, especially as there are practical actions that we can
all undertake to improve many aspects of our health. Our growing understanding of the science of circadian rhythms is providing
multiple opportunities to make a real difference across many
domains of health, and highlighting this knowledge and encouraging each of us to embrace this information will be the main
theme of the following chapters.

## Questions and Answers

**1. I have read that our internal body clock can change grad-
ually, and adapt to night shift work, like adapting to flying
across time zones. Is this true?**
Sadly, this is not true except for a few exceptional circumstances,
such as night shift workers on oil platforms in the North Sea
(chapter 3). A study on night shift workers showed that 97 per
cent of them do not adapt to the demands of working at night.[188]
The only solution is to hide from natural light during the day and
increase the amount of light during the night shift. But for most
of us this is not possible.

**2. Should I get rid of my alarm clock?**
Ideally, we should all wake naturally from REM sleep and not be
dependent upon an alarm clock. Rather like our ancestors, whose
rhythms were set by sunset and sunrise. However, if you are
working this is often completely impractical. When, or if, I retire,
I may just give my alarm clock to a teenager!

### 3. I like to catch up with errands immediately following the night shift. Is this a good idea?

It is best to try to sleep immediately after you get back from the night shift. Your sleep pressure is very high and the circadian drive for wake is low in the morning, but rises throughout the day making it more difficult to sleep later. As you finish the graveyard shift – and it is not called this for nothing (Table 1) – the people around you are waking up, and you may be tempted to join them and run errands, catch up on recorded TV programmes, or speak to friends on the phone. Try not to do this because these activities will alert you, making it less likely that you will sleep.

### 4. How can I maintain a relationship as a night shift worker?

It is certainly more difficult, as the higher divorce rate testifies, but the following might be worth trying.

- Think before you speak – tiredness drives irritability, impulsivity and lack of empathy. Knowing this, think carefully before speaking and wait until you are both fully rested to tackle serious domestic issues.
- Discuss positive experiences and not negative problems and issues when tired.
- Schedule time together – nine-to-five working allows evenings to be spent together. This cannot be assumed if one or both of you are on the night shift, so work out when you will both be free and schedule time and activities that the two of you enjoy.
- Schedule domestic chores – again this is a problem. The weekly shop, house cleaning, meal preparation and even filling the car with petrol should be timetabled to avoid friction or last-minute and stressful 'chore catch-up'.
- Keep channels of communication open – you may not see much of each other. So keep in regular contact via

whatever social media works best, and keep the exchanges jolly.

- Maintain your health – vulnerability to illness as a night shift worker increases and this can have an impact on your relationship.
- Try and exercise together when possible, and if not together, then schedule exercise for yourself. This will also aid relaxation.
- Plan regular holidays – get away and spend time together in a relaxing stress-free environment. Again, planning ahead is important as you both have something to look forward to.

### 5. What is 'lucid dreaming' and should I worry about it?

A lucid dream is a dream in which the individual is aware that they are dreaming, and sometimes can take some control over the dream content in terms of the characters, the form of the dream or the environment where the dream is taking place. Lucid dreams seem to occur most often in REM sleep, and there is some suggestion that they represent an 'in-between state' where you are not fully awake but not asleep either. Some individuals try to encourage lucid dreaming by waking from REM sleep, focusing on the dream experience and then going back to sleep hoping to re-enter REM sleep. Any benefits of lucid dreaming are unclear, but a potential danger is that if you are susceptible to mental health problems such as schizophrenia, lucid dreams may blur the line between what is real and what is a dream experience and cause added confusion. Recent studies have linked activity in the prefrontal cortex of the brain (Figure 2, p. 24) with the generation of lucid dreams.[346]

# 7.

# The Rhythm of Life

## Circadian rhythms and sex

Physics is like sex; sure, it may give some practical results,
but that's not why we do it.

Richard Feynman

If you thought courtship was complex as a human, being a male praying mantis is significantly more demanding. When a male praying mantis wants to attract a mate, his mating dance includes a vigorous flapping of his wings with broad swings of his abdomen. Should he attract a female she responds by allowing him to copulate. So far so good, but success is bitter sweet. The downside is that while still copulating she will bite off his head. The male continues to copulate in this decapitated condition, but at some point the female will decide that enough is enough by devouring the remains of her still-mating partner. Conception is the act of conceiving offspring, and across the biological world it is a complex, and sometimes dangerous, business. It is all about getting the correct materials in the right place, in the right amount, at the right time. Conception is a spectacular example of finely tuned orchestration, and it is rather humbling to think about what our parents had to do to get us started in life. But . . . what did they do? I assume that the basic facts of life will be well known to you. So I will skip over this rudimentary knowledge and jump straight into the glories of the menstrual cycle and ovulation – which is the release of an egg from the ovary into the fallopian tube, where it may, or may not, then get fertilized.

## When to Have Sex

Ovulation is the release of an egg from one of the ovaries, and the timing surrounding this event is truly extraordinary, involving circadian rhythms and an interaction with multiple hormonal systems which then stimulate oestrogen and progesterone release from the ovary. The menstrual cycle begins with the first day of bleeding, or menstruation, which involves the shedding of the lining of the uterus (endometrium) that had been all set to nurture a fertilized egg (zygote) – but no zygote arrived (Figure 5).

Menstrual cycles can range from around 21 to 40 days, with only about 15 per cent of women having so-called 'normal' cycles of 28 days (Figure 5). This is a good example of why an 'average value' can be very misleading, and can even cause worry. It is in fact *normal* to have menstrual cycles different from 28 days. In about 20 per cent of women, cycles are irregular due to a range of factors, but, as we shall discuss, some of this irregularity may be linked to sleep and circadian rhythm disruption (SCRD).[347–8] Menstrual bleeding lasts 3–7 days, averaging five days. The menstrual cycle (Figure 5) has three phases: the ***follicular phase*** – getting the ovary and uterus ready before release of the egg; the ***ovulatory phase*** – release of the mature egg; then fertilization of the mature egg by sperm in the fallopian tube.[349] Following ovulation, sperm are capable of fertilizing an egg in the human female reproductive tract (fallopian tube) for around four days, although fertilization is most likely when sperm are already in the fallopian tube, with pregnancy most likely when intercourse occurs around three days *before* ovulation. By the way, there is no evidence that the timing of sexual intercourse in relation to ovulation has any influence on the sex of the baby.[350] So the first key point to make is that there is a relatively narrow time window of a few days when fertilization is most likely to take place, and ideally sperm should be in the female reproductive tract three days before ovulation.

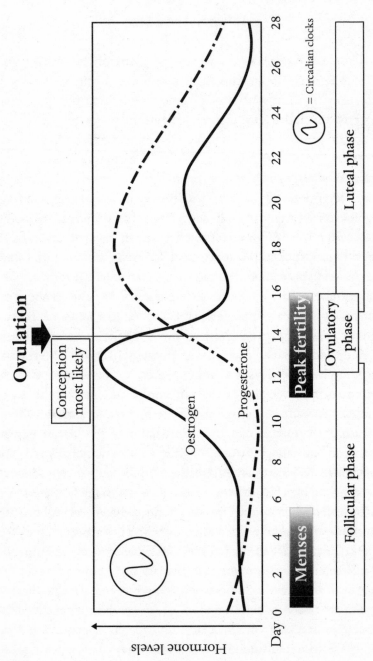

**Figure 5.**

**Figure 5. The changes in oestrogen and progesterone across the female menstrual cycle leading to ovulation**. The timing of ovulation (release of a mature egg that may be fertilized) involves a complex set of interactions involving the hypothalamus, pituitary gland, ovaries and circadian clocks located in all these tissues and organs. The activity of these peripheral circadian clocks is coordinated by the 'master clock' located in the suprachiasmatic nuclei (SCN). Synchronization of these complex systems is essential for successful reproduction. Conception (fertilization of the egg) is most likely around three to four days before ovulation. Circadian disruption in women, as in night shift workers, is associated with irregular and longer menstrual cycles, reduced fertility and a higher risk of miscarriage.[347-8]

Next is the *luteal phase*, getting the uterus ready to receive the fertilized egg (zygote) – or not. If the egg is fertilized, the zygote implants into the uterus wall (endometrium) and starts to develop. If the egg is not fertilized, the endometrium is shed and menstrual bleeding (menses) occurs and a new menstrual cycle begins. If the zygote is implanted, the cells around the developing embryo produce a hormone called human chorionic gonadotropin (hCG). Pregnancy tests are based upon detecting rising levels of hCG.

The circadian system is involved at every stage of this process, from the timing of the release of all the key hormones and the responses to those hormones by the various target tissues such as the ovary.[351-2] There is circadian synchronization between the master clock in the SCN and the release of gonadotropin-releasing hormone from GnRH cells in the hypothalamus, which then stimulates the pituitary gland to release luteinizing hormone (LH) and follicle-stimulating hormone (FSH). LH and FSH travel to the ovary in the blood and then stimulate the release of oestrogen and progesterone. Critically, disruption or lack of synchrony between the different circadian clocks in the hypothalamus, pituitary gland or ovaries can contribute to reproductive problems. The importance of the circadian timing system in reproduction has been demonstrated in 'clock mutant mice' whose circadian clocks either don't work or run at different times (chapter 1). Such clock mutants have markedly disrupted ovulation and reproductive cycles, along with reduced fertility and fewer young.[353] This discovery in mice might help explain observations in women who experience SCRD as a result of night shift work or repeated jet lag. Such disruption has been associated with a significant increase in irregular and extended menstrual cycles, abnormal levels of reproductive hormones and reduced fertility. For example, irregular and long (more than 40 days) menstrual cycles have been noted in long-term night shift workers along with a lowered chance of pregnancy.[347, 354-5] It is important to stress that while the effects of shift work on fertility vary greatly between individuals, it is probably important to consider shift work and jet lag as potential risk factors for fertility in some women. Many clinical practitioners advise *not* to undertake shift work or multiple long-haul flights if undergoing *in vitro* fertilization (IVF).

Before we consider pregnancy, and because this is one of the most frequently asked questions I get, some mention has to be made about the influence of the Moon on the menstrual cycle. In folklore, a woman's monthly cycle has been tied to the lunar cycle. The probable reason for this is that the average menstrual cycle lasts around 28 days and a lunar cycle is about 29.53 days. But, as mentioned above, menstrual cycles range from 21 to 40 days, with only about 15 per cent of women having cycles of 28 days. As a pragmatist, I would argue that if there were a strong influence of the Moon, then one might reasonably expect more women to have cycles closer to 29 days. In addition, folklore suggests that the menstrual cycle is synchronized to the phase of the Moon, but this is not the case. Early reports rejecting a link between the phase of the Moon and the menstrual cycle date back as far as 1806.[356] But articles continue to surface from time to time claiming to demonstrate a connection. For example, in the 1980s one study suggested that women were more likely to menstruate during the full moon, while another argued that women were more likely to ovulate during the new moon! Longer-term and more detailed research has failed to find any correlation. A paper published in 2013 reported the results on 74 women studied over a year, and found no correlation between their menstrual cycles and the phases of the Moon.[357] A recent study by Clue, a menstrual health app developed by the Berlin-based technology company BioWink, also found no correlation. The Clue team analysed 7.5 million cycles from 1.5 million women who used the app to track their periods and 'found no correlation between the lunar phases and the menstrual cycle or period start date'. Menstrual period start dates fell randomly across the month, irrespective of the lunar phase. Sadly, the results of this major study have only been made available on a blog post and have not yet been peer reviewed. But, with these caveats, the conclusion overall is that there is no predictable influence of the lunar cycle on the menstrual cycle. Finally, and in contradiction of the studies outlined above, a very recent report suggested that women with menstrual cycles longer than 27 days could be 'intermittently synchronous' with the Moon's phase.[358] Maybe this is the case, or, perhaps like other proposed links between the Moon and our biology, the relationship will dissolve with greater scrutiny and a different set of statistics.[25, 359] There are also no links between the phase of the Moon and the frequency of baby deliveries, delivery complications or the gender of the baby.[360] And while we are on the topic of the Moon, there is probably no clear relationship between the phase of the Moon and its influence on sleep in modern humans.[359] But, as I shall discuss later in this chapter, the menstrual cycle has an influence on sleep.

While there is no robust evidence that human reproduction is influenced by the Moon, this is not the case in other animals. Perhaps the best documented example is the palolo worm (*Eunice viridis*) which is found on several coral islands near to Samoa and the Fiji Islands, and lives in crevices and cavities in coral reefs. These worms time their reproduction by swarming at a particular phase of the Moon in October and November. Rather dramatically, during the last quarter of the Moon, the worms break in half; the long tail section (the 'epitoke'), which contains both eggs and sperm, swims to the surface, possibly attracted upwards by the light of the Moon, where it releases eggs and sperm. Tens of thousands of epitokes swarm at the same time. The front section of the worm (the 'atoke') remains below and grows a new epitoke for the following year. The peoples of the Samoan Islands have known this for centuries and predict the date and time of day when the emergence occurs, collecting the worm tails for food. Palolo worms have been kept in isolation from the environment and the Moon, yet still show a Moon-phase-related rhythm in behaviour, suggesting that these worms have an internal clock that can predict lunar time, just like we have a circadian clock that is used to predict daily time. In fact, such circa-lunar clocks have been found in many animals that live in the intertidal zone by the sea.[361-2]

Back to us humans. For successful pregnancy, timing is every-thing, and while the beautifully timed production of a mature egg is essential, it is not sufficient. The egg needs to be fertilized – which brings us back to conception. The best time for conception is when sex takes place shortly before ovulation (Figure 5), with pregnancy most likely when intercourse occurs around three days before ovulation. A lot of biological effort has gone into producing a nicely timed ovulation, and so the question arises: 'Is sex similarly timed to maximize fertilization?' Two studies, one in 1982 in 48 young married couples[363] and a more recent study in 2005 on 38 university students[364] examined when and why humans have sex. Although these studies were fairly small, the results from both were similar. Sex occurred across the day, but most sex-ual encounters occurred around bedtime (11 p.m.–1 a.m.), with a smaller peak around wake time in the morning. In the 2005 study, participants were also asked: 'Why do you have sex at these times?' The responses were as follows: 23 per cent said because their partner was available; 33 per cent responded that this was when work schedules permitted the time; 16 per cent said it was because they were already in bed; while only 28 per cent responded by saying it was when they 'felt sexual'. The 1982 study showed that the weekend was when couples had sex the most, with an increase in bedtime and morning sex over the weekend. All this suggests that sex is driven, primarily, by strong environmental factors based upon work schedules and partner availability, rather than any internal biological drivers.

Such findings have led to the conclusion that sex occurs basically any time with no link to ovulation. But this conclusion is too simplistic. Once again, pregnancy is most likely when intercourse occurs around three days before ovulation; and by 12–24 hours after ovulation fertilization is unlikely. With this narrow window for successful fertilization, it would be surprising if our biology did not, in some way, prepare us for action. Indeed, there is now good evidence that human sexual behaviour *does* indeed change around the time of peak fertility. Unconsciously, both heterosexual male attractiveness to women and the attractiveness of women to men varies across the menstrual cycle, and brave researchers have provided the evidence. One study showed that heterosexual women are more attracted to strong male features during peak fertility, including strong male facial features, a deeper voice, overtly dominant behaviour and increased height. Women also experience greater sexual desire and are also more likely to have an affair during their peak fertility.[365-8] The physiological basis for these changes remains unclear. Heterosexual men, however, seem to be affected by odorants secreted by women called 'copulins'. Copulins increase in concentration during the follicular phase (anticipating ovulation) and decrease in concentration during the luteal phase (Figure 5). Men exposed to copulins produce more testosterone, are less interested in the attractiveness of women's faces (become less discriminating) and behave less cooperatively. The effects are not large but they are significant.[369] For example, heterosexual men were asked to rate the sexual attractiveness and intensity of the odours 'sniffed' from T-shirts worn previously by women at different stages of their menstrual cycle. The odours were rated as most attractive by the men when the T-shirts had been worn by women during mid-cycle, when they would have been most fertile. In this study, the controls were straight women, who also sniffed the T-shirts, and reported no change in odour attraction. The results suggest that heterosexual men can use odour cues to distinguish between ovulating and non-ovulating women, and that this can potentially alter their behaviour.[370] Such observations are not new. Back in 1975 a study showed that vaginal secretions from preovulatory and ovulatory phases were considered less unpleasant in odour than those from vaginal secretions in the luteal phases.[371] So the timing of ovulation does influence the timing of human hetero-sexual behaviour. To date, no studies have been undertaken in homosexual attractiveness in either men or women.

In addition to altered behaviours, there are also physiological changes that promote fertilization during peak fertility. Cervical mucus changes over the menstrual cycle and just prior to ovula-tion it resembles a raw egg white, and this is the best time to have intercourse to become pregnant. It seems that the mucus at this stage helps sperm move up the cervix to fertilize an egg, and it also keeps the sperm healthy during their long journey – so a bit like a hearty packed lunch during a vigorous hiking trip. Another physio-logical indicator of ovulation is that a woman's body temperature increases by about 0.5°C at ovulation. This is driven by the rise in the hormone progesterone, which is from the cells of the follicle in the ovary that released the egg. Whether this rise in tempera-ture helps fertilization is unknown, but it has been used to track ovulation for the planning of intercourse and achieving fertiliza-tion. Various studies suggest it is an unreliable technique to determine ovulation, with reliability as low as only 22 per cent.[372]

Interestingly, there is a circadian rhythm of testosterone release from the testes of men, rising from about midnight and peaking in the morning just before wake (Figure 1, p. 20),[373] with levels of testosterone in young men about 25–50 per cent higher at this time than for the rest of the day. As there is a link between male sexual drive and testosterone,[4] this morning peak in testosterone may contribute to the increase in sex observed in the morning. Another study showed that semen collected in the early morning before 7.30 a.m. showed the highest levels of sperm concentration compared to other times of the day.[374] These findings suggest that men may increase their chances of fertilizing an egg by producing top-notch sperm at a particular time of day. So sex – and I really mean successful fertilization – is like everything else. It's about getting the correct materials in the right place, in the right amount, at the right time. A big round of applause for our parents – whose biology clearly got it right!

## A Time to be Born

The relatively high risk of death for a child during childbirth has led this day to be called 'the most dangerous day of our lives'. While I appreciate the logic of this statement, I feel obliged to point out that statistically the most dangerous day of our lives is, in fact, the day we die. Anyway, around the world, the peak hours for natural childbirth, i.e. that has not been induced by drugs or surgery, fall between 1 a.m. and 7 a.m., peaking at around 4–5 a.m.[375-6] Clearly, babies can be born at any time of the day or night, but this clustering of births in the early morning strongly implicates some level of circadian regulation and has raised the question, why? What might be the evolutionary advantage of this timing? Suggestions include that birth overnight would have been useful for mothers in a hunter-gatherer society, because the group would be reunited at night, providing protection and social support, which would not be available during the day when the group was dispersed to forage. In addition, there would be reduced predator activity and reduced heat in the middle of the night. So the peak in human births during the early-morning hours could be a vestige of our evolutionary history, when survival was increased if birth occurred at this time.[376]

These evolutionary explanations have been coupled to key changes in hormone levels. Melatonin is released by the pineal gland at night (Figure 1, p. 20) and could drive a nocturnal signal for the production of hormones that increase the contractions of the uterus, such as oxytocin. Support for this idea comes from studies which show melatonin levels are higher in late pregnancy and that melatonin may increase the sensitivity of the uterus to oxytocin, which in turn stimulates powerful contractions that move the baby down and out of the birth canal.[377] Although the original evolutionary selection pressure to give birth in the early hours of the morning has gone in modern humans, the physiology driving timed birth remains. If melatonin was originally co-opted as the biological marker of the night, then childbirth in modern humans may still remain locked to this nocturnal melatonin signal.

This all makes sense, but my colleague in Oxford, Professor Alastair Buchan, who helped 'fact-check' this book, is not persuaded by these arguments. He makes the point that it is a lack of oxygen during birth, called neonatal hypoxia-ischemia, that is the most common cause of death and disability in human babies.[378] And as a result we should investigate the links between night-time birth and the possible protective mechanisms that reduce the chances of hypoxia in newborn babies at this time. It's a really interesting idea. Perhaps a slightly cooler temperature at night reduces the chance of neonatal hypoxia?

## Circadian Sexual Dimorphism

Men and women are biologically different – they are broadly 'sexually dimorphic', and this extends to several notable differences in our circadian system. The next section is all about the different circadian patterns in heterosexual men and women. I apologize in advance to the LGBTQ communities, where scientific studies exploring potentially different patterns of circadian behaviour are lacking. The first clear evidence, from both human and animal studies, was for a difference in 'chronotype' and the tendency to be either a more morning or evening person (chapter 1). The finding that chronotype differed between males and females originated with studies on rodents. One of the earliest studies examined male and female hamsters who were placed in individual cages and provided with a running wheel. The lights were turned off, and in the absence of a light/dark cycle no entrainment was possible. Under these conditions the internal freerunning circadian rhythms of activity and rest were recorded. Female hamsters had shorter freerunning rhythms, and hence 'faster' clocks, compared to male hamsters. Other studies in rodents confirmed this finding.[379] Remarkably, similar results have been found in us. In one study, circadian rhythms of core body temperature, sleep/wake and alertness were compared between women and men. All these rhythms occurred earlier in women.[380] Diary data from the American Time Use Survey between 2003 and 2014 showed that men are typically later chronotypes, with the greatest differences between men and women between the ages of 15 and 25 years.[381] After the age of 40, men and women have more similar chronotypes, but men show much greater variability.[382] These findings have also been confirmed in a very recent study that examined 53,000 individuals. On average, women were again more likely to be morning types, while men had a later chronotype.[379] This sexual dimorphism in chronotype has been linked, perhaps unsurprisingly, to the sex hormones oestrogen (from the ovary) and testosterone (from the testes). Why there should be this difference in chronotype remains unclear, but the bottom line is that women will tend to want to get up earlier in the morning than men.

The influence of oestrogen and testosterone extends beyond just chronotype. For example, oestrogen in females has been linked to more consolidated circadian rhythms with a greater amplitude (higher peak to trough).[379] In short, oestrogen is associated with more robust circadian rhythms. Interestingly, as women age, oestrogen levels decline, and this may be a contributing factor to the age-related increase of insomnia reported by many women (chapter 8).

In male mice, and humans, testosterone has been linked to a reduced sensitivity to light for circadian entrainment. For example, female mice adapt faster to simulated jet lag (shifting the light/dark cycle by eight hours) compared to male mice, with females adapting after six days but males taking 10 days.[383] And there may even be an explanation for this aspect of circadian sexual dimorphism. The distribution of oestrogen and testosterone receptors in the brain, and especially in the suprachiasmatic nuclei (SCN) is different between males and females. The SCN is divided into two broad parts, the 'core' and 'shell'. The core of the SCN receives the main projection from the eye (the retinohypothalamic tract), which is used to entrain the SCN to the light/dark cycle (chapter 3). The core of the male SCN has a large number of testosterone receptors, so perhaps these receptors act to desensitize the clock to light in males. As females lack these testosterone detector mechanisms in the SCN and/or produce much less testosterone, they can respond more quickly. In addition, the shell of the SCN sends circadian 'output' messages to the rest of the body. In females, this region has many oestrogen receptors. As oestrogen is associated with driving more robust circadian rhythms with a greater amplitude, perhaps the oestrogen receptors in the shell of the SCN promote closer coupling between individual SCN neurons, producing a more robust output rhythm from the clock.[379] This mechanism might explain why oestrogen is associated with producing 'stronger' circadian rhythms in females.

## Impact of the Menstrual Cycle

Compared to men, women are reported to have double the lifetime risk of developing mood disorders (e.g. depression) and a 25 per cent greater overall chance of SCRD.[384] The assumption has been that changes in female reproductive hormones (primarily oestrogen and progesterone), across the menstrual cycle (Figure 5, p. 146) and during the menopause, are responsible for these differences in mood and depression between the sexes. This idea may have its origins in 'folk wisdom', or patriarchal attitudes, but there is now increasing evidence that changes in female reproductive hormones are indeed an important contributing factor for the development of SCRD, mood changes and depression. Let's start with mood changes across the menstrual cycle.

## Mood changes

Estimates vary but suggest that between 20 and 80 per cent of women experience some form of emotional change such as mood swings or irritability, along with poorer sleep, during the premenstrual phase of the menstrual cycle (second half of the luteal phase and just before menstrual bleeding). These changes fade after the onset of menstrual bleeding (menses). Emotional changes that cause distress or discomfort have been classified as 'premenstrual syndrome' (PMS). We do not fully understand the causes of PMS. However, key structures in the brain that are associated with emotion and mood, including the hypothalamus, amygdala and hippocampus (Figure 2, p. 24), have receptors that detect oestrogen and progesterone. Overall, oestrogen is associated with a more positive mood, with one of its actions to promote brain levels of serotonin, the hormone most associated with happiness. High levels of progesterone are associated with anti-anxiety effects and sleepiness.[385] These actions of oestrogen and progesterone fit neatly with the dynamic changes seen in these hormones across the menstrual cycle and the timing of PMS (Figure 5, p. 146). Oestrogen rises during the follicular phase prior to ovulation and is linked to a more positive mood, and following ovulation progesterone levels are high for the first half of the luteal phase, promoting sleep and reducing anxiety. But during the second half of the luteal phase progesterone levels fall. The falling level of progesterone, combined with already low levels of oestrogen, is thought to be responsible for the poorer mood and reduced sleep associated with the later luteal (premenstrual) phase of the menstrual cycle, and hence premenstrual syndrome.

## Circadian and sleep changes

In addition to altered mood, there is strong evidence that circadian rhythms are directly altered during the premenstrual phase of the menstrual cycle in women. As discussed above, oestrogen is associated with delivering more consolidated circadian rhythms, and, as oestrogen is much lower during the premenstrual phase of the menstrual cycle (Figure 5), this may contribute to a weaker circadian drive for sleep, and hence worse sleep.[386-7] Interestingly, there are reports that circadian responses to morning light in women are smaller during the premenstrual phase.[388] Whether this is the result of increased progesterone or decreased oestrogen remains unclear. But the net result would be a weaker entrainment signal, and hence a greater vulnerability to sleep/wake disruption, especially in the absence of natural light exposure (chapter 3). Additional factors for SCRD during the premenstrual phase are the reduced levels of progesterone. High levels of progesterone reduce anxiety and promote sleep, while low progesterone during the premenstrual phase will have the opposite effect.[385]

## Mood, circadian and sleep interactions

In 3–8 per cent of women, premenstrual syndrome (PMS) is sufficiently severe to warrant the status of a depressive diagnosis termed premenstrual dysphoric disorder (PMDD), in which women experience irritability, anger, depression, anxiety and marked insomnia. Indeed, there is a strong association between insomnia, combined with daytime sleepiness, and premenstrual mood changes in PMDD. The worse the insomnia, the worse the mood changes.[354, 389] Clearly, SCRD and mood are linked during the late luteal phase of the menstrual cycle. But, in addition, SCRD can activate the stress axis, and the greater the stress responses the greater the risk of SCRD. To recap, SCRD is associated with poor mood[390-91] and depression[392]; low levels of progesterone and oestrogen occur during the premenstrual phase and are likely to combine to promote both SCRD and poor mood; SCRD activates the stress axis, which promotes both poor mood and an exacerbation of SCRD. This triangle of interactions during the premenstrual phase between hormones, SCRD and stress probably explains why the late luteal phase of the menstrual cycle can be such a challenge for many women. And when menstrual cycles stop, the menopause surfaces as another potential agent of SCRD.

## The Impact of Menopause

For women, menopause represents the end of menses and menstrual cycles (Figure 5, p. 146). But this is not an abrupt change, there is a 'menopausal transition' which starts some 4–6 years before the end of menses, and occurs at an average age of 51. This transition is associated with fluctuating levels of oestrogen and progesterone from the ovary. Changes in these, and other hormones, have been associated with a variety of conditions, including sleep disturbance, hot flushes (also known as hot flashes) and fluctuations in mood. Specific links between hormone levels and sleep disturbance have been difficult to define precisely, but lowered levels of oestrogen and progesterone have been strongly implicated in a number of studies.[393-4] In addition, a history of sleep disturbance will probably increase the chances of SCRD during menopause.[395] Sleep disturbances can be severe in many women, with associated daytime sleepiness and alterations in mood.[396]

Self-reported sleep difficulties during the menopausal transition range between 40 and 56 per cent compared to pre-menopausal women at 31 per cent.[397] The most commonly reported form of insomnia is multiple awakenings at night, difficulty getting to sleep and waking up early,[398] a textbook description of insomnia (Figure 4, p. 103).

Hot flushes are a unique feature of the menopausal transition and are reported in up to 80 per cent of women. Feeling hot, sweating with associated anxiety, and then chills can last between three and 10 minutes and can occur during the day or night (night sweats).[399] Hot flushes have been associated with a decline in oestrogen, but this is not the whole story. Recent studies have shown that changes in hypothalamic neurotransmitters, including noradrenaline and serotonin, may also be linked to hot flushes.[400] The key point is that hot flushes at night are almost always associated with insomnia, and particularly waking up at night.[401] Significantly, treatment of hot flushes with hormone replacement therapy (HRT), which is usually a combination of oestrogen and progesterone, improves sleep.[402] Such data suggest strongly a clear hormonal link between declining levels of oestrogen and progesterone, increased hot flushes and poorer sleep. SCRD varies widely between individuals across the menopausal transition, but if insomnia is chronic, depression, anxiety and poor physical health increase and cognitive abilities decline.[400]

Sadly, hot flushes driving insomnia are only part of the problem. Sleep-disordered breathing (SDB), as discussed in chapter 5, is more likely during the menopausal transition, along with periodic limb movement disorder and obstructive sleep apnoea (chapter 5).[403] In a cohort of women with sleep complaints, 53 per cent had SDB and/or periodic limb movement disorder. The decline in oestrogen and progesterone has again been associated to SDB, with both hormones linked to improved breathing during sleep,[404] reminding us that so many hormones perform not just one but multiple roles – beyond their most recognized function. The causes of insomnia during the menopausal transition are clearly complex, but HRT to reduce hot flushes is effective for some.[405] While HRT, in consultation with your medical practitioner, could be considered as a treatment option, it should be used in parallel with cognitive behavioural treatment for insomnia (CBTi) (chapter 6). Indeed, CBTi alone has been shown to be effective in helping to improve SCRD during the menopausal transition.[406]

If we return to the question raised at the beginning of this section: 'Are the higher rates of depression and insomnia in women across the lifespan really related to changed hormone levels during the menstrual cycle and menopause?', the evidence reviewed above would certainly seem to suggest a yes. This conclusion is reinforced by the finding that the higher rates of insomnia and depression in women appear from puberty onwards,[407] suggesting that the dynamic changes in oestrogen and progesterone account for the higher rates of depression and sleep problems in females than in males. And just in case male readers are feeling left out, lowered testosterone levels with age have been associated with sleep problems such as nocturnal awakenings, poorer sleep and symptoms of depression.[408] However, cause and effect have not been clarified.

### Studying the circadian biology of men versus women

The topic of circadian sexual dimorphism has been recognized by circadian and sleep researchers, but until recently this area has been poorly studied. In most circadian experiments that use mice, only male mice are studied. This is because the four-day mouse oestrus (reproductive) cycle slightly alters the circadian timing of activity, due to changed levels of oestrogen and progesterone. This makes it more difficult to study the impact of drugs or other agents on the circadian system – especially if the drug effects are small. This is also true for humans. It is sometimes difficult to isolate subtle changes in the human circadian system from the effects of the female menstrual cycle. However, there is an increasing realization that by only studying men, and usually just young male university students, you get a highly selective, and potentially very misleading, understanding of human biology. And while it is more difficult to design experiments that consider the more complex biology of women, this has to be done. Indeed, some funding bodies now require that human circadian studies include both males and females. It is also worth mentioning that although circadian biology and sex differences (male versus female) are now being explored, the influence of circadian biology on gender identity or sexual orientation remains entirely unstudied at a behavioural level – such as chronotype (Appendix I). The only studies undertaken have been on brain anatomy. For example, in a sample of post-mortem brains of homosexual men the suprachiasmatic nucleus (SCN) was found to contain twice as many cells as the SCN of heterosexual men.[409] However, there are problems with these studies. Not least because men were classified as being homosexual on the basis that they had died of AIDS, and the size of the SCN in 'homosexual' men varied greatly, overlapping very markedly with the heterosexual men.[410]

### Impact of pregnancy and early parenthood

It is a sober reminder that the arrival of new human life is closely associated with SCRD. If severe, this can lead to pre-clinical and clinical levels of depression. From the first trimester 40 per cent of pregnant women experience some form of insomnia, increasing to 60 per cent by the third trimester.[411] Following birth, irrespective of the type of delivery, sleep deteriorates further, with major sleep loss at night due to newborn feeding and sleep patterns. Daytime naps can increase total sleep time, but fragmented sleep at night almost always leads to increased daytime sleepiness.[242] Three months after birth, sleep for the mother has begun to improve but usually does not return to pre-pregnancy sleep.

I don't have the data to support the assertion I am about to make, so consider what follows as a topic for discussion. Many societies have moved from living in an extended family to the nuclear family – of just two parents with children. This shift has arisen from increased economic wealth and the freedom to choose how we organize our lives. But an unintended consequence has been to make the sleep of new mothers even worse. In previous generations, with many more members of the same family sharing the same domestic space, childcare could be distributed across the family, allowing new mothers to more easily catch up on sleep. New mothers often feel guilty today because they 'cannot cope' with the sleep loss. But as a species we did not evolve to manage with just one parent doing essentially all the childcare. In humans, and in our closest relatives, the apes and monkeys, childcare is invariably spread across other family members. So it is emphatically not a failure for mothers to reach out for help when young babies arrive. This is especially important if there is a personal history of SCRD and mental health problems.[412] Depression, mania, anxiety, suicidal thoughts, psychosis and obsessive-compulsive disorder (OCD) are all reported in mothers with newborn babies.[242] At three months after birth, mothers with symptoms of depression reported high levels of sleep disturbance, difficulty falling asleep, waking earlier and daytime sleepiness.[413] In addition to the demands of the new child, sudden hormonal changes immediately after birth, and particularly the decline in progesterone, may be important. Progesterone helps maintain pregnancy, and during pregnancy it has a mildly relaxing and sleep-promoting effect. Under normal circumstances, the decline in progesterone may be compensated for by prolactin, which is released during breastfeeding and may help promote sleep. Studies suggest that women who breastfeed exclusively averaged around 30 minutes more sleep at night than women who used formula at night.[414] While SCRD is worse in mothers,[415-16] fathers do not escape completely. New fathers report increased levels of daytime sleepiness during the first month after birth.[416] Parents with different chronotypes, such as a lark with an owl partner, could certainly be useful during these early months or even years.

Despite the prevalence of SCRD and depression in new mothers, evidence-based approaches to treat these conditions are limited. As already mentioned, breastfeeding seems to help improve sleep[414], as does CBTi[417], and napping while the baby sleeps during the day can help. For women at a high risk of depression, one report showed that a longer stay in hospital of up to five days, combined with staff bottle-feeding the baby at night with breast milk pumped by the mother during the day, reduced the chances of postpartum depression.[418] Education prior to giving birth, to raise aware-ness of sleep loss during early motherhood, has also been shown to be helpful to new mothers.[419] Increasingly, in some enlightened societies, couples are genuinely able to share the demands of post-partum childcare. Working out in advance who will do what and how often, how household chores will be managed, and who else might be involved in helping with baby care all act to reduce anxiety and stress. But there is no getting away from the fact that early par-enthood, with its associated lack of sleep and its causal relationship with mental illness vulnerability, is a significant issue for many. And currently robust treatment options are few. Young mothers in par-ticular should not be afraid to seek the help of others during this critical period.

## Questions and Answers

### 1. Should pregnant women sleep on their left side?

I am often asked this question and the advice goes back and forth. A recent study concluded that going-to-sleep on your left *or* right side was equally safe in terms of any risk of stillbirth. However, lying flat on your back was a contributing factor for late stillbirth in pregnancy. Specifically, if pregnant women, more than 28 weeks' gestation, slept on their side rather than their back there was a 5.8% *reduction* in the chances of stillbirth.[420]

## 2. Do humans show seasonal breeding?

We do show annual rhythms in many aspects of our biology, including suicide, heart disease, some cancers and birth rate.[25] For birth rate, the range from most to least births in pre-industrialized societies was recorded to be 60 per cent or greater. Today, this is far less marked, being either non-detectable or of very low amplitude (around 5 per cent). The mechanisms that produced this seasonality in the past, and why there has been a marked drop in seasonal births today, are unclear. Changes in societal customs, the local economy, perhaps the lack of exposure to seasonal cycles such as day length, and, most recently, efficient birth control all seem to be important.[25, 421]

## 3. Does being on the pill help with premenstrual syndrome (PMS)?

You would think that this would be a clear yes, because oestrogen and progesterone or progesterone-only birth control pills prevent ovulation-related hormonal changes. However, some women experience worse symptoms and others report relief.[422, 423] A worsened mood is often reported as the reason why women have stopped using a birth control pill.[424] The explanation is not clear but may relate to the use of the synthetic *progestin*-based hormonal birth control pills, which, instead of promoting sleep and relaxation, can actually increase depression[425] and worsen emotions[426].

## 4. Because men and women differ in their chronotypes, with men later than women, does this affect partner relationships?

A recent study suggested that the frequency of sexual intercourse was generally unrelated to partners' chronotype. However, in the same study, women reported being happier in their relationship when they and their partner shared the same chronotype.[427] But the links between chronotype, sexual relationships and marriage are, at best, very complicated and highly variable, depending upon interacting social, economic and personality factors.[428]

## 5. What is the difference between the menstrual cycle seen in women and the oestrus cycle in other mammals, like mice?

Oestrous cycles are named after the cyclic occurrence of sexual activity (oestrus) that occurs in all mammals except for higher primates. Menstrual cycles, which occur only in higher primates, are named for the regular appearance of menses (menstrual bleeding) due to the shedding of the endometrial lining of the uterus. In mammals with an oestrus cycle, the endometrial lining is absorbed and not shed. Most mammals show an oestrus cycle, and females are generally only prepared to mate around the time of ovulation. This is sometimes referred to as being 'in heat'. By contrast, higher-primate females, including us, can be sexually active at any time in their cycle. Intercourse outside the optimum time for fertilization has been linked to strengthening male/female partnerships.

## 6. Do men produce oestrogen?

Testosterone and oestrogen have been considered to be either male or female sex hormones, but this is incorrect. A form of oestrogen (oestradiol) is produced from testosterone, and the enzyme that converts testosterone to oestradiol is abundant in the brain, penis and testes of men. In the brain, oestradiol production from testosterone is increased in areas related to sexual arousal. Oestradiol helps regulate sexual drive (libido), the erection of the penis and sperm production.[429] So, and perhaps surprisingly, testosterone made in the testes drives key aspects of male physiology and behaviour *after* it has been converted to oestradiol locally in the brain, penis and testes.[429] In the recent past gay men were treated with a synthetic form of the female hormone oestrogen known as Stilbestrol, or Stilboestrol, in order to suppress their sexuality – this has been called 'chemical castration'. This synthetic form of oestrogen acts on the pituitary gland to suppress the hormones (LH and FSH) that would normally stimulate the production of testosterone by the testes. As a result, male libido, penile erection or sperm production are suppressed by Stilboestrol, along with a host of other unpleasant side effects, including breast enlargement.

## 7. The circadian clock is important for the timing of ovulation in women, but are circadian rhythms important in sperm production?

A major study examined a total of 12,245 semen samples from 7,068 men. The samples were examined for sperm concentration, total sperm count, motility and normal morphology. The semen samples collected in the early morning before 7.30 a.m. showed the highest levels of sperm concentration, although motility did not show a diurnal rhythm.[374] Why sperm production should peak in the early morning remains unclear.

# 8.

# The Seven Ages of Sleep

## How circadian rhythms and sleep change as we age

Change is inevitable – except from a vending machine.

Robert C. Gallagher

Even if the Antarctic glass sponge possessed our degree of consciousness, and I'm confident it does not, it wouldn't have to worry too much about getting old. The Antarctic glass sponge is thought to be the 'oldest living animal' in the world, with an estimated life span of 15,000 years. Some sponges alive today would have been around when the Sahara was wet and fertile. But this sponge is not immune from change. It lives in the Antarctic shallows in waters less than 300 metres deep, where until recently there was extensive seasonal sea ice blocking out sunlight. The sponges feed on small bacteria and plankton that they filter from the surrounding water.

A recent study has suggested this ancient sponge may be one of the few species to benefit from global warming. The recent collapse of Antarctic ice shelves due to regional warming around parts of the Antarctic has exposed huge areas of sea floor to sunlight, which has allowed an explosion of algal growth. The algae are the primary source of food for the Antarctic ecosystem. Over a four-year study period, in areas where the Arctic ice has disappeared there has been a two- to threefold increase in Antarctic glass sponges. They are doing well, presumably due to all the extra algae.[430] We live far shorter lives than the Antarctic glass sponge.

Methuselah, a figure in Judaism, Christianity and Islam is thought to have died at the age of 969 and is reported to be the longest lived of all humans, at least according to the Book of Genesis. Sadly, this report has not been independently verified. According to Wikipedia (July 2021), and I appreciate some may feel the Book of Genesis has greater authority than Wikipedia, the oldest person recorded is Jeanne Calment (1875–1997) of France, who lived to the age of 122 years. The oldest man is Jiroemon Kimura (1897–2013) of Japan, who lived to the age of 116 years. The bottom line is that most of us living in the advanced economies hope, and indeed expect, to live close to 100 years, and what is absolutely certain is that we will experience every sort of change: social, political, environmental and, of course, biological.

Our circadian rhythms and sleep change profoundly as we age, and while variation can be very marked between individuals, some trends can be considered universal. The amount of time we spend asleep (sleep duration) shortens as we age; our circadian rhythms become less 'robust' and provide a weaker drive for our 24-hour biology, including sleep, which can become more fragmented; there is a change in our circadian timing, with a tendency to show a later chronotype as we progress through the teenage years and into early adulthood, and then shifting increasingly to an earlier chronotype, beginning in our twenties and continuing into old age. As we age many of us feel we don't get the sleep we want or need.

While age-related changes in our sleep and biological timing are inevitable, altered patterns do not have to mean worse patterns. It's all about managing our individual expectations. Knowing what is likely to happen, and preparing (physically and emotionally) in advance is critical. This chapter is about taking this information on-board. For ease of consumption I have divided the chapter into three 'bite-sized' chunks. 'Sleep during the early years': some of the greatest, and certainly the most rapid, changes in sleep and circadian rhythms occur between birth and adolescence, with important consequences for education and wellbeing. 'Sleep after adolescence': following adolescence, there are also manifest changes in sleep. These tend to occur more slowly and are the complex product of our biology, along with the varied demands of work, stress, parenthood and potential illness. And 'Sleep and the impact of neurodegenerative diseases': as we age, we are more vulnerable to key diseases such as Alzheimer's and Parkinson's disease. While such conditions are not an inevitable consequence of old age, they are more frequent and have a major impact upon our personal sleep and the sleep of those who share our lives.


## Sleep During the Early Years (Infants to Adolescents)

### Sleep during pregnancy

As far as we can tell, babies spend most of their time in the womb sleeping. At the age of 38 and 40 weeks of gestation (birth is around 40 weeks), babies are thought to spend 95 per cent of the time asleep.[431] By contrast, sleep for the mother can be a challenge during pregnancy, especially in the last trimester. Physical discomfort, combined with hormonal changes, the need to pee at night as the baby presses on the bladder, leg cramps, acid reflux and kicks from baby all promote increasing levels of insomnia (chapter 5). During the first trimester, expectant mums may sleep more. However, this has little impact on perceived sleep, with increased reports of sleepiness and fatigue. The high levels of human chorionic gonadotropin (hCG) (chapter 7) and progesterone required to maintain pregnancy induce mild sleepiness. However, hCG and progesterone may also induce a slight increase in body temperature, which is often not helpful for sleep.[432] From the second to the third trimester, almost 50 per cent of women report poor sleep.[433] Snoring and obstructive sleep apnoea (OSA) (chapter 5) increase in women who are at risk of developing these conditions,[433] and should be treated because of the potential risks. Restless-legs syndrome (RLS) and periodic limb movements also increase during pregnancy and are found in about 20 per cent of pregnant women. In consultation with your general practitioner, RLS and periodic limb movements may be treated with iron supplements[278] (chapter 5). When pregnant, my mother was told to drink Guinness every day to get the extra iron, but it turns out that Guinness contains very little iron. For this, and more importantly due to the alcohol, this beverage would not be recommended during pregnancy today. Luckily for me, my mother said Guinness made her feel sick and she would not drink it.

### Infant (0–1 year) and parent sleep

As I discussed in chapter 7, before the advent of nuclear families, childcare was shared across the family. Now, all of that responsibility is usually devolved to the mother, with varying degrees of help from the father or a partner. The assumption is that the chronically tired mother should manage. But we have not evolved to undertake these duties alone,[434] and mothers should seek help when it is needed. When a child is born, sleep dominates every aspect of the new family's life. At birth, children lack an established circadian rhythm and so will sleep at varied intervals throughout the day and night in short bouts, probably related to their high feeding needs at this time. Around 10–12 weeks of age, the first signs of a circadian rhythm begin to develop, with a progressive increase in sleep throughout the night. Over this period, sleep duration decreases from around 16 to 17 hours in newborns, to 14–15 hours at 16 weeks of age, and 13–14 hours by six months of age.[415, 435] The need for day sleep decreases, and night sleep increases through the first year of life, resulting in mostly night-time sleep with little daytime sleep by the first year of age.[436] However, around 20–30 per cent of all children experience some form of night awakenings throughout the first two years of their lives.[437] Although robust 24-hour sleep/wake patterns take 6–12 months to fully develop, a good sleeping environment should be maintained from the outset both to encourage sleep and eventually to help synchronize the circadian system. Babies should be exposed to a stable and sufficiently bright light/dark cycle. Based upon my family photographs, in the old days babies were often left outside in their prams during much of the day! At night the bedroom should be as dark as possible, using blackout curtains, and the room should also be as quiet as you can make it. As the baby gets older, ensure there is a stable mealtime schedule, and again a strong 24-hour pattern of light during the day and dark at night.[438] Although 'cause and effect' remain unclear, studies support an important link between good infant sleep and increased cognition and physical growth during this phase of early development.[437] However, it also needs to be stressed that humans, like most other animals, show marked 'developmental plasticity' and the ability to develop in a variety of ways.[439] So poor sleep in the infant years does not necessarily mean there will be permanent problems in cognition or growth later on. Highly disrupted sleep in an infant may be an indication of neurological problems, and advice from your medical practitioner should be sought if you are concerned.

Parents are bombarded with advice from family, friends and the media about what to do to help babies get to sleep and stay asleep. The best general advice is to do what works, and you may need to try different approaches. For our three children, we deployed 'self-soothing' from about 4–6 months, which meant that we did not immediately go to the child when crying started, allowing them to soothe themselves for longer periods of time, starting with short bursts of several minutes and then longer.[440] Friends told us this was too stressful for both parent and child, and instead of 'self-soothing' practised 'parental soothing', gently rocking or singing their child back to sleep when it woke. The key point is that you should always do what works for you, promoting both your sleep and the child's sleep. Of course, there are boundaries, and the eighteenth-century practice of letting your baby suck on a 'gin-soaked rag' is now considered to be a bad idea.

New parents experience substantial sleep and circadian rhythm disruption (SCRD) and need to be aware of this fact, not least in considering the potential risks of accidents at home and work. If you are tired, don't drive. Get family and friends to visit *you* and at a time that is good for you! Insomnia will also affect emotional responses, placing relationships under pressure. In addition, cognition and decision-making skills will be impaired (chapter 9). Under these circumstances, coping strategies by parents could include a regular and early bedtime, and certainly don't feel guilty about getting an early night. Take naps when the baby naps, and request that family and friends babysit for a few hours while you sleep. Basically, as new parents, prioritize sleep wherever and whenever you can. Parents often find it useful to discuss sleep issues, and various coping strategies, before the baby arrives. Developing strategies when tired is always more of a challenge.

### Children (1–10 years)

Many parents get anxious about how much sleep children should get. And the answer is 'as much as they need'. Again, sleep should be given the highest priority. Between 15 and 35 per cent of children are reported to have some type of sleep disturbance in the first five years of life which usually disappears with age.[441] As in adults, sleep in children is very important for health and cognition.[442] For example, sleep problems in children have been clearly associated with worse academic performance[443], and there are strong links to obesity[444]. The long-term consequences of insufficient sleep during childhood remain a matter of much debate, and currently we don't really know what problems might turn up later. Total sleep time decreases through childhood, from around 16 hours in young children to an average of 8–9 hours into the teenage years. The non-rapid eye movement/rapid eye movement (NREM/REM) cycle (chapter 2) increases in duration from about 60 minutes in newborns to 75 minutes at two years, and reaches about 90 minutes by around six years of age, which is similar to most adults. What this change in NREM/REM sleep actually means again remains unclear. Perhaps, as we age, less time is needed to process information and consolidate memories, because fewer experiences are novel. This makes intuitive sense, but remains unproven.

Difficulty falling asleep, and an unwillingness to go to bed, is a problem for both children and those who care for them.[445] Good sleep practices are essential. Bedtime routines such as bathing, reading, singing/lullabies and cuddling/rocking psychologically prepare the child to go to sleep.[446] Lighting conditions before and during sleep are again critically important. Bright light immediately before bedtime will alert the brain[447] and possibly delay the circadian system[448]. Both of these modulators make it more difficult for the child to go to sleep. The pre-bedtime ritual should occur under dim light and ideally all light should be removed during sleep. If a child is anxious and is comforted by a dim low light intensity night-light in the 5 lux range or less, this is unlikely to be a problem. Children, like adults, should also avoid caffeine, and any other alerting activities.[447]

How do you know if children are not getting enough sleep? A reliable measure is if they are unruly, recalcitrant or capricious. There may be other signals. Higher rates of obesity and obstructive sleep apnoea (OSA) in children over the past 20 years have been associated with childhood insomnia, which, as discussed, leads to daytime sleepiness. Daytime sleepiness can cause increased aggression, anxiety, depression, hyperactivity, learning and memory difficulties. These are all important signals that may underpin a sleep problem.[445]

## *Adolescents (10–18 years)*

Adolescence begins with the onset of puberty and ends with adulthood. Having said that, the ages at which these changes occur differ between individuals and even the concept of adolescence varies between cultures.[449] As I mentioned earlier, the amount of sleep at night declines from childhood to late adolescence; however, it is not clear that sleep need declines in parallel.[450] Adolescents, therefore, get less sleep but probably need just as much sleep as before puberty. Adolescents seem to be sleeping less today than in previous generations, with a notable decline over the past 20–30 years[451], perhaps as much as an hour less per night over the previous 100 years[452]. But how much sleep do adolescents actually need? There will be individual variation (chapter 5), but the National Sleep Foundation recommends between 8 and 10 hours per night for teenagers aged between 14 and 17 years.[453] The American Academy of Sleep Medicine also concluded that between 8 and 10 hours per night is optimal for 13- to 18-year-olds.[454] However, many adolescents get much less sleep than these recommendations.[183] For example, a major survey of teenage sleep concluded that adolescents were typically getting significantly less than eight hours on school nights[455], and in some cases as little as five hours or less on a school night[456].

Sleep loss in adolescence has been highlighted as an 'epidemic' in the USA[457] and a matter of public concern in the UK. The reason for taking this issue so seriously is that not enough sleep in adolescents comes with important consequences, in terms of poorer physical and mental health.[458] Less than eight hours of sleep on a school night has been associated with a variety of adverse behaviours, including smoking tobacco or marijuana, the consumption of alcohol, physical fighting, feelings of sadness and even the serious consideration of suicide.[459] Short sleep is also associated with an increased risk of obesity[444] (chapters 12 and 13). A consistent finding has also been that shortened sleep results in poorer academic and school performance.[460-61] Laboratory studies have compared academic performance in adolescents allowed to sleep 10 hours in bed against 6.5 hours over five nights. Those sleeping only 6.5 hours showed a significantly worse academic performance.[462] Importantly, adolescents are generally aware that not enough sleep reduces their mood, concentration and ability to make decisions.[463] In a recent review, it was shown that 75 per cent of adolescents with a diagnosis of SCRD also had mental health problems.[464] Clearly, an important issue for parents, carers and society in general is to convince adolescents that they need to do something about their poor sleep (chapter 14). But first let us consider some of the reasons why sleep in adolescents can be difficult.

## THE BIOLOGICAL DRIVERS OF ADOLESCENT SLEEP

There are changes in the biological drivers of sleep that result in a delayed sleep/wake cycle, or a later chronotype, in adolescents. The result is that bedtimes occur later into the night and wake times, especially on free days, are pushed into the late morning or even afternoon. Maximum lateness occurs around 19.5 years for women and 21 years for men, which is approximately two hours later than individuals in their mid-fifties and early sixties.[381] Such differences often generate conflict with parental and societal expectations and accusations of 'laziness'. As we learnt in chapters 1, 2 and 3, our chronotype depends upon our genes, our development and when we get light exposure. More evening light exposure acts to delay the clock and drive a later chronotype. And there is good evidence that adolescents with a later chronotype get more evening compared to morning light.[92] This, of course, can be corrected by trying to get more morning light (chapter 3). In addition, the progressively later chronotype in adolescents correlates very closely with the hormonal changes that occur during puberty. It seems likely that the sex hormones (oestrogen, progesterone and testosterone) interact with the master clock in the suprachiasmatic nuclei (SCN) at some level and alter sleep timing. As discussed in chapter 7, there is evidence that oestrogen and progesterone can interact with the circadian system across the menstrual cycle and during pregnancy, so it seems likely that these hormones will influence the clock during female puberty. Recent studies also provide good evidence that testosterone levels in males act to drive the clock to a later chronotype.[465]

In addition to the circadian changes during puberty, sleep pressure (chapter 2) is also altered. Late-stage adolescents have been shown to have a slower build-up of sleep pressure than pre- or early pubertal children,[466] suggesting that later-stage adolescents may be able to stay awake longer without feeling tired. These findings are supported by a study that measured the time it took to fall asleep after a period of wakefulness of 14.5, 16.5 and 18.5 hours. Pre-pubertal adolescents fell asleep much faster than mature adolescents,[467] again suggesting that the response to sleep pressure is diminished at this age. As a result, the data suggest that late adolescents can stay up longer 'biologically' in the late evening. However, sleep pressure and the circadian drivers of sleep during adolescence do not act alone, and this 'biological predisposition' for a late chronotype must be considered alongside the varied environmental modulators of sleep, some of which are considered here:

ENVIRONMENTAL REGULATORS OF SLEEP IN ADOLESCENTS
There are key ways in which adolescent sleep can be altered. These include:

**The impact of caffeine.** As discussed in chapter 2, caffeine is frequently consumed as an 'antagonist' of sleep. It blocks the receptors that detect the neurochemical adenosine, which increases within the brain as a result of wakefulness. Adenosine is considered as one of the key drivers of sleep pressure.[468] Caffeine in drinks like coffee lasts in the body for a considerable time before it is broken down, so caffeine in the late afternoon or early evening will act to delay sleep.[468] Adolescents are targeted by advertising campaigns to consume 'energy' drinks which contain between 70 and 240mg of caffeine in a standard drink, or 'energy shot'. In the UK, more than 70 per cent of 10- to 17-year-olds are reported to consume such drinks.[469] The assumption is that adolescents are consuming caffeine-rich drinks to increase alertness, and combat daytime sleepiness. This may indeed be the case, but because caffeine can last in the body for many hours, an alerting afternoon or early-evening caffeine drink will act to delay sleep. This caffeine-induced delay will reinforce the delayed biological drivers of sleep.

**The use of social media.** A major concern in recent years has been the growing use of electronic devices as a factor in delayed sleep (Figure 4, p. 103). Such devices have been considered to act both to replace the time available for sleep and as a cognitive and emotional arousal mechanism, delaying sleep.[470] And the data support these concerns. For example, a large survey in the USA investigated whether shorter sleep in adolescents was correlated with electronic-device use, social media and screen time. This detailed study concluded that it was.[471] In another survey, gaming and the use of mobile phones, computers and the Internet were linked to a significant delay in adolescent bedtimes.[472] Mobile phone use alone has also been shown to drive poorer sleep behaviours[473], including delayed sleep[474] and increased emotional arousal[475].

**School start times.** Most schools do not consider the later chronotype of adolescents when planning the school curriculum. Many adolescents are forced to wake up earlier than the start of their biological day to get to school on time. This misalignment between the biological day and societal demands has been termed 'social jet lag'.[195] In practical terms, social jet lag represents the difference between when an adolescent wants to wake up (as on free days) and when the adolescent is forced to wake up (as on a school day). Sleep need (sleep debt) builds over the school week, and the response is to catch up with extended sleep at the weekend, getting up much later. The result is to further delay sleep, not least because of the loss of morning light exposure, and the start of the school week begins with shortened sleep.[241] School start times work against later chronotypes, but favour earlier chronotypes who, no surprise, show higher levels of academic performance and attention in class.[476]

One approach to address the delayed chronotype of adolescents, and to counter the impact of social jet lag, has been to adopt a delayed start time to the school day. This has been embraced with some enthusiasm in the USA, driven by advocacy groups such as Start School Later, which recommends that middle and high schools should not start the school day earlier than 8.30 a.m. Note that the majority of middle and high schools in the USA start well before 8.30 a.m., and closer to 7 a.m.[477] The findings from the USA show consistently that a delayed school start time increases sleep duration and reduces daytime sleepiness, depression and caffeine use, improves punctuality, absenteeism and grades, and results in fewer motor vehicle crashes.[478-80] A delayed school start time seems to be highly beneficial in situations where schools normally start before 8.30 a.m., such as the USA, Singapore and Germany. However, many countries such as the UK traditionally start the school day much later than 7 a.m., and closer to 9 a.m. Under these circumstances it remains unclear if an even later start time will be beneficial or whether 'sleep education' may be the answer. Interestingly, many private schools in the UK have decided to start the school day at 10 a.m. or later.

**Sleep education and cognitive behavioural therapy for insomnia.** Sleep education aims to address the social and lifestyle factors that drive sleep to a later time. Sleep education, or good 'sleep hygiene', can be very beneficial. Combining a regular sleep schedule, a bedtime routine that promotes sleep, and morning light exposure really helps if adolescents can be persuaded to do it (chapter 6 ).[481] However, adolescents as a group are notoriously resistant to acting upon advice[482], and sleep education programmes have shown that while sleep knowledge can be improved, changes in sleep behaviour don't always follow[483–4]. But prioritizing and nurturing sleep in the home environment, by encouraging and discussing good sleep practice, can play an incredibly useful role in adolescent health and wellbeing.[485] Simple approaches to increase adolescent sleep have been shown to work, such as the introduction of earlier bedtimes.[486] Resolving the problems of adolescent sleep has to be achieved through a partnership between adolescents and sleep education at school, along with gentle reinforcement in the home environment by parents and/or carers. Sadly, such partnerships are rare, not least because the information needed to guide adolescent sleep behaviours is not standardized or readily accessible. I will return to this topic in chapter 14.

## *Sleep After Adolescence*
### *(Adulthood to the Healthy Aged)*

Benjamin Franklin said: 'In this world nothing can be said to be certain, except death and taxes.' Another certainty of life is that with age, and after youth, our patterns of sleep and circadian rhythms will change again. These changes can worry many, but change does not always have to mean worse. It's all about how you respond to these altered sleep/wake patterns. Globally, the population is ageing and the World Health Organization (WHO) predicts that, by 2050, the population aged 60 years or more will double, while those aged 80 years or more will number 400 million. Increased life expectancy has led to a change in the categories of what constitutes middle and old age, and very recently the leading medical journal the *Lancet* defined 'mid-life' as being 45–65 years.[487] So, by definition, old age would be older than 65 years. As things stand, half of the general population of older adults' report changed or disturbed sleep.[254] My point is that there are going to be a *lot* of people on our planet who will experience marked changes in their sleep and circadian rhythms, and we all need to know what to expect.

The changes in sleep as we progress from youth into middle age and then beyond are:

- shorter night time sleep (reduced total sleep time, or TST)
- a reduced amount of REM sleep
- an increase in the lighter stage of sleep (NREM stages 1 and 2)
- a corresponding decrease in deeper sleep (NREM stages 3/SWS)
- a longer time to get to sleep (longer sleep latency)[488]
- more awakenings during the night with excessive daytime sleepiness.

One third of older adults report early morning awakening and/ or difficulty staying asleep on a regular basis (several times per week).[489] Changes in sleep provide an important reason for the increased use of sleeping pills in the ageing population.[490] The 2003 National Sleep Foundation 'Sleep in America' poll found that 15 per cent of people over the age of 55 said they have daytime sleepiness, several days a week, so severe that it interferes with their daytime activities. In the same survey, and very alarmingly, 27 per cent of the individuals interviewed aged between 55 and 64 reported driving a car while sleepy in the previous year, 8 per cent reported actually falling asleep at the wheel, and 1 per cent reported having a traffic accident after falling asleep while driving. This raises the important point that the age-related increase in SCRD not only affects the health and the quality of life of the individual, but also the safety of the broader community.

Progressive changes in the circadian drive for sleep, along with altered sleep pressure, represent the key drivers for the changed patterns of sleep as we age. These include:

## Circadian timing (phase)

The most obvious change in the circadian system as we age is an advance in sleep/wake behaviour, driving the timing of sleep onset to an earlier hour. The timing of the circadian rhythm of core body temperature is earlier in both middle-aged and older adults compared to young adults (20–30 years).[491] The circadian rhythm of melatonin also seems to move earlier with age[492], as does the timing of the cortisol rhythm[493] (Figure 1, p. 20). However, these circadian rhythms do not all move to an earlier time in the same way. The timing of the rhythms of core body temperature and melatonin lag behind the sleep/wake cycle.[492] This means that in older individuals there is a greater tendency for 'internal desynchrony', and that getting the correct materials in the right place, in the right amount, at the right time of day becomes less precise. Unsurprisingly, this earlier wake and sleep time is not considered a positive experience by most aged individuals.[494]

## Circadian 'amplitude'

There is also good evidence that the amplitude, or robustness, of circadian rhythms diminishes with age. For example, the circadian rhythm in temperature cycles flattens with age[491], along with hormonal cycles[495], but again there is significant individual variation[496]. As discussed in chapter 7, in women this may be due to reduced oestrogen, and in men reduced testosterone. There is also evidence that the activity of our SCN might change with age, perhaps because the individual SCN clock cells are not so tightly connected (coupled), leading to a flattened circadian output rhythm.[497] The SCN might also lose neurones, and so be less capable of maintaining the appropriate levels of circadian drive.[498]

There are some fascinating new studies that have examined the circadian clock properties of skin cells (peripheral clocks) collected from young and older subjects and then studied in fresh culture media. The length, amplitude and timing of the clock cells were all identical in the two age groups, even though the actual sleep/wake behaviour recorded in these individuals was very different, with aged individuals showing earlier and flattened circadian rhythms. This suggested that the basic clock properties of peripheral clocks do not change with age, as the circadian rhythms of isolated young and old cells responded in the same way in culture. Remarkably, if 'young' cells were then cultured in the presence of blood serum from older individuals, rather than fresh artificial serum, the skin clocks resembled 'old' clocks with earlier and reduced amplitude rhythms. These findings suggested that something in the blood of older individuals was changing the circadian properties of the cells.[499] This is truly extraordinary, and reminded me of Countess Elizabeth Báthory de Ecsed (1560–1614), a Hungarian noblewoman who is said to have enjoyed drinking the blood of virgins, believing that it would preserve her beauty and youthfulness. To make my position absolutely clear on this, such a practice is a very bad idea.

## Altered circadian responses to light

There is evidence that young adolescents show increased sensitivity to evening light, and that this helps shift adolescent clocks to a later time.[500] By contrast, elderly people show decreased circadian photosensitivity to dusk light,[501] which would result in an earlier clock time. One suggestion for this decline in photosensitivity is that it could be due to eye problems such as cataracts, which might filter out light, and especially blue light, for circadian entrainment (chapter 3).[502] We tested this idea by examining sleep/wake cycles before and after cataract surgery using either a clear replacement lens which only blocks potentially damaging UV light or blue-light-blocking lens replacements. Six months after surgery, sleep quality was improved in both patient groups. These findings suggest that decreased light through the lens, as a result of cataracts, might indeed contribute to decreased circadian photosensitivity. But the reduction in blue-light transmission into the eye by blue-light-blocking lenses is not sufficient to affect circadian entrainment.[503] So at least in terms of the circadian system, and despite some alarming stories in the media, don't worry about what type of lens you get if you have cataract surgery.

### *The circadian regulation of sleep*

As discussed in chapter 2, the timing of sleep depends upon an interaction between the circadian system and the 'homeostatic drive for sleep', or sleep pressure. These two biological timers need to be properly aligned for stable patterns of sleep and wake. Under normal circumstances, sleep pressure builds through the day, while the circadian drive for wakefulness increases, reaching its maximum in the evening hours when sleep pressure is at its highest. This interaction to keep us awake is called the 'wake-maintenance zone'. The circadian system not only keeps us awake late in the day, but also provides us with an active drive for sleep at night, which reaches its maximum during the early morning just prior to our awakening, a time when sleep pressure is low. Ideally, the circadian system and sleep pressure interact to drive consolidated sleep and wake, but it seems that changes in the circadian system as we age reduce the robustness of this interaction. For example, with an earlier circadian clock because of age, the circadian drive for sleep in the early morning will be reduced, leading to an earlier wake time. In the same way, in the evening, the circadian drive for wake and sleep will occur earlier promoting earlier sleep. In addition, if the amplitude of the circadian drives for sleep and wake have been flattened with age, then the maintenance of sleep and wake will be less effective, leading to greater daytime sleepiness (napping) and night-time awakenings. Finally, the circadian timing system is also thought to influence the timing and duration of both REM and NREM sleep.[504] Which also helps explain the altered patterns of NREM/REM sleep as we age.

With the above drivers and mechanisms in mind, let's look at post-adolescence changes in sleep, and how they relate to SCRD in more detail. Overall, the lower the SCRD experienced, the better the chances of good mental, cognitive and physical health as we age.[505]

## Adulthood/middle age (19–65 years)

There are multiple societal and biological causes of SCRD, but during the middle years there are some particular drivers that need to be highlighted. As we try to balance family life with career aspirations, sleep is often squeezed out of our priorities. We also experience the increased risk of clinical sleep disorders, particularly associated with weight gain (obstructive sleep apnoea) or increased and chronic stress. As we age, we also become more morning types and our sleep duration reduces. It seems that the circadian driver of sleep and the processes that give rise to our sleep pressure become 'sloppier'[506] and less able to control the sleep/wake cycle with its earlier precision. Gender differences in sleep also become more apparent as we age, not least the menopause, which has a major effect upon sleep (see chapter 7), inducing night sweats, mood changes and difficulty falling asleep. Post-menopausal women have nearly double the rate of reported insomnia compared to pre-menopausal women. However, when sleep is measured objectively, sleep would seem to be worse in pre-menopausal women. This has led to the suggestion that hormonal changes may affect the perception of sleep. The risk of OSA is three times higher after the menopause, partly due to the redistribution of fat, as a result of the hormonal changes during the menopause. Significantly, some women report a reduction in OSA after hormone replacement therapy (HRT).[507]

*The healthy elderly (65–100 years) and disturbed sleep*

Many people experience noticeably altered patterns of sleep as they get older and regard this as 'poorer sleep' and an inevitable consequence of ageing. However, different sleep is not necessarily 'bad'. Freed of the constraints of work and other pressures, the elderly can relax and just stop worrying about their sleep, and enjoy the sleep they get. I know several people in their eighties who feel they have never slept better, and friends and family are told firmly *not* to call before noon. Noon has become the new breakfast time! There is the assumption that older people need less sleep, or are incapable of getting good sleep, but neither of these assumptions may be true.[238] Aged individuals generally take longer to get to sleep, and experience more broken sleep, and shorter night-time sleep. All this leads to an increased likelihood of daytime naps. But this is not a problem unless it affects your daytime ability to function.[238] There has been a lot of discussion around older people having more sleep problems because they produce less of the pineal hormone melatonin. While it is true that we do produce less melatonin as we age, melatonin 'replacement therapy', giving melatonin as a sleeping aid to older people, is not successful at improving sleep.[254] This suggests strongly that lower melatonin is not the cause for the sleep changes observed as we age. This reinforces the argument that melatonin is not a 'sleep hormone' (chapter 2). Another potential problem in the healthy aged is temperature regulation. We need to experience a slight decrease in core body temperature to promote sleep. If you have poor circulation with cold hands and feet, you cannot lose as much heat from the extremities. Warming the hands and feet, which causes vasodilation and increases body heat loss from the body[508], increases sleepiness and the likelihood of falling asleep[509]. So my Grandma Rose was right: bedsocks and mittens help you get a better night's sleep!

## *The need to pee at night*

Without doubt, the question that I am most often asked when speaking to senior citizens about sleep is 'Why do I need to get up at night and go for a pee?', or expressed more formally, to void urine (nocturia). The kidneys typically produce 250–300ml of urine overnight, and a normal bladder can usually hold up to 350ml of urine. So, ideally, emptying the bladder before going to sleep should prevent the need to get up and pee at night. Sadly, this ability declines with age. For a long time nocturia was thought to be only a problem for men and occurred as a result of an enlarged prostate gland (benign prostatic hypertrophy). However, several recent studies have shown that nocturia is an issue for both men and women.[510–11] In terms of age, self-reports of nocturia found that less than 5 per cent of young adults experience nocturia; this figure rises to about 50 per cent for people in their sixties, and around 80 per cent in their late seventies[512], and is considered a major cause of sleep disruption and daytime sleepiness[513].

Because sleep is not as deep as we age, sleep disturbance is much more likely to occur. Following awakening we are more likely to become aware of our bladder 'status' and then be prompted to go for a pee. Light sleep will also make the individual more aware of the signals coming from the stretch receptors in the bladder, which will drive awakening. Indeed, if older individuals are given sleep medication, nocturia is reduced.[514] In addition to lighter sleep, several additional causes of nocturia have been proposed:

### REDUCED BLADDER CAPACITY

One study showed that the bladder capacity in older men and women was almost half of the capacity of younger individuals.[515] This can be due to a variety of causes, including obstruction, inflammation or cancer. However, as mentioned, a full bladder may not be the actual cause of awakening, and nocturia is a secondary consequence of sleep disturbance and the decision to pee.[513]

### BENIGN PROSTATIC HYPERPLASIA (BPH)

Twenty years ago, BPH was considered *the* cause of nocturia, but is now recognized as only a contributing factor in men.[516] The problem arises because in most men their prostate continues to grow throughout life, and in many this continued growth enlarges the prostate enough to cause a significant block in urine flow. The prostate gland is located beneath the bladder, and the tube (urethra) that transports urine from the bladder passes through the prostate. When the prostate enlarges, it begins to block urine flow. Because the urethra is being squeezed, the bladder needs to apply more pressure to get the urine out. This leads to a thickening of the bladder walls, which makes the walls less flexible, and less able to contract and fully empty the bladder.[516]

### CIRCADIAN REGULATION OF VASOPRESSIN

Under controlled laboratory conditions, older adults produce more urine at night compared to younger adults, and it seems there is a 'flattening' in the circadian rhythm of urine production.[513] Urine production is regulated by two key hormonal drivers, vasopressin and atrial natriuretic peptide. Let's first consider vasopressin, which is also called arginine vasopressin, and usually abbreviated as AVP. AVP is released from the posterior pituitary gland and into the general circulation. A key function is to cause the kidneys to reabsorb water from the blood and return it to the circulation. The consequences of this action are to make the urine more concentrated and reduce urine production. This protects the body against dehydration during sleep. A circadian rhythm of urine output, low during the night (10 p.m.–8 a.m.) and higher during the day (8 a.m.–10 p.m.), is established by the age of five years. In young adults, AVP shows a diurnal rhythm with a peak during the night-time hours, promoting lower urine production. However, there is some suggestion that in older individuals the rhythm in AVP is either missing or flattened.[517] A synthetic replacement for AVP called desmopressin, or DDAVP, can be taken before bedtime as a treatment for nocturia and has been shown to be effective in reducing night-time urine production and interrupted sleep.[518-19] So a flattened circadian rhythm in AVP may be another contributor to nocturia.

### URINE PRODUCTION AND ATRIAL NATRIURETIC PEPTIDE (ANP)

The second key hormone involved in urine production is atrial natriuretic peptide. ANP is secreted by heart muscle cells, which detect increased stretching of the heart wall due to increased blood volume/pressure. ANP acts on the systems of the body that increase the excretion of sodium and water from the kidneys. This increases urine production (diuresis), and so lowers blood volume and hence blood pressure. During the day, and due to inactivity, fluid from the blood accumulates in the legs and ankles. At night, when lying down to sleep, the fluid in the legs and ankles is reabsorbed by the body. Blood pressure then increases, ANP is released in response and urine is produced. Some individuals can make over 1,000ml of urine once they lie down for the night, and as the normal bladder can only hold about 350ml this will promote nocturia. Another link between blood pressure and nocturia is seen in individuals with obstructive sleep apnoea (OSA), a condition that can lead to high blood pressure. With increased apnoea events there was a direct correlation with increased ANP secretion at night. This leads to increased urine output and nocturia.[520] Treatment of OSA with continuous positive airway pressure (CPAP) has been demonstrated to reduce nocturia[521] (chapter 5). In this regard, OSA should be viewed as part of the diagnosis, and indeed a target for treatment for nocturia.

### URINE PRODUCTION, ANP AND ALDOSTERONE

An important action of ANP is to inhibit aldosterone release from the adrenal cortex in the adrenal gland. Aldosterone acts at the kidneys, primarily, to reabsorb sodium into the blood, leading to an increase in blood pressure. Normally, aldosterone secretion is closely tied to the sleep/wake cycle, with high levels of aldosterone during sleep, reducing urine production. This day/night difference seems to be driven primarily by sleep itself, and not the circadian system, as sleep deprivation leads to a blunted rhythm in aldosterone.[522] So collectively ANP, aldosterone, high blood pressure and sleep disruption can all contribute to drive nocturia. Irrespective of the causes of nocturia, it can have a profound impact upon both the actual and the perceived health and quality of life of the individual. Nocturia is highly correlated with excessive daytime sleepiness, an increased risk of injury as a result of falling at night[523], depression[524] and perhaps even early death[525]. It is worth pointing out that there are multiple 'pee bottles' on the market for both men and women that can be kept close to the bed and that can be used to minimize awakening trips to the lavatory at night. And at the risk of going 'off-piste' (sorry), don't throw your urine away if you have a garden. It can be emptied onto your compost heap and, diluted, onto the soil, providing plant nutrients like nitrogen. You will find extensive advice about such practices online. Somewhat surprisingly, I discovered that there are even clubs you can join.

### ANTIHYPERTENSIVES AND URINE PRODUCTION

I will discuss the benefits of taking blood pressure medications called antihypertensives before bedtime to reduce the chances of stroke in chapter 10. But there is a downside. Some antihypertensives increase urine production.[526] Diuretics, sometimes called 'water pills', stimulate the kidneys to remove water and salt from the blood into the urine. This will reduce blood volume, and so reduce blood pressure, but it will increase urine production if diuretics like furosemide (Lasix®) are taken before bedtime. Calcium channel blockers (e.g. Amlodipine) also increase nocturia.[527] Calcium channel blockers relax the blood vessels and lower blood pressure. But they will also act to inhibit bladder contraction and reduce bladder emptying, making it necessary to pee more and pee at night.[528]

## Sleep and the Impact of Neurodegenerative Diseases

### The elderly, dementia and Alzheimer's disease (65–100 years)

For clarity, dementia is a general term for a decline in mental ability severe enough to interfere with daily life. The most common form of dementia is Alzheimer's disease (below). Dementia is not an inevitable consequence of ageing, but 50 per cent of those aged 85 years or older will be affected to some degree,[254] and in terms of its links with SCRD the facts are truly brutal. Some form of SCRD is reported in at least 70 per cent of patients in early-stage dementia[529], and SCRD in individuals with dementia is highly predictive of poorer outcomes in terms of more severe cognitive and neuropsychiatric symptoms, along with poorer quality of life[530]. Remarkably, an estimated 70–80 per cent of people with dementia have a sleep-related breathing disorder (SBD) such as obstructive sleep apnoea (OSA) (chapter 5), and the worse the SBD the greater the dementia's severity.[254] This raises the possibility that SBD may drive the progression of dementing illnesses and, in turn, that dementia may exacerbate SBD.[531] This seems likely as sleep-related breathing disorders have been linked to poorer attention, cognition and reaction speeds. SBDs also carry a 2–6 times greater risk of mild cognitive impairment and an earlier onset of dementia.[532] How SBDs like OSA lead to dementia is unclear, but it seems likely that the lack of oxygen to the brain (hypoxia) is a probable contributing, or even driving, factor. It is also worth being aware that SBDs, and shortened sleep in general, are linked to a thinning of the cortex of the brain and brain ventricle enlargement; both are features of cognitive decline and dementia.[254] A really important point is that many SBDs can be treated, and so the detection and treatment of conditions like OSA could have a big impact upon cognitive decline and dementia in the elderly. Significantly, individuals with mild dementia and who have SBD tolerate continuous positive airway pressure (CPAP) treatment no differently from non-demented SBD patients, and so this treatment should be strongly encouraged; although individuals with full-blown dementia along with neuropsychiatric symptoms do *not* cope well with CPAP.[533]

### Alzheimer's disease

As I mentioned, Alzheimer's disease is by far the most common form of dementia, accounting for around 80 per cent of all cases. In 2020 approximately 5.5 million US citizens were thought to have Alzheimer's disease, and by 2050 the number is estimated to rise to 13.8 million. In terms of care, the financial costs to families and the state are estimated in the hundreds of billions of dollars in the USA alone.[534] The exact cause of Alzheimer's disease is still debated but is characterized by 'plaques' and 'tangles' in the brain due to two proteins called amyloid (plaques) and tau (tangles).[535] These proteins 'clog up' the workings of brain cells and cause their death. One consequence is the death of neurones in the basal forebrain which produce acetylcholine. These neurones normally project to and stimulate the hippocampus and the cerebral cortex (Figure 2), which are involved in generating memories and cognition.[536-7] This observation is consistent with the fact that Alzheimer's disease is characterized by the loss of acetylcholine, with the progressive symptoms of: difficulty remembering recent events while having a good memory for past events; poor concentration; difficulty recognizing people or objects; poor organizational skills; confusion; disorientation; slow, muddled or repetitive speech; withdrawal from family and friends; and problems with decision making, problem solving, and planning and ordering tasks. In addition, acetylcholine is one of the key wake-promoting neurotransmitters, and its loss causes daytime sleepiness, along with cognitive decline. Drugs such as Donepezil, Rivastigmine and Galantamine all act to block an enzyme called acetylcholinesterase from breaking down acetylcholine in the brain. This results in higher levels of acetylcholine, which can lead to a slight but significant improvement in cognition and the sleep/wake cycle. In some individuals, acetylcholinesterase inhibitors can slow cognitive decline for several months. However, there is an important issue with these drugs that is often overlooked. Acetylcholinesterase inhibitors increase REM sleep and nightmares in patients with Alzheimer's disease. As a result, these drugs must be taken in the morning rather than at night.[538] Sadly, this is frequently not appreciated and acetylcholinesterases are often prescribed by a clinician to be taken before bed, leading to severely disrupted sleep and vivid dreams. Such vivid dreams are the primary reason why patients will stop taking these drugs.

Around 70 per cent of individuals with Alzheimer's disease are reported to have disrupted and fragmented circadian rhythms with night-time wakefulness, delayed rhythms, disrupted rhythms of core body temperature and frequent daytime napping (Figure 4, p. 103). Many will also show 'sundowning' characterized by agitation and disruptive behaviours in the evening or night (Figure 8, p. 226). Circadian disruption begins early in the course of the disease and progresses until death.[539] Not surprisingly, this high degree of SCRD is associated with poorer daytime cognitive performance, along with aggression and agitation.[254] Such disruption has been linked to a degenerate SCN in Alzheimer's patients, suggesting that the degeneration of the SCN may contribute directly to the SCRD.[254]

SCRD earlier in life seems to be a good predictor of an increased risk of developing Alzheimer's disease.[540] In one study, a short sleep duration of five hours or less and a long sleep duration of nine hours or more was associated with an increased risk of dementia.[541] In another study, markedly fragmented circadian patterns of sleep/wake behaviour were associated with a 50 per cent greater risk of developing Alzheimer's disease.[542] However, until recently, cause and effect have been difficult to establish. SCRD often occurs before a clinical diagnosis of Alzheimer's disease, and so disentangling whether SCRD promotes the development of Alzheimer's, or whether SCRD is the consequence of early Alzheimer's, has been a problem. Recent and robust links are emerging between SCRD and increased amyloid ($A\beta$) levels in the cerebrospinal fluid and brain.[543] Plaques are formed from aggregates of these misfolded proteins and build up in the spaces between nerve cells. SCRD is associated with a disruption of the recently discovered 'glymphatic system'. The glymphatic system is a type of waste clearance system that removes toxic agents from the cerebrospinal fluid, including $A\beta$. Importantly, the glymphatic system is most active during sleep[544], and this suggested that sleep might actually help prevent $A\beta$ accumulation and, by extension, Alzheimer's disease[545]. A very recent study using brain-scanning approaches showed that just one night of sleep deprivation in healthy human subjects increased $A\beta$ deposition in the brain.[62] And, even more recently, another study showed that OSA *is* associated with higher levels of $A\beta$ within the brain.[546] Significantly, abnormalities in some of the key genes that generate the circadian molecular clockwork have been linked to increased levels of $A\beta$ and cognitive impairment in mouse models of Alzheimer's disease.[547] So mechanisms are emerging between SCRD, the glymphatic system, the molecular clock and dementia. If these links turn out to be correct, and the evidence is already strong, then treating SCRD as soon as it is detected could be used to slow the progression of Alzheimer's.

*Parkinson's disease (PD)*

Parkinson's disease is a neurodegenerative disease caused by a deficiency of the brain neurotransmitter dopamine. The main symptoms of PD are involuntary shaking of parts of the body (tremor), along with slow movements and stiff and inflexible muscles. Individuals with PD can also experience other symptoms, including: depression and anxiety; balance problems; loss of a sense of smell; memory problems; and very commonly SCRD – affecting 60–95 per cent of patients.[548] Most people with Parkinson's start to develop symptoms when they are older than 50 years, and the condition affects around 2 per cent of adults over the age of 65.[549] PD can progress to a form of dementia due to the build-up of a misfolded protein called α-synuclein that aggregates to form 'Lewy bodies', which are considered to be an important mechanism in brain degeneration. In addition, amyloid (Aβ) deposition forming plaques, like Alzheimer's disease, is linked to cognitive decline in PD.

The SCRD features in Parkinson's disease include: daytime sleepiness, insomnia and REM sleep behaviour disorder (RBD). As with dementia, SCRD exacerbates the symptoms of PD, and the main symptoms of PD can make SCRD worse. Because dopamine is a key neurotransmitter that plays a critical role in maintaining the sleep / wake cycle, the decline in dopamine in PD is almost certainly the cause of many of the sleep problems associated with the condition. RBD (chapter 5) is strongly associated with PD and dementia. The normal muscle atonia (paralysis) during REM sleep (chapter 2) is impaired or lost in PD subjects and results in considerable physical activity during sleep, such as acting out violent dreams during REM sleep. RBD can serve as an early 'biomarker' that predicts the onset of PD. This could be very important because future drugs to prevent the onset of this neurodegenerative condition need to be given before symptoms start. The diagnosis of RBD is associated with a five-year 20 per cent risk, 10-year 40 per cent risk, and 12-year 52 per cent risk of developing a neurodegenerative disease such as PD.[40] It has been suggested that Lewy body formation in the SCN may reduce the capacity of the SCN to provide a strong circadian drive to regulate multiple aspects of sleep, including REM sleep.[550-51]

### Can I do anything about SCRD in dementia and Parkinson's disease?

Because SCRD can exacerbate the symptoms of dementia and PD, it is important to think of SCRD stabilization as a therapeutic target. We have discussed what to do about SCRD in general terms in chapter 6, but let's review some of the current specific approaches for tackling SCRD for these devastating neurodegenerative diseases. As with the general population, including the healthy aged, individuals with dementia can improve the symptoms of SCRD and daytime sleepiness by exercise outside, such as walking.[552] This is one of the first activities to be diminished or lost with disease progression. It is perhaps worth mentioning that in care homes the routine is to keep individuals in bed during the day for ease of care. The result will be the loss of any kind of exercise and hence the removal of an intervention that could improve SCRD. Also, reducing or eliminating caffeine intake, especially close to bedtime, and reducing alcohol consumption can help.[254] Two further areas warrant a detailed mention.

### Sleep medications in dementia and Parkinson's disease

Medication, in the form of sleeping tablets, should not, ideally, be the first approach to help SCRD in dementia. The benzodiazepine drugs are not recommended for dementia or Parkinson's because they can increase the decline in cognition and mood, promote daytime sleepiness and increase the risk of falls. There are also the added problems of addiction and drug–drug interactions.

A recent review of the use of the newer 'Z drugs' (zopiclone, zaleplon, zolpidem, etc.) to treat insomnia in dementia concluded that any benefit was considerably outweighed by the risks.[553] Low doses of antidepressants such as selective serotonin reuptake inhibitors (SSRIs) or trazodone have also been used frequently to treat insomnia in individuals with neurodegenerative diseases, but there are few data to support their use, and again these drugs produce the side effects of daytime sleepiness, dizziness and weight gain – leading to obstructive sleep apnoea (OSA).[554] Antihistamines are found widely in many over-the-counter sleep aids (e.g. Benadryl), but should be avoided in Alzheimer's because they can reduce acetylcholine in the brain, which makes the cognitive impairment worse.[555] Melatonin has also been used to try to improve sleep in dementia. However, the effects are either small or absent.[254] In view of the findings to date, the treatment of SCRD in dementia and Parkinson's should probably avoid the existing sleep medications, or at least consider them as a very last resort.

### Light exposure in dementia and Parkinson's disease

Exposure to natural bright light or the use of bright-light therapy (chapter 6) seems to be a particularly useful approach for two reasons: older adults appear to show reduced circadian light sensitivity;[501] and individuals with dementia or Parkinson's do not get outside very much and so experience very little natural light, especially in the nursing home environment. In one early study, individuals within a nursing home were found to be exposed to an average daily light exposure of only 54 lux, with only 10.5 minutes spent over 1,000 lux.[556] Have a quick look at Figure 3 (p. 61) to give you a sense of natural-light exposure. In a recent report, elderly individuals in a nursing home were compared to non-institutionalized elderly individuals. Those in the nursing home showed worse overall sleep, higher levels of daytime sleepiness and higher depressive symptoms.[557] In a key study, long-term daily treatment with either bright light (~1,000 lux) or dim (~300 lux) light in a nursing home environment over a period of several years showed that bright light partially consolidated sleep/wake cycles, reduced daytime sleepiness, and reduced both cognitive decline and depressive symptoms in individuals with dementia.[558] So bright-light therapy, in some form, has now been shown to consolidate sleep/wake cycles, increase daytime wakefulness, reduce evening agitation and improve cognition.[559] In contrast to drug-based medications, bright-light therapy represents a potentially very powerful therapeutic intervention to improve the health and wellbeing of the elderly and those suffering dementias. Importantly, this information should be incorporated into the design and construction of the next generation of nursing homes.

## Questions and Answers

### 1. Should children nap?

Sleeping during the day could reduce sleep at night, as in adults, but this seems unlikely in most children as they need more sleep. Preventing napping may deny children much-needed recovery sleep if they are not fulfilling their full sleep need at night. Nevertheless, napping should be age-appropriate, and a reduction in daytime naps may be necessary if getting to sleep at bedtime is a problem.

### 2. What is the link between night-time light exposure and myopia?

This is a somewhat controversial area. Myopia, or nearsightedness, is a significant public health concern in many developed and developing countries. It is estimated that by 2050 approximately 50 per cent of the world population will be affected by myopia. Comparisons of children with and without myopia have strongly suggested that lack of time spent outdoors and exposure to bright natural light is a potential risk factor for the development of myopia. So what about the use of dim bedroom night lights that can give children an extra feeling of security? It has been reported that exposure to artificial lighting at night during the first two years of life is associated with the development of myopia. However, there are no strong data to support this. One large study showed that there were no differences in the incidence of myopia in children who had slept from 0 to 2 years in a bedroom in darkness or with a night light.[560] A later study confirmed this.[561] The consensus seems to be that night light use will not cause myopia, but the failure, when young, to be exposed to bright natural light during the day could be a risk factor.

### 3. Do weighted blankets promote sleep in children with autism?
Autism is a complex neurodevelopmental condition with features that include atypical communication skills, poor social interaction, impaired information processing and motor skills, and very often poor sleep. Estimates suggest that 44–83 per cent of individuals (adults and children) with autism experience sleep disturbances. Weighted blankets have been used as an intervention strategy to improve sleep in children with autism who have sleep problems. However, a very recent study[562] supports earlier findings that use of a weighted blanket had little influence on sleep in children with autism.

### 4. I have noticed that as I have got older I wake up with a wet patch on my pillow. Should I worry about this?
Drooling (or sialorrhea) while sleeping is quite common. Saliva can build up in your mouth while asleep. Normally, saliva collects at the back of your throat and this triggers an automatic swallowing response. But if you are lying on your side, the saliva trickles out of the corner of your mouth and onto the pillow. A bit unpleasant, perhaps, but harmless.

# 9.

# Time Out of Mind

## The effects of time on cognition, mood and mental illness

We are not interested in the fact that the brain has the
consistency of cold porridge.

Alan Turing

Albert Einstein is the poster boy for both genius and sleep. He is
*the* example I often use in my lectures of intellect fuelled by regular and extended sleep. His general theory of relativity and Nobel
Prize in Physics were the product of a nightly sleep of around 10
hours followed by a highly structured day. Often, at this point in
the lecture, a hand will be raised and someone will ask: 'Well,
what about Salvador Dalí, he didn't sleep and you can't say that
he wasn't a genius?' 'Good point,' I respond! The story goes that
Dalí would sit with a key or spoon in one hand, and some dis-
tance below this hand he would place a metal plate on the floor.
When he fell asleep the spoon would fall from his relaxed hand,
hit the plate and the noise would wake him. Dalí felt sleep was a
waste of time. And for Dalí, and especially for the creation of his
art, it probably was. This is because sustained lack of sleep induces
paranoia, hallucinations and an altered state of consciousness.
Such vivid perceptions are usually abstract images or sounds,
although they can also be other sensations such as smell or taste.
Hallucinations are illusions, and Dalí claimed that the melting
watches in his famous painting, *The Persistence of Memory* came
from self-induced hallucinations induced by lack of sleep. In

contrast to Dalí, Einstein needed critical and forensic thinking to understand the nature of the universe, and sleep helped him achieve this crystal-clear reality. Dalí wanted a surrealist vision of the universe, and for him the distorting lens of sleep deprivation and altered consciousness helped achieve his singular perspective. So, time permitting, that's how I deal with the 'no sleep and Dalí genius' question. However, if time is short, I might just quote George Orwell, who expressed the view in 1944 that Dalí's art was 'diseased and disgusting' and the man 'cruel and repellent'. Indeed, reading Dalí's autobiography can be a very shocking experience.

Consciousness has been variously defined as our individual awareness of our unique thoughts, memories, feelings, sensations and environments: essentially, our awareness of self and the world that surrounds us. For most of us, our consciousness sits somewhere between the extremes of Einstein and Dalí, and precisely where will be profoundly influenced by the time of day and the level of SCRD we are forced to endure. I have divided the following discussion into two parts, considering first 'Cognition, Time of Day and SCRD', and then 'Mood, Mental Illness and SCRD'. My aims are to provide you with some insight into your own consciousness and the consciousness of others, and how you might want to improve both.

## Cognition, Time of Day and SCRD

Let's start with a definition for the much-used, but poorly understood, word 'cognition' – as in the phrase 'your cognitive abilities', and of course cognition contributes greatly to our consciousness. Cognition describes a broad range of the processes within the brain that are needed to gather information and then understand, store and respond appropriately. In many instances actions will draw upon memories and previous experiences. There are three key elements to cognition: **attention**, whereby we notice key features in the environment and filter out 'irrelevant' information; **memory**, which describes our ability to retain and retrieve information, initially as a transient memory before it becomes 'set' into long-term memory; and **executive function**, which relates to our brain's ability to plan, monitor and then control complex behaviours that achieve specific goals or enable us to complete specific tasks. Essentially, our 'executive functions' are the processes in the brain that allow us to solve problems – like Einstein's $E = mc^2$, or for most of us how to turn what we find in the fridge into dinner. Collectively, our cognitive processes can be conscious, when we deliberately set out to address an issue or problem, or unconscious, when the brain attends to features in the environment or draws upon memories to solve a problem, while we are unaware of this activity until a solution suddenly emerges in the form of a 'flash of inspiration' or 'insight' (see below).

The first point to make is that our overall cognitive abilities, which can be assessed using a whole host of different tests, vary markedly across the 24-hour day. This daily change is determined by an interaction between our circadian system, chronotype, sleep need and age. It is often difficult to disentangle the individual contribution of each of these processes, but the final result is that the cognitive abilities of most adults rise rapidly from waking and peak in the late morning and early afternoon (Figure 6). In a famous study, from an Australian researcher, Drew Dawson, our cognitive abilities in the early morning between 4 a.m. and 6 a.m. were shown to be worse than the cognitive impairment produced by a level of alcohol consumption that would make us legally drunk.[563] This time-of-day effect makes any activity, especially driving, particularly dangerous in the early hours of the morning. But this change in cognition across the day is a bit different for teenagers and young adults, who tend to have a delayed chronotype. Cognitive abilities tend to rise and peak later in the day. On average they are delayed by about two hours, peaking in the mid-afternoon (Figure 6).[3] Such findings have been used to argue that teenagers, and particularly those teenagers with a very late chronotype, should be allowed to sit their school exams in the afternoon, rather than the usual practice of first thing in the morning.[564-6] These findings also highlight an interesting dilemma. Generally, because of age, the teachers of adolescents and teenagers will be most cognitively alert in the morning, but their teenage pupils will be far less alert at that time. By mid-afternoon, however, teacher cognition will be declining, while pupil cognition will be peaking. Such a mismatch prevents an optimal teaching experience for most teachers and their pupils, unless of course, the teacher is particularly young or has a delayed chronotype. This all illustrates the point that our cognitive abilities are not fixed and vary across the day. Importantly, there will be times in our lives when we need to engage the full cognitive resources that our brain can muster. We all have to make choices and sometimes, as an adult, it may be a good idea to make those choices in the late morning rather than the late afternoon.

Let's now move from the daily changes in our cognitive abilities and consider the impact of sleep and circadian rhythm disruption (SCRD) on the three key elements of cognition I introduced earlier: attention, memory and executive function:

## Attention and SCRD

Attention is particularly vulnerable to sleep loss. The 1986 catastrophe at the Chernobyl nuclear plant began at 1.23 a.m., precipitated entirely by human error. It was discovered later that the operators responsible had been working with greatly reduced sleep and failed to recognize that the powerplant was headed for disaster. Sleep loss over time will greatly decrease our ability to stay focused, alert and 'on task'. For example, after seven days, individuals who had the opportunity to sleep nine hours each

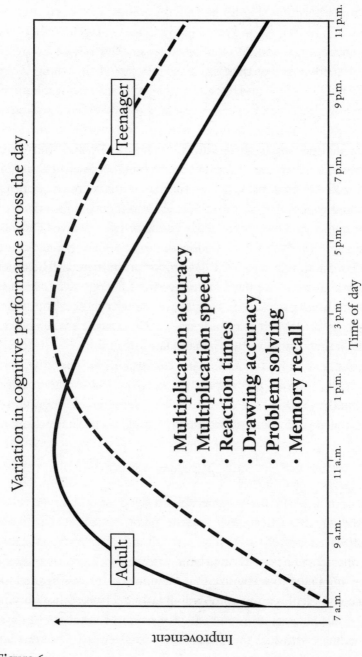

**Figure 6.**

**Figure 6. Cognitive performance across the day for adults and teenagers**. On average, cognition rises sharply from waking to peak in the late morning/early afternoon for adults, and mid-afternoon for teenagers. Unless chronically sleep deprived, teenagers usually show greater cognitive abilities throughout the afternoon compared to adults. The measurement of our cognitive abilities involves a range of different tests that assess the speed and accuracy of multiplication tasks, the speed of responses or reactions to an alert, the time taken to copy a drawing, solving problems such as rearranging shapes to make new ones, and the ability to recall memories.[3] Interestingly, our mood shows a similar variation across the day.

night had no lapses in attention. However, the opportunity to sleep seven hours per night, over seven days, led to five episodes of lost attention in the study period; five hours of sleep over seven days led to seven episodes of lost attention; and three hours of sleep per night over seven days produced seventeen episodes of lost attention.[567-8] These findings make the point that even a small reduction in optimum sleep time can accumulate over time to cause a problem with attention. The Space Shuttle *Challenger* explosion in 1986 is a classic example of the failure to pay full attention to a complex problem. The inquiry that followed the disaster stated that accumulated lack of sleep and sleep deprivation due to night shift work (resulting in loss of attention combined with impaired executive function) contributed to the poor judgements made when launching the shuttle. A similar set of problems led to the *Exxon Valdez* oil spill. The final report of the Alaska Oil Spill Commission published in 1990 stated that overwork and extended sleep deprivation were major contributing factors that caused the grounding of the *Exxon Valdez* oil tanker in 1989, and the resultant massive oil spill in Prince William Sound, Alaska.

Simple, repetitive tasks requiring sustained vigilance are badly affected by sleep loss. Multiple studies have shown that individuals have more lapses of attention on monotonous tasks than they do when completing more complicated tasks.[569] These lapses in vigilance can be due to brief involuntary episodes of 'microsleep', when there is a temporary lapse in consciousness that can last between three and 30 seconds, during which time the individual is largely unresponsive and unaware they have experienced a micro-sleep.[570] Tragically, one of my remarkable undergraduate professors, Thomas Thompson, died as a result of a microsleep driving home on a motorway. Sleep loss, combined with a boring task leading to a microsleep, probably accounts for the very many sleep-related motorway accidents and deaths worldwide.[571] For example, the Selby train crash in 2001 in the UK is a terrible example of the dangers of having a microsleep. Gary Hart fell asleep at the wheel of his Land Rover and veered off the motorway and onto railway tracks at Great Heck near Selby, North Yorkshire. This microsleep caused a 125 mph London express train and a 1,800-tonne freight train to collide, killing six passengers and four railway staff, and injuring more than 80 others. Hart was sleep deprived, and had little sleep the night before the morning crash. He was found guilty of 10 counts of causing death by dangerous driving. Hart was sentenced to five years in prison. Microsleeps are almost always preceded by head nods and drooping eyelids and these movements form the basis of some 'eye-video detection' devices to warn drivers that they are experiencing microsleeps. It seems likely that most new cars will incorporate this technology in the next few years.

The mechanisms in the brain that drive our attention are strongly linked to our levels of alertness, which originate from the release of 'excitatory neurotransmitters' within the brain. Alertness is a measure of brain wakefulness. Normally, during the day, excitatory neurotransmitter release is increased by the circadian system, which provides an increasing drive for wake throughout the day. This circadian drive for wakefulness is opposed by a sleep drive and the build-up of sleep pressure throughout the day, which on its own would decrease alertness and promote sleep (chapter 2). Normally these two drivers are in balance to help generate a sleep/wake cycle. Sleep loss, however, increases sleep pressure, and the pressure for sleep can then build and exceed the circadian drive for wake, leading to a marked reduction in alertness. The bottom line is that sleep loss reduces our attention and vigilance, and leads to cognitive impairment.

In addition to reduced attention, alertness and overall vigilance, sleep loss makes our cognitive performance more variable and erratic, and, importantly, this variability increases as we approach night-time and normal sleep. We sense we are sleepy and try to keep ourselves alert by an activation of the stress axis and the increased release of excitatory neurotransmitters within the brain. But in the evening sleep pressure is high and this acts to flip us from wake into sleep more quickly.[572] This variability can be particularly dangerous. Our attention will seem fine at one point, and we fool ourselves that we can cope, only to be followed by a sudden, potentially catastrophic loss of attention.[197] This seems to have been the main problem in 1979, at the Three Mile Island nuclear plant in Pennsylvania. The near calamity occurred when shift workers, working between 4 and 6 a.m. failed to notice a serious change in the power plant that very nearly resulted in the meltdown of the nuclear reactor the following day.

## Memory and SCRD

Memory involves the ability to both learn and retain experiences over days, months and years. Sleep is critical for the consolidation (establishing) of new memories, and an area of the brain called the hippocampus (Figure 2, p. 24) has been shown to be very important for the initial formation and organization of new memories. It seems that the pattern and sequence of neural activity generated initially within the hippocampus when information is gathered (acquired) is then 'replayed' in part during sleep. This replaying allows a strengthening of the connections between nerve cells to develop the first stage of memory formation (consolidation).[573] Sleep deprivation reduces the activation of the hippocampus, and this is linked directly with the failure to recall new events the following day.[574] By contrast, sleep promotes activity in the hippocampus, along with the formation of memories[575] (chapter 2).

The development of memory is divided into three processes: **acquisition**, or encoding, whereby a new memory is formed, but it is susceptible to being forgotten; **consolidation**, when new memories are gradually turned into a long-term, stable memory; and the **retrieval**, or recall, of a consolidated memory. Consolidated, or long-term, memories are divided into two types: **declarative memories** are those memories that are under conscious control and which we recall as facts and concepts, often called 'common knowledge', such as understanding the difference between a dog and a cat or knowing that Richard Wagner wrote *Der Ring des Nibelungen* (The Ring of the Nibelung); and **procedural memories** are those memories relating to how we perform different actions and skills. Basically, memories of how you do certain things such as riding a bike, tying your shoelaces or cooking a Beef Wellington. Although the difference between declarative and procedural memories can get a bit blurred, this distinction is not just semantic; these two different types of memory seem to be encoded differently during sleep. Multiple studies have shown that declarative memories are more associated with slow-wave sleep (SWS) and are stored long term in the temporal lobes of the brain, whereas procedural memories are more associated with REM sleep and are stored within an area of the brain called the cerebellum (Figure 2).[576] To be clear, REM sleep is not *just* associated with procedural memories, but also emotional memories, notably the memories associated with post-traumatic stress disorder (PTSD)[338] (see chapter 6, p. 134).

After a new experience, one of the key functions of sleep is to acquire and consolidate new memories. Interestingly, we know that sleep loss, and the sleep status of the brain, can influence the *type* of declarative memories that are retained. In a classic study, individuals who had either slept normally or who had been sleep deprived for 36 hours were asked to remember words with a different emotional association that were either negative (e.g. hate, war, murder), positive (joy, happiness, love) or neutral (e.g. cotton). After two subsequent nights of sleep, the subjects were asked what words they could remember. The sleep-deprived group showed an overall 40 per cent reduction in the recollection of all words, showing the importance of sleep deprivation in preventing memory acquisition. But the striking finding was that when the results were separated into the three emotional categories (positive, negative or neutral), there was a highly significant and major reduction in the retention of positive words, but only a trend to forget words with a negative or neutral association.[63] These data show that the tired brain is much more likely to remember negative, rather than positive, associations. To be clear, these and other data support the idea that sleep loss promotes the acquisition and the retention of negative, rather than positive, memories. Sleep loss drives a 'negative salience' of the world.

So why forget positive associations and remember negative ones? Overall, we humans seem to be programmed to expect that encounters with other people, and our experiences in general, will be pleasant or at worst neutral. As a result, negative behaviours or experiences are more unexpected and therefore more noteworthy (more salient). As a result, we ordinarily pay more attention to negative experiences or events. The key point is that the tired brain has an even greater preference to remember negative, over positive, experiences. This means that negative memories will play a greater role in our judgement process. Normally this could be useful, as negative experiences are more likely to do us harm, so it is worth remembering them. A difficulty arises when such negative experiences dominate our overall 'view of the world'. Indeed, a negative salience is a key feature of many mental health problems.

There is also very strong evidence that sleep not only helps in the retention of declarative memories (recall of facts and memories), but is also involved in the retention of procedural memories, such as learning a particular task. This has been shown in very many studies.[577] In one such study, participants were asked to learn a specific sequence of keystrokes on a keyboard, and each key press was associated with a particular sound. After the acquisition period, participants were left to consolidate this memory either by allowing them to sleep or by keeping them awake. While this consolidation occurred, the sounds associated with the correct sequence of keystrokes were played to the participants, reactivating the learnt sequence in the brain. Those who had slept were far better at recalling the learnt sequence.[578] Many studies have also shown that sleep deprivation inhibits procedural memory formation and task learning. A recent study showed that total sleep deprivation after learning a task resulted in much poorer task performance, and, significantly, performance was not improved by an afternoon nap, or by increased practice of the task.[579] This highlights the fact that a lost night of sleep greatly impairs the learning of a task, to an extent that a nap or extra revision cannot compensate for. And you know who I am talking to now – yes, you adolescents out there!

## Executive function and SCRD

Beyond attention and memory, the third element of cognition is executive function, which describes our ability to solve problems. This goes way beyond the notion that sleep is just there to help us form memories, but that sleep helps us to come up with novel solutions to complex problems. How often have we been told to 'sleep on it tonight, and make the decision tomorrow'? I can hear my grandmother saying this to me, and her intuitive wisdom was absolutely correct. Remember – Grandma Rose was also right about bed socks . . . Solving a problem after a night of sleep has been widely reported and is called gaining 'insight', and there are many famous examples. The Nobel Prize winner Otto Loewi reported that he woke with the idea of how to show that his theory of chemical neurotransmission within the brain was correct; the man who laid out the periodic table of chemical elements, Dmitri Mendeleev, reported that his insight for this achievement came after a night's sleep; August Kekulé had been pondering how the atoms in the benzene molecule are arranged – unable to find a solution, he slept and then woke remembering a dream of a snake biting its own tail, and realized that the benzene molecule is made of a ring of carbon atoms. And it is not just in the sciences that insight follows sleep. Beatles legend Sir Paul McCartney woke after a night of sleep in 1964 with the melody for 'Yesterday' fully formed. 'Yesterday' is one of the most recorded songs of all time with more than 2,000 versions to date. One wonders how much money that particular insight has generated . . .

But can such anecdotal experiences be verified in the laboratory? Does sleep really help us to solve a problem or allow us to generate a new idea? This was tested in what is now considered a 'classic' study – which means I teach it to my students. Subjects were introduced to a complex cognitive task in which there was a hidden pattern, and discovery of this 'rule' (gaining insight) allowed the participants to complete the task quickly and easily. Subjects received an initial training for several hours in the morning, and were then divided into three groups. Group 1 performed the task the same afternoon, and about 20 per cent discovered the hidden pattern; group 2 performed the task the following afternoon but were not allowed to sleep, and again around 20 per cent discovered the hidden pattern; group 3 performed the task the following afternoon but were allowed to sleep normally, and more than 60 per cent discovered the hidden pattern – they had gained 'insight' following sleep.[580] This wonderful study showed that sleep enables the extraction of knowledge and promotes insightful behaviour.

So collectively our cognitive abilities allow us to gather information, retain that information and respond appropriately. To be absolutely clear, cognition comprises selective attention, memory formation and an appropriate executive action. Sleep is essential for each of these processes to take place within the brain, and various forms of SCRD massively impair our overall cognition. It should also be clear by now that sleep is the best, and certainly the safest, 'cognitive enhancer' available to humanity.

## Mood, Mental Illness and SCRD

The old expression 'Did you get out on the wrong side of the bed this morning?' indicates, to my generation at least, that someone is in a bad mood, and alludes to our everyday experience that sleep and mood are intimately connected. Our 'mood' refers to a temporary state of mind or emotion that may be good, bad or neutral, which varies across the day. Generally, in healthy individuals, mood rises rapidly from waking through the morning, peaking in the early afternoon, declining slowly in the evening and then worsening at night.[581-2] In this regard, mood has a similar daily profile to cognitive performance (Figure 6). The observations that both cognition and mood are lowered in the evening and before bed provide additional reasons (chapter 6) why important discussions should wait until the following day!

Mood swings refer to changes in our mood from good to bad or vice versa. While everyone experiences mood swings to a certain degree, extreme mood swings can be characteristic of mental illnesses such as bipolar disorder, and are a symptom of other mental illnesses, including schizophrenia. Strikingly, SCRD in some form (Figure 4, p. 103) is a ubiquitous feature of mental illness.[247] This is an immense subject, and only a brief overview can be provided here. I think it is worthwhile introducing a few definitions before we start to distinguish between mood disorders and psychotic disorders, although the boundaries between these conditions are becoming increasingly blurred as we learn more about these states. In brief:

## MOOD DISORDERS

These are mental health problems that affect an individual's emotional state, such that a person experiences long periods of extreme happiness or extreme sadness, or flips between both mood states. There are multiple types of mood disorder: the most familiar is **depression** (also called major or clinical depression), in which grief or sadness occurs in response to a life event such as the death of a loved one, job loss or a major illness. In such circumstances, if the depression continues after the traumatic event is over, and there is no additional cause, for a period of two weeks or more, this is generally classified as clinical depression. Interestingly, circadian rhythms in core body temperature, melatonin and cortisol are flattened (reduced amplitude) in individuals with depression. Although the circadian rhythm in cortisol release is flattened, overall levels are higher.[581] Such findings have suggested that flattened circadian rhythms are a symptom and perhaps a contributing factor in depression. As circadian amplitudes appear to be decreased in depression, perhaps approaches that enhance the circadian drive might provide a useful therapeutic target (chapter 14). **Postpartum depression**, which is discussed below, is associated with the arrival of a newborn. **Seasonal affective disorder (SAD)** occurs during particular seasons of the year, usually starting in the late autumn and lasting until the spring or summer. **Psychotic depression** is a type of severe depression that occurs with psychotic episodes, such as hallucinations (seeing or hearing things that others do not) or delusions (having strongly held but false beliefs). **Bipolar disorder** (manic-depressive disorder) is defined by swings in mood from depression to mania. When in a low mood, symptoms resemble those of clinical depression. During a manic episode, a person may feel elated or can also feel irritable or have increased levels of activity.

### PSYCHOTIC DISORDERS

SCRD is also a feature of psychotic disorders, which are conditions that make it hard for someone to think clearly, make suitable judgements, respond with the appropriate emotions and actions, communicate coherently, assess reality and behave appropriately. When symptoms are severe, individuals are unable to handle daily life. An important element for the development of a psychotic disorder is considered to be 'aberrant salience', where abnormal patterns of brain neurotransmitter release lead to undue significance (cognitive attention) being paid to environmental stimuli that would normally be considered irrelevant.[583] An example of this would be when you fleetingly make eye contact with another passenger on public transport. We normally ignore this, but with psychosis such accidental eye contact can be interpreted as threatening and that you are being followed. There are different types of psychotic disorders, including the following. **Schizophrenia**, where individuals exhibit delusions and hallucinations with episodes lasting longer than six months; the impact on work or school, as well as maintaining personal relationships, can be very serious. **Schizoaffective disorder** is a condition in which individuals have symptoms of both schizophrenia and a mood disorder, such as depression or bipolar disorder. **Brief psychotic disorder** describes short periods of psychotic behaviour, often in response to stressful events, such as the death of a loved one, and in this case recovery is quick, usually less than a month. **Delusional disorder** describes a state where false and fixed beliefs are held that are plausible, but not true, such as being followed or being plotted against. **Substance-induced psychotic disorder** results from the use of, or withdrawal from, drugs such as hallucinogens and crack cocaine.[584]

There is a high chance that at some point in our lives we will all personally experience a mood or psychotic disorder, and almost a certainty that we will know someone with one of these conditions.[585] The statistics reported by several mental health charities are striking. For example, Mental Health First Aid (MHFA) England state that: 1 in 4 people experience mental health issues each year; 792 million people are affected by mental health issues worldwide; at any given time, 1 in 6 working-age adults have symptoms associated with mental ill health; mental illness is the second-largest source of burden of disease in England; mental illness is responsible for 72 million working days lost and costs £34.9bn each year. Self-reported depression is increasing, and there is debate as to whether this is a genuine increase, or due to increased societal awareness of mental illness, which has led to a changed threshold regarding what constitutes 'depression'. In short, our perception of how we feel has changed. For many the expectation is that we should feel 'happy', and if we don't feel happy we are by default 'depressed'. I am not trying to dismiss lack of happiness, but it is worth reminding ourselves that depression is not just lack of happiness, but a constant feeling of sadness and loss of interest which prevents us doing some or many of our normal activities.

Despite the diversity of mental illness states, *all* of these conditions are linked to some form of SCRD, and in some cases SCRD can be very severe.[246] The links between SCRD and mental illness have been known for a long time. For schizophrenia, SCRD was first described in the late nineteenth century by the German psychiatrist Emil Kraepelin (1856–1926).[586] Kraepelin is often called the 'Father of Modern Psychiatry' as he believed that psychiatric disease had a biological and genetic basis and so it would be possible to identify the cause of major psychiatric disorders and eventually find ways to treat these conditions. He was also a champion against the cruel treatments that psychiatric patients experienced in the asylums of the day. In these ways he was a pioneer, but, as with so many scientists, artists and policy makers in the late nineteenth and early twentieth centuries, he was also a strong proponent of eugenics and racial hygiene. Which leads us to the inevitable question, how can the same person hold simultaneously ideas that are brilliant and even kind, while other ideas are utterly repugnant? Intelligence in one domain is clearly no guarantee of decency.

Today, a severe level of SCRD is reported in more than 80 per cent of patients with schizophrenia, and SCRD is recognized as a characteristic feature of the disorder, although SCRD is seldom treated.[587] The nature of SCRD in psychosis and mood disorders is very variable, and all the sleep/wake patterns illustrated in Figure 4 (p. 103) have been observed. This notable diversity in sleep/wake patterns almost certainly reflects the diverse mechanisms that generate mental illness, arising from a complex interaction between genetics and the environment (chapter 5), including the demands of work, emotional pressures and physical illness.[586–91] Very importantly, the SCRD associated with mental illness contributes in a major way to the poor health, poor quality of life and social isolation experienced by the majority of individuals suffering from mental illness. The World Health Organization states that there is a 10- to 25-year life expectancy *reduction* in patients with severe mental disorders, highlighting the human cost of SCRD.[587, 592–3] Importantly, schizophrenia patients often comment that an improvement in sleep is one of their highest priorities during treatment.[594] It is also becoming clear that SCRD impacts upon the triggering of mental illness, its progression, relapse and chances of remission.[595–6]

The links between mental illness and SCRD are clear, but the mechanisms linking SCRD and mental illness have remained a puzzle. When considered at all, SCRD was thought to be an unfortunate side effect of mental illness, due to external factors such as social isolation and the lack of employment. In the absence of any social structure, such as a fixed pattern of work, sleep fell apart in conditions like schizophrenia. In addition, some psychiatrists thought SCRD arose exclusively from the use of anti-psychotic medication.[247] This is a particularly baffling explanation as SCRD has been reported in mental illness, such as schizophrenia, for nearly 150 years and long before the introduction of anti-psychotics in the 1970s. Such explanations are easier to understand when viewed in the context of the 'mind versus body' debate that dominated the thinking in psychiatry for decades. The 'mind' is all about mental thought and consciousness, while the 'body' is considered to be the physical processes underpinning brain function. As a neuroscientist, it is beyond my comprehension that such a distinction should be made in the first place. For me and, to be fair, many psychiatrists these days and starting with Kraepelin, the mind is the product of neural circuits and the physical structures of the brain, and there is no mystical place for the mind outside the brain. In psychiatry, sleep was considered to arise from the mind, and so explanations for SCRD in mental illness were sought from the environment and not from within the brain. Baffling!

A fresh look was needed. While it seemed likely that the lack of a social routine and/or the use of anti-psychotic medication might contribute to SCRD, it did not make sense that these factors were the direct and only cause of SCRD. Our team in Oxford decided to explore the idea that SCRD arose from a lack of social constraints by examining the sleep/wake patterns of individuals with schizophrenia and comparing the levels of SCRD in these individuals with unemployed control subjects.[246] The findings showed that severe SCRD exists in individuals with schizophrenia, but sleep/wake patterns in unemployed subjects are stable and essentially normal. As a result, SCRD in schizophrenia, at least, could not be explained on the basis of lack of employment. Importantly, SCRD in schizophrenia was also not linked to anti-psychotic medication.[246] These findings, along with an increased understanding of how sleep and circadian rhythms are generated and regulated within the brain, led us to an alternative idea, which suggested that mental illness and SCRD share common and overlapping pathways in the brain.[597] As discussed in chapter 2, the sleep/wake cycle ('flip/flop') arises from a complex interaction between multiple genes, brain regions, all the key brain neurotransmitters and multiple hormones. So it follows that changes in any of the pathways that give rise to mental illness will almost certainly have an effect, at some level, upon the sleep/ wake cycle. In fact, we have been able to show that brain circuits involved in mood and psychotic disorders overlap with the circuits associated with normal sleep/wake generation and regulation.[597-9] As a result, it is no surprise that SCRD is very common in mental illness; the pathways in the brain *are* linked! In addition, SCRD will act to exacerbate the severity of mental illness, and the mental illness will exacerbate the level of SCRD. The overlapping relationships between mental illness and SCRD are illustrated in Figure 7. It is also worth noting that many of the illnesses caused by short- and long-term SCRD (Table 1, p. 77) are very common in neuropsychiatric illness. So much of the poor health seen in individuals with long-term mental illness may arise from, or be made worse by, SCRD. Sadly, such poor health is rarely linked to

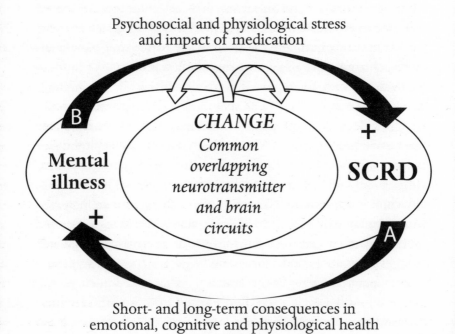

Figure 7.

**Figure 7. Model showing the relationships between mental illness and SCRD.** The model illustrates the emerging view that mental illness and SCRD share common and overlapping pathways within the brain. As a result, an altered pattern of brain neurotransmitter release that predisposes an individual to mental illness will result in a parallel impact upon the sleep and circadian systems. Disruption of sleep (shown as **A**) will, likewise, impact upon multiple aspects of brain function with both short- and long-term consequences in emotional, cognitive and physiological health (see Table 1, p. 77), and in the young may even affect brain development. The consequences of mental illness (shown as **B**), which give rise to psychosocial (e.g. social isolation) and physiological stress (e.g. altered stress hormone release – see chapter 4), along with the possible impact of medication, will intrude upon the sleep and circadian systems. Once started, the positive-feedback loop can rapidly develop, whereby a small change in neurotransmitter release could be amplified as a result of positive-feedback loops into more pronounced SCRD and then poorer mental health.

SCRD, or indeed treated, but is dismissed as some unidentified by-product of mental illness.[600-601]

Do we have good evidence for the model illustrated in Figure 7? The short answer is a very clear yes! As mentioned, we now know that genes associated with mental illness *also* play a role in sleep and circadian rhythms,[597, 602] and genes previously linked to sleep and circadian rhythms are now known to be involved in various forms of mental illness[597] (Figure 7). What about the other links? The assumption has long been that mental illness causes SCRD. But if the circuits between SCRD and mental illness overlap, then you would predict that SCRD may occur prior to a clinical diagnosis of mental illness. This is indeed the case.[602] Individuals at risk of developing bipolar disorder show elements of SCRD *before* any clinical diagnosis of bipolar. And, finally, the model predicts that reducing the level of SCRD should reduce the severity of mental illness. In a fairly recent study from the team in Oxford, led by my colleague Dan Freeman, experiments were set up to determine whether treating SCRD would indeed reduce levels of paranoia and hallucinations in individuals with SCRD. The trial involved a randomized controlled trial across 26 UK universities, and students with insomnia were randomly assigned to receive either digital cognitive behavioural therapy for insomnia (CBTi)

(1,891 individuals) or no intervention (1,864 individuals). As discussed in chapter 6, the CBTi helps identify those thoughts, feelings and behaviours that are contributing to the symptoms of insomnia and suggests ways to correct these issues. The study measured levels of SCRD, paranoia and hallucinatory experiences. The results showed that a reduction in SCRD, by using digital CBTi, was correlated with a highly significant reduction in paranoia and hallucinations over the study period. The study concluded that SCRD is a causal factor for the occurrence of psychotic experiences and other mental health problems.[603] I see these findings as highly significant as they demonstrate that treatments for SCRD represent a potentially new and powerful therapeutic target for the reduction of symptoms in mental illness.

## Questions and Answers

### 1. How are anxiety, depression and sleep loss linked?
This is thought to be a very important set of interactions. In brief, the connections would be as follows. Anxiety will increase the release of stress hormones such as cortisol and adrenaline. Stress will act to disrupt sleep and circadian rhythms and this will impact upon two key areas: SCRD will alter cognition and promote a negative 'salience' (increased prominence). The world will appear worse than it is, predisposing to depression; and SCRD will also disrupt multiple neurotransmitter pathways in the brain that regulate mental health. Such disruption will predispose to the worsening of mental health states.[604]

### 2. Are there links between suicide and SCRD in adolescents?
Half of all mental health conditions start by 14 years of age, and suicide is the third leading cause of death in 15- to 19-year-olds in the USA. A series of studies have observed notable levels of sleep disturbance in the weeks before a suicide attempt or completed

suicide. These findings indicate that sleep difficulties should be carefully monitored in adolescents for the prevention of suicide.[605] Interestingly, there is a 'time of day' effect on suicide, with more suicides tending to occur in the late afternoon and evening[606] (Figure 8, p. 226).

### 3. What happens to sleep in bipolar disorder?

The mania phase of bipolar disorder involves a very high level of energy and activity. It is common for people in this phase to experience racing thoughts and have difficulty concentrating. Individuals with mania may have trouble sleeping or feel that they need less sleep. Some individuals may stay awake for more than 24 hours, or only sleep three hours a night, but report that they have slept well. This sleep loss may contribute to the emotional and cognitive changes seen during mania (Table 1), including increased impulsive, and potentially dangerous, behaviours such as drug use, unprotected sex, excessive spending or reckless driving. Such a change in sleep patterns is a characteristic symptom of bipolar disorder, but shortened sleep can also trigger the condition. For example, shift workers, individuals who work long hours or who travel across multiple time zones, and students who lack sleep during exams are all at risk of having a recurrence of a mood episode, which is why vulnerable individuals should make every effort to protect their sleep[607] (chapter 6). If you think you are at risk of developing a mental health condition such as bipolar disorder, you should contact your medical practitioner immediately.

### 4. How is sleep affected by grief and the loss of a loved one?

Grief is complex and poorly understood. And, as many of us know, grief is often associated with SCRD. SCRD can then lead to poor physical and mental health. While mental and emotional symptoms such as depression might be expected, many physical symptoms can also occur. For example, some individuals experience daytime aches, pains and fatigue. Dry mouth, difficulty breathing and anxiety may also occur. Eating habits may change and there can be heightened sensitivity to noises. All of these could occur as a consequence of SCRD – but also make SCRD worse. A vicious cycle often develops. Loss of a loved one is almost inevitable at some point during our lives, and the normal response to such loss is 'resilience' and the ability to both withstand immediate loss and bounce back over time following loss. Most people eventually get through such times without requiring any outside intervention. However, as SCRD has such a big impact upon mental and physical health, paying attention to sleep disturbances following bereavement is important. Sleeping tablets are often prescribed, and may be useful short term, but, in older adults especially, falls and other side effects, including daytime cognitive impairment, represent a potential concern. Alternatively, the appropriate behavioural approaches outlined in chapter 6 have been shown to be helpful in improving both night-time sleep and daytime sleepiness following bereavement.[608] Grief and SCRD are certainly linked, and every effort should be taken to minimize SCRD during this vulnerable time.

### 5. How is lithium treatment involved in mental illness and SCRD?

This is a really interesting question. Lithium acts as a mood stabilizer and is used to treat mood disorders such as mania (feeling highly excited, overactive or distracted), regular periods of depression, and bipolar disorder, in which mood changes vary between feeling very positive (mania) and very low (depression). How lithium improves these conditions is not clear, but, interestingly, lithium is known to lengthen the period and increase amplitude of circadian rhythms in cells[609] by acting on signalling pathways involving two key proteins, GSK3B and IMPA1. So one suggestion is that lithium acts by correcting circadian rhythm disruption in conditions like bipolar disorder. Although lithium is an effective mood stabilizer in many individuals, it does not work for all.

A very recent study from our team has suggested that lithium treatment is more likely to fail in individuals who have variable or long circadian periods.[610] The conclusion from these and other studies suggests that the chronotype of an individual may influence their response to lithium. The precise links are still being worked out.

# 10

# When to Take Drugs

## Stroke, heart attack, headaches, pain and cancer

I want to die peacefully in my sleep, like my grandfather . . .
Not screaming and yelling like the passengers in his car.

Will Rogers

The United States Food and Drug Administration (FDA) has approved over 20,000 prescription drugs for sale, and a similar number of drugs have been permitted for use by the European Medicines Agency (EMA). Currently, around 50 new drugs are registered for use each year, but in terms of amount consumed and length of use Aspirin breaks most of the drug records. Around 100 billion standard Aspirin tablets are produced each year, and estimates suggest that over one trillion Aspirin tablets have been consumed in the past 100 years. Aspirin, also known as acetylsalicylic acid (ASA), is the synthetic version of salicylic acid originally extracted from willow (*Salix*) bark and other salicylic acid-rich plants such as myrtle and meadow sweet. The history of Aspirin use is truly remarkable. Clay tablets from the Sumerians dating to around 4,000 years ago describe how willow leaves can be used for joint pain (rheumatic disease). The Egyptians describe the use of willow leaves or myrtle for joint pain, and Hippocrates (460–377 BC) recommended an extract of willow bark for fever, pain and childbirth. Ancient Chinese, Roman and Native American civilizations have long recognized the benefits of plants containing salicylic acid for their medicinal benefits. Jumping forward to the modern era, in 1758 the Reverend Edward Stone (1702–68), from Chipping Norton in Oxfordshire, not far from where I live, explored the properties of willow bark, looking for a cheaper remedy than the really expensive Peruvian cinchona bark for treating 'the agues', which means fever and shivering. As we will discuss in chapter 11, cinchona bark contains quinine, which is a medicine used to treat malaria, and is different from salicylic acid. Stone gave extract of willow bark to his parishioners with fever and he showed that this caused a marked reduction in fever. He presented his findings to the Royal Society of London in 1763, and willow bark became the 'go to' remedy for fever. The purification of salicylic acid from willow bark followed in 1828. It was made synthetically in 1859, and the first clinical trials were in 1876, by Thomas MacLagan, who used salicylic acid to treat joint inflammation in patients with acute rheumatism. A pure and stable form of salicylic acid, in the form of ASA, was made by the Bayer drug company in 1897, and with this step Aspirin dominated the world and the modern pharmaceutical industry was born.[611] Today the total global pharmaceutical market is valued at about 1.27 trillion US dollars. My prediction is that by taking circadian biology into account this global industry is about to enter another phase of growth and development.

Because our physiology and biochemistry change so profoundly over the 24-hour day, it will not be a surprise that various conditions and the symptoms of different diseases change across the day. The peaks in some of these disease symptoms are summarized in Figure 8.

In view of the fact that disease symptoms vary across the day, it should also come as no surprise that many of the medicines we take show variable levels of effectiveness across the day/night cycle. Ideally, medications and treatments should be delivered at a time of day when they are most needed and most effective, anticipating the disease severity illustrated in Figure 8. This is the concept of 'chronopharmacology'. However, most medications

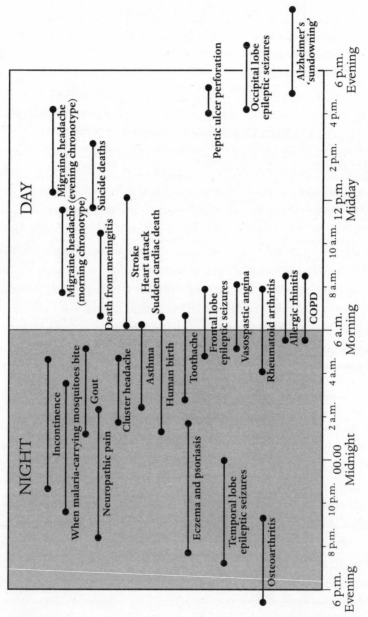

**Figure 8.**

**Figure 8. Circadian changes in disease events and disease severity**. The timing of disease states varies across the day. There will be differences between individuals, and there is not always agreement between the scientific studies, but, with this in mind, shown are the average peak times when these events and disease severity are most likely to occur. **Osteoarthritis** pain and joint stiffness peaks at the end of the day in the early evening.[612] **Eczema and psoriasis** induce intense itching, which peaks during the late evening and overnight, which can disturb sleep. Itching is less severe towards dawn.[613] **Neuropathic pain** feels like a burning or shooting pain and is most painful in the late evening and early morning.[614] **Biting by mosquitos carrying malaria** varies between different species of mosquito, but usually occurs at night and is characterized by looking for human victims after dusk, increasing to a peak bite rate around midnight, with 60–80 per cent of bites estimated to occur between 9 p.m. and 3 a.m.[615] **Toothache** peaks between 3 and 7 a.m.[616] **Epileptic seizures** vary in type and peak at different times: **frontal lobe** seizures between 5 and 7.30 a.m., **occipital lobe** epileptic seizures between 4 and 7 p.m., and **temporal lobe** epileptic seizures from 7 p.m. to midnight.[616] **Peptic ulcer perforation** is the rupture of the stomach wall and the escape of digestive fluids into the abdominal cavity. Rupture has a main peak time between 4 and 5 p.m., with a smaller chance of rupture between 10 a.m. and 12 noon, and then at 10 p.m.[617] **Incontinence** (nocturnal enuresis) in elderly adults occurs throughout the night and the early hours of the morning.[618] **Cluster headaches** start at around 2 a.m.[619] **Gout pain**, arising from an accumulation of uric acid crystals in joints such as the big toe, ankle, knee and wrist, peaks in the middle of the night, around 3–4 a.m.[616] **Asthma**'s worst symptoms occur around 4 a.m., with sudden death due to asthma also occurring around this time.[620] This will also depend upon the degree of exposure to triggering allergens in the bedroom. **Human birth** occurs mainly between 1 and 7 a.m., peaking around 4–5 a.m.[375-6] **Vasospastic angina** is a type of angina (chest pain) that usually occurs in the early morning, around 6 a.m., and feels like a constriction or tightening in the chest.[621] **Chronic obstructive pulmonary disease (COPD)** includes chronic bronchitis and emphysema, which are chronic inflammatory lung diseases, often the result of heavy smoking, that cause restricted airflow in the lungs, causing breathing difficulty, coughing, the production of mucus (sputum) and wheezing. Symptoms are worse during the early morning hours.[622] **Rheumatoid arthritis** pain peaks in the morning.[623-4] **Allergic rhinitis** is inflammation of the inside of the nose caused by an allergen, such as pollen, dust, mould or flakes of skin from certain animals. Cough frequency and handkerchief use is most frequent during the initial hours after awakening from sleep.[625] **Meningitis** is an infection of the protective membranes that surround the brain and spinal cord (meninges). Most deaths occur between 7 and 11 a.m.[616] **Migraine** onset varies with chronotype: morning types tend to have migraine in the morning while evening types have migraines in the afternoon / evening.[626] **Stroke** peaks between 6 a.m. and noon.[627] **Heart attack** and **sudden cardiac death**, like stroke, peaks between 6 a.m. and noon.[628-9] **Alzheimer's 'sundowning'**, which has symptoms such as agitation, confusion, anxiety and aggressiveness, happens usually in the late afternoon and evening but can occur at night.[630] **Suicide deaths** occur most frequently between late morning and early afternoon.[631] For further details see these reviews.[616, 632-3] are not taken at a time that optimizes their impact, but instead at a time that helps us remember to take them. And this may not be the best time for an optimal effect. Different drugs have a different 'half-life'. The half-life of a drug, as the name suggests, is the time it takes for the amount of drug in your body to be reduced by half, and this depends upon how your body processes and then excretes the drug. This is called the 'pharmacokinetics' of the drug, which basically means what the body does to the drug. The half-life of different drugs can vary from a few hours to a few days. I stress that the half-life of a drug does not tell you when it has stopped working, but the time taken for the level of the drug to reach one half its original concentration.

In addition to the time of day and the half-life of a drug, drug effectiveness can vary significantly between individuals. These differences are due to changes in how we process our medications. Individually, we unwittingly change the pharmacokinetics of a drug. With age our kidney and liver function can change, altering drug processing. Fat deposits can absorb fat(lipid)-soluble drugs and extend their half-life within the body. Also, our sensitivity to drugs can change, especially if we have been taking the same drug over a long period of time.[634] It is also important to note that drugs with a long half-life, and taken daily, can build to high concentrations in the body. And more does not always mean better. Counterintuitively, some drugs become less effective at higher levels. In addition, higher concentrations can lead to unwanted side effects like nausea, an upset stomach, allergies such as skin rashes, or worse.[635] Well-documented side effects occur with the anti-allergy drug Benadryl (chemical name diphenhydramine). Peak levels occur in the blood after two hours, with a half-life of between 3.5 and around nine hours. Diphenhydramine is used to ease allergic reactions, but it also reduces the action of the wake-promoting neurochemical acetylcholine (chapter 2), which can lead to drowsiness. And for this reason Benadryl is not recommended for use in individuals with dementia or other neurodegenerative diseases (chapter 9).

Additionally, diphenhydramine produces other side effects, such as a dry mouth, which, if taken before sleep, can drive you out of bed at night to get a glass of water. Side effects can also occur if different drugs interact. Such 'drug interactions' occur with alcohol and certain painkillers (narcotics such as morphine and codeine), which can cause an accidental overdose leading to death. Surprisingly, grapefruit juice alters blood concentrations of certain drugs because it prevents drug breakdown by blocking a key enzyme in the liver. So instead of being metabolized, more of the drug enters the blood and stays in the body longer, resulting in too much in circulation. The action of grapefruit juice can alter some key medications, including those that regulate blood pressure (some antihypertensives, e.g. amlodipine) and cholesterol-lowering medicines (some statins, e.g. simvastatin). For these reasons it is *always* critically important to check the information that comes with your drugs. Many people don't. In addition, some supplements (e.g. St John's Wort or Goldenseal) can interact with prescribed drugs. So before taking any supplements, discuss with your medical practitioner if there are interactions with the drugs you are taking.[636]

Given all these varied interactions, it is truly remarkable that circadian-driven changes in drug pharmacokinetics remain an important aspect of drug efficacy, and that any effects are not lost in the 'noise' of all the other interactions going on. This illustrates that circadian-driven changes in pharmacokinetics are robust and important and should be incorporated into the healthcare advice provided to patients. In fact, circadian changes in more than 100 different drugs have been recognized, and this has led to the development of guidelines concerning when to take a drug for a particular illness, not least for cancer or cardiovascular disease.[637] Sadly, this information is not always acted upon.

It can also be difficult for each of us to work out when we should take our medications, and ideally this should be undertaken in consultation with your healthcare practitioner. Hopefully the examples below can act as a guide. To illustrate the importance of circadian-driven changes in drug pharmacokinetics I want to consider the importance of drug timing in three key areas of health: stroke and heart attack; pain, migraine and headaches; and, finally, cancer.

## Stroke and Heart Attack

Let's start with the head and a leading cause of death and disability. A stroke occurs when a blood vessel in the brain ruptures and bleeds (haemorrhagic stroke), or when there's a blockage in the blood supply to the brain (ischemic stroke). For clarity, a transient ischemic attack (TIA), is a brief loss of blood flow (ischemia) to the brain which does not progress to a stroke. Large numbers of politicians seem to die, or be affected, by a stroke or TIA. Ten of the 46 presidents of the USA have suffered strokes during their presidencies or soon after leaving office. The Allied war leaders, Roosevelt, Stalin and Churchill, all ultimately died from some form of stroke in 1945, 1953 and 1965 respectively. On 8 April 2013, the former British prime minister Margaret Thatcher died of a stroke at the age of 87. And, as I shall discuss below, it is likely that sleep and circadian rhythm disruption (SCRD) played, and continues to play, a role in the high frequency of stroke in politicians. The term 'stroke' has its origins as a medical description from the 1500s when it became a shortened form of 'Struck by God's hand'. Which brings me to the communist theorist and revolutionary Leon Trotsky. Trotsky had high blood pressure and feared he would die from a stroke. He did, in fact, die from blood loss from the brain. However, the cause of his 'stroke' was being 'struck', not by the hand of God, but by an ice axe driven into his skull in 1940 under the orders of Stalin in Mexico City. And if you are puzzled by what an ice axe was doing in Mexico, there are glaciers on Mexico's tallest mountain peaks, and Trotsky's murderer, Ramón Mercader, claimed to be an experienced mountaineer.

Coronary heart disease is when your coronary arteries (blood vessels that supply the heart) become narrowed by a build-up of fatty material. Coronary heart disease is sometimes called ischemic heart disease and a heart attack (myocardial infarction) occurs when the supply of blood to the heart is suddenly blocked. The blockage or rupture of a blood vessel in the brain or heart prevents glucose and oxygen from sustaining these highly meta-bolically active organs. Such life-changing events show a daily circadian variation. In a major review of 31 studies, and based upon 11,816 stroke patients, all subtypes of strokes showed a significant time of day effect. There was a 49 per cent increase in the chance of a stroke, for all types, between 6 a.m. and noon compared to the rest of the day. The three subtypes of stroke showed a risk of 55 per cent for ischemic strokes, 34 per cent for haemorrhagic strokes and 50 per cent for transient ischemic attacks.[627] Similar findings have been documented repeatedly for heart attacks.[638] Collectively, these data show unambiguously that you are most likely to die from a stroke or ischemic heart disease first thing in the morning (Figure 8). Perhaps each of us can take some comfort when we get to 12.01 p.m., appreciating that we have survived one of the most dangerous times of the day!

The reasons for this 6 a.m.–12 noon 'window' in risk and death arise from a combination of events. A key contributor is the circadian-driven rise in heart rate and blood pressure. This anticipates the demands of activity – the switch from sleep to consciousness, and the increased need for oxygen and nutrients. The rise in blood pressure is in large part mediated by the autonomic nervous system, that part of the nervous system responsible for the unconscious control of bodily functions. The autonomic nervous system consists of two parts – the sympathetic and parasympathetic branches. The sympathetic branch drives an increase in heart rate, while the parasympathetic system has the opposite effect and reduces heart rate. Both are regulated by the circadian system with parasympathetic activity higher at night and the sympathetic system peaking in the morning, leading to the rise in blood pressure and heart rate. Altered behaviour, in the form of a major change in activity and posture following wake, also acts to increase blood pressure.[639] Increased physical activity is accompanied by physiological changes, including a circadian-driven rise in cortisol, testosterone (Figure 1, p. 20), insulin and glucose release, all of which help drive a higher metabolic rate and increased activity. Raised metabolism requires more oxygen and glucose, and increased blood pressure delivers these essentials. Very significantly, there is also an increase in pro-clotting factors in the blood in the morning, including the activation of platelets.[640] Platelets normally act to clump together to form blood clots and prevent blood loss following injury. But they can also work against us by producing clots that block blood vessels, resulting in an ischemic stroke. Blood clotting peaks in the morning hours – corresponding to the greatest chance of having a stroke or heart attack (6 a.m.–12 noon, Figure 8).[641] If you are healthy, these dynamic changes are not a problem, but poor health combined with SCRD can make them lethal.

Anything that prevents our rhythmic physiology getting the correct materials in the right place, in the right amount, at the right time of day will increase our health risks. Sustained night shift work, jet lag or marked sleep/wake disruption (chapter 4) is associated with an increased chance of a stroke and heart attack. SCRD is linked to clinically elevated blood pressure and higher triglyceride levels. High triglycerides contribute to the hardening and thickening of the artery walls, which increases the risk of stroke, heart attack and heart disease. SCRD-driven inflammatory responses and a greater risk of Type 2 diabetes are, again, risk factors for stroke and heart disease.[642-3] And I stress, in addition to stroke, there is also an increase in the number of other heart-related problems during the morning hours (6 a.m.–12 noon), such as ventricular arrhythmias and sudden cardiac arrest (Figure 8).[644-6] Importantly, individuals who have had a heart attack during the 'morning risk window' experience more heart damage and a poorer chance of recovery compared with heart attacks at other times of the day.[628] My colleagues in Germany refer to this window in time as the *Todesstreifen* (*To·des·strei·fen*), or 'death zone' – a word originally used to describe the buffer zone between West and East Germany, and the strong likelihood of death from sniper fire if you entered this zone.

It also seems that SCRD can have an effect on the recovery from a stroke or heart attack. Allow me to expand upon this. Sleep disorders such as insomnia, obstructive sleep apnoea (OSA) and restless-legs syndrome (RLS) (chapter 5) are all associated with worse recovery and death following a stroke.[647] Importantly, in terms of treatment, sleep and sleep/wake stability after a stroke or heart attack aid recovery.[648–51] So along with the medications discussed below, the reduction of SCRD should be part of both the preventative *and* recovery management programmes for stroke and heart attack.

These daily changes in stroke and cardiovascular risk are now being considered in both the development of new drugs and the delivery of existing medications. For example, drugs that are used to treat high blood pressure (hypertension) are called antihypertensives. Significantly, when antihypertensives are taken before bedtime, rather than in the morning, they are more effective in regulating blood pressure levels and in reducing stroke and heart attack.[652] Also taking Aspirin before bedtime reduces platelets clumping together to form unwanted blood clots and helps prevent heart attacks and stroke.[653] By taking Aspirin in the evening, compared to taking Aspirin in the morning, platelet activation is greatly reduced in the morning.[654–5] In the most extensive trial to date, taking antihypertensive medicine at bedtime was associated with both improved blood pressure regulation and a near halving of cardiovascular deaths and cardiovascular problems when compared with morning medication. It is worth looking at this study in a bit of detail. The trial randomly assigned nearly 20,000 hypertensive individuals with an average age of 60.5 years to take their entire daily dose of one or more anti-hypertensives at bedtime or on waking in the morning. The patients were followed up yearly, with detailed health checks, for over six years. The patients who took their anti-hypertensives in the evening had a considerably lower risk (nearly halved) of cardiovascular death, including heart failure and stroke.[656] The lead scientist of this study, Ramón Hermida, said:

> Current guidelines on the treatment of hypertension do not mention or recommend any preferred treatment time. Morning ingestion has been the most common recommendation by physicians based on the misleading goal of reducing morning blood pressure levels. The results of this study show that patients who routinely take their antihypertensive medication at bedtime, as opposed to when they wake up, have better controlled blood pressure and, most importantly, a significantly decreased risk of death or illness from heart and blood vessel problems.[657]

Currently, there are no formal guidelines regarding when you should take your blood pressure tablets, or indeed Aspirin. New studies are underway in an attempt to support the findings of Ramón Hermida and colleagues. If confirmed, I hope guidelines to general practitioners will be implemented quickly. Of course, each of us can be guided by the current evidence, and given the existing data my personal choice would be to take this class of antihypertensive medication before bed. My colleagues in Australia also tell me that this is the usual advice on Australian prescriptions. Which prompts the questions:

*Why take anti-hypertension medication at night if the danger is most acute in the morning?*

This relates to the pharmacokinetics of the antihypertensive medications, in terms of how they are absorbed and distribution throughout the body; how the drugs are metabolized; and then finally broken down and excreted. These processes all take time.[658] By taking antihypertensive medication at bedtime the drug levels rise and remain in the body at relatively high levels (long half-life), acting to reduce blood pressure across the time window when the sharp rise in blood pressure normally occurs, between 6 a.m. and 12 noon. If antihypertension medication is taken in the morning, drug effectiveness will have risen and then peaked *after* the critical surge in blood pressure.

As discussed in chapter 8, some antihypertensives can increase urine production.[526] Antihypertensive diuretics stimulate the kidneys to remove water and salt from the blood into the urine. This reduces blood volume, and so reduces blood pressure, but it increases urine production. Calcium channel blockers relax the blood vessels and lower blood pressure. But they will also act to inhibit bladder contraction and reduce bladder emptying, making it necessary to pee more and pee at night.[528] So, if nocturia is a problem, then it may be a good idea to discuss which antihypertensives may be best for you with your health care provider.

## What about Aspirin?

Aspirin inhibits the pro-clotting factors in the blood and reduces platelet activation – Aspirin 'thins the blood'. But Aspirin levels in the blood rise rapidly and then decline fairly quickly (short half-life), certainly within a few hours. So how does bedtime Aspirin reduce platelet 'stickiness' in the morning? Interestingly, Aspirin prevents platelets clumping together to form clots for the lifetime of the platelet, which is around 10 days. So platelets are permanently 'turned off' after exposure to Aspirin.[659] However, a hundred billion new platelets are made freshly each day, and they are made in the evening.[640] Taking Aspirin in the evening will ensure that new platelets will be effectively deactivated before the dangerous window for stroke the following morning – between 6 a.m. and 12 noon. Aspirin taken in the morning will be much less effective because the new platelets made the previous evening will already be promoting clotting, before the Aspirin can get to the platelets and deactivate them. In addition, Aspirin will be metabolized and cleared from the body before the next wave of new platelet production occurs in the evening. Again, there is a downside. Taking Aspirin before bed may increase the chances of damaging the stomach and/or intestinal lining, leading to the development of small and large sores (ulcers), which can bleed or perforate. However, the use of proton-pump inhibitors (PPIs) and drugs that protect the gastrointestinal system can resolve this problem.[660]

### When should I take my statins?

Another risk factor for a heart attack and stroke is increased levels of cholesterol in the blood. Cholesterol is an essential building material to make healthy cells and key hormones, but high levels of cholesterol can lead to the development of fatty deposits on the walls of blood vessels. These deposits can grow, making it difficult to get enough blood to the organs of the body, including the heart and brain. This is called a 'stenosis'. Such a stenosis prevents oxygen-rich blood reaching the heart or brain, leading to a heart attack or stroke. Cholesterol is transported in the blood in the form of a protein complex (lipoprotein), and two types of lipoproteins combine with and carry cholesterol to and from cells. One is low-density lipoprotein, or LDL. The other is high-density lipoprotein, or HDL. LDL cholesterol is considered the 'bad' cholesterol, because this lipoprotein–cholesterol complex contributes to fatty build-up in arteries (atherosclerosis). This narrows the arteries and increases the risk for heart attack and stroke. The HDL lipoprotein–cholesterol complex is thought of as 'good' because it transports LDL (bad) cholesterol away from the arteries and back to the liver, where the LDL–cholesterol complex is broken down and excreted. But HDL cholesterol doesn't completely eliminate LDL cholesterol. Only about 30 per cent of the LDL cholesterol is carried to the liver by HDL. But this is where drugs can help. Drugs called 'statins' (e.g. simvastatin, lovastatin, pravastatin, atorvastatin) are particularly good at lowering LDL cholesterol. Statins work by slowing down the production of LDL cholesterol in the liver, where it's made, resulting in a fall of blood cholesterol levels. Statins slow down LDL cholesterol production by blocking an enzyme called HMG-CoA-Reductase (HMG-CoA Reductase inhibitors). So is there a best time to take statins?

Levels of cholesterol in the blood follow a circadian rhythm and it is normally produced at night, between 12 midnight and 6 a.m. Statins remain active in the blood for many hours. Some statins are effective for 4–6 hours, but others remain effective for 20 or even 30 hours. And this is the crucial point, if the statin you take is short-acting in the 4–6 hours range (e.g. Simvastatin), then these statins should be taken around bedtime to 'hit' the night-time production of cholesterol. However, if you take statins that are effective for much longer (20–30 hours; e.g. Atorvastatin), these statins can be taken any time, as their effectiveness will always overlap with higher levels of cholesterol production at night.[661] If you are uncertain about the effective lifetime of the statins you take, consult your general practitioner for advice and find out if they are short- or long-acting.

These three examples (antihypertensives, Aspirin, and statins) illustrate how circadian variation in our physiology and drug pharmacokinetics must be considered when timing medications. But there is also one other critical factor that must be considered – the extrapolation from animal studies to the human condition.

*Drug development and the extrapolation from*
*animal studies to the human condition*

For very good reasons, mice have become the animal of choice in medical research. We understand their genetics, they are easy and relatively cheap to look after, and their basic biology closely resembles that of humans. Except, of course, in one critical and often overlooked respect. Mice are nocturnal (night active) and we are diurnal (day active). During the daytime a mouse is normally inactive or asleep, but because animal facilities are generally operated between 7 a.m. and 5 p.m., experimental results, such as the effects of drugs, are collected from mice when we are awake but the mice are biologically prepared for sleep.[662] The mouse results are then extrapolated to the human condition, though not to human sleep but incorrectly to human wake. Mice show huge circadian-driven changes between their sleep versus wake biology, not least in their drug effectiveness and toxicity.[663] This has been known for many decades but is often forgotten in the early stages of drug testing using mice. This important issue was investigated recently by looking at the effect of three different drugs in treating the effects of a stroke in mice, and the treatments were delivered at different times of the day. In mice, the stroke was simulated by reducing the blood supply to the brain. All three drugs reduced tissue death if given during the day – when the mouse was biologically prepared for sleep – but failed when given at night-time – when the mouse was awake.[664] Such findings explain why treatments that were found to be successful in mice subsequently failed in human trials. They were given during the day, like the mice, but at the wrong biological time. They should have been given prior to bed, during our sleep time at night! These findings show that time of day really matters when taking preventative stroke medication and that biological time in mice must be aligned to the appropriate biological time in humans to ensure that 'like is being compared to like'.[664] Researchers have considered developing a 'good' diurnal rodent model, such as the 'Nile grass rat' (*Arvicanthis niloticus*),[665] but the challenge has proved overwhelming.

## Pain, Migraine and Headaches

A whole range of disorders that generate pain show 24-hour patterns of pain intensity.[666] Rheumatoid arthritis is an autoimmune disorder, which means your body attacks itself, and the result is joint stiffness and pain. The pain from rheumatoid arthritis occurs in the morning (Figure 8). By contrast, osteoarthritis is a degenerative joint disorder causing a breakdown of the cartilage that cushions the joints. Joint pain from osteoarthritis occurs in the evening and at night (Figure 8). We have known this for many decades, and morning joint stiffness/pain has been used as a way to distinguish rheumatoid arthritis (morning) from osteoarthritis (evening).[623] Most recently, evening therapy for rheumatoid arthritis using glucocorticoids (a type of corticosteroid that acts as an immunosuppressant) resulted in reduced morning stiffness and pain compared with the same glucocorticoid dose taken in the morning.[667] Such findings and chronopharmacological approaches are now being used to both understand and treat other pain disorders such as headaches and neuropathic pain (pain after nerve damage). Let's first consider headaches. These nasty little demons come in a variety of forms.

### Headache

A headache is a general term to describe pain from any region of the head, and can occur on either side of the head, have a specific location, or spread across the head from one point to another. The pain may be sharp, throbbing or a dull ache. Headaches were originally thought to be related to blood vessels swelling or increased blood flow to parts of the brain. Today, most headaches are thought to be due to changes in the nervous system. The two most common types of headache, with strong links to the circadian system, are cluster headaches and migraine:

### CLUSTER HEADACHE

Cluster headaches can be very unpleasant. Indeed, they are sometimes called 'suicide headaches', as over 50 per cent of individuals
with these severe headaches have contemplated suicide.[668] They
occur in around 1 in 1,000 people, in more men than women, with
a first occurrence between 20 and 40 years of age.[669] The pain is
usually on one side of the head and lasts for around 15 minutes to
three hours. Most (90 per cent) will have the same sort of intense
pain every day for several weeks or months and then nothing for
several months. The trigger for a cluster headache is not well
understood, but seems to arise from the activation of key pathways
in the brain and involving nerves from the trigeminal nucleus (Figure 2, p. 24) and the autonomic nervous system, regulated by parts
of the hypothalamus[670] and possibly the suprachiasmatic nuclei
(SCN) (chapter 1). This has been suggested because a key feature
of cluster headache is that in many individuals headaches occur at
precisely the same time every day, and often at the same time every
year. A large study showed that 82 per cent of individuals have cluster headaches at the same time every day[671], with the most common
time for the onset of an attack at 2 a.m. (Figure 8)[619]. Interestingly,
several studies have suggested that the circadian rhythms of individuals with cluster headaches are abnormal, with multiple
hormones and brain neurotransmitters desynchronized, including
melatonin, testosterone, prolactin and growth hormone. In addition, abnormalities in the genes associated with the molecular
clockwork and the 'transcription / translation feedback loop' (chapter 1) have also been linked to cluster headaches.[669] So a picture is
emerging that links cluster headaches with dysregulation of the
circadian system.

### MIGRAINE

A migraine attack induces a throbbing, moderate or severe one-
sided headache pain which lasts between four and 72 hours and is

often accompanied by nausea, vomiting, and sensitivity to light, noise and motion.[672] It's a common condition with almost 18 per cent of women and 6 per cent of men experiencing at least one migraine per year.[673] Friedrich Nietzsche, the German philosopher, cultural critic and writer, suffered from bad migraine attacks from his childhood.[674] As an aside, Nietzsche's writings have often been misappropriated after his death, and, for clarification, it was Nietzsche who was the first to say: 'That which does not kill us makes us stronger', and not Arnold Schwarzenegger in the 1982 film *Conan the Barbarian*. The brain pathways that trigger a migraine appear similar to cluster headaches, involving nerves from the trigeminal nucleus (trigeminovascular system) and activation by the hypothalamus (Figure 2), and again like cluster headaches, migraines are rhythmic events. In a recent study, migraine occurred in 'morning types' (go to bed and wake early) in the morning, while 'evening types' (go to bed late and wake late) were more likely to have migraines in the afternoon (Figure 8).[626] Migraine has also been linked to the menstrual cycle, and lower levels of oestrogen during the luteal phase (Figure 5, p. 146).[675] Triggers include stress, abnormal meal timing, the menstrual cycle, abnormal light exposure and SCRD.[676] In addition, a key circadian gene that leads to advanced sleep phase syndrome is also linked to migraine (Figure 4)[248], along with other genes known to affect circadian regulation[669]. Because circadian rhythm disruption is associated with migraine, the expectation was that night shift workers will have higher rates of migraine. However, a recent review of the literature suggested that there were no clear links with migraine and night shift work.[677] Perhaps this is not too surprising after all. As migraine is so debilitating, vulnerable individuals will almost certainly self-select *not* to do night shift work. A more interesting approach would be to survey individuals with migraine and ask if they had ever undertaken night shift work, and whether this experience resulted in worse attacks.

## *Chronotherapy for headache disorders*

The strategies discussed in chapter 6 for the stabilization of sleep and circadian rhythms disruption (SCRD) have been shown to help with both cluster headaches and migraine, including: morning light and stable meal timing.[678] Interestingly, several recent drugs used to treat headaches have known actions on the clock, including: Valproate, which shifts the circadian clock[679], as does Baclofen[680]; Verapamil is commonly used in migraine and cluster headache, and has been shown to alter circadian rhythms[681]. How the 'circadian actions' of these drugs influence therapeutic outcomes remains unclear. So, although the data are not extensive, there is a growing body of evidence that headaches are influenced by SCRD, and that SCRD stabilization will have an important role in reducing the occurrence of headaches. Indeed, new drugs designed to stabilize the circadian system are being developed to alleviate the pain of headaches.[669] The National Health Service (NHS) in the UK has just introduced a device called gammaCore™ for headache and migraine therapy, a handheld medical device that enables patients to self-administer vagal nerve stimulation (nVNS) by placing the contact electrodes on the neck.[682] Presumably such stimulation interacts with the trigeminal nucleus (Figure 2) and the trigeminovascular system. The timing of nVNS, combined with SCRD stabilization, may provide help to those individuals suffering from headache and migraine.

## *Neuropathic pain*

Neuropathic pain is caused by damage or disease of the sensory neurones and pathways that detect changes such as touch, pressure, pain, temperature, vibration, etc. outside or inside the body. This is called the 'somatosensory system'. Neuropathic pain feels like a burning or shooting pain. Daily, 24-hour patterns in nerve pain are well documented. For example, in experiments, electrical stimulation of a nerve in the calf region (sural nerve) is most painful in the late evening and early morning.[614] The same is also true for nerves damaged by disease (e.g. diabetic neuropathy), with pain increasing throughout the day and peaking at night[683] (Figure 8). Such changes are thought to be driven by the circadian modulation of pain receptors at a variety of levels.[669] In a key set of experiments, the molecular clock has been shown to directly alter 24-hour patterns of a gene which produces a pain-signalling molecule called 'substance P' which regulates the intensity of neuropathic pain.[684] So there is a clear link between neuropathic pain intensity and the circadian system. The key point is that knowing when neuropathic pain is worse offers the opportunity to deliver painkillers at a strength and time of day when they will be most useful in reducing pain; and, importantly, reducing the impact of pain on sleep disruption. Overall, understanding the mechanisms that drive daily rhythms in pain intensity provides the opportunity to target these pathways with new drugs to block, or lower, pain thresholds. Again, this is an area of active research and I suspect will provide important breakthroughs in the next few years.

## Cancer

Circadian rhythms regulate multiple processes that are linked to both the protection from and development of cancer, including cell division, programmed cell death (apoptosis), DNA repair and immune function. Indeed, 'immunotherapy' is a treatment for some types of cancer which enhances the ability of the immune system to find and kill cancer cells.[685] Many cancer-related genes are under circadian control, and, importantly, tumour development and growth are much faster if the circadian system is disrupted. This has been shown many times in the laboratory. For example, one approach has been to shift the light/dark cycle of mice with tumours every few days, simulating repeated jet lag. Mice with jet lag showed much faster tumour growth compared to normally treated mice.[686] Other experiments have looked at mice in which the molecular clock has been disrupted. For example, key elements of the molecular clock (Figure 2D, p. 24) are the PER proteins. Mice lacking PER1 and PER2 show both very disrupted circadian rhythms and higher rates of cancer.[687] Mutations in PER2 in the mouse liver have been shown to dramatically increase liver cancer.[688] However, restoring PER2 to cells in which PER2 is defective reduces tumour growth. Another study disrupted the molecular clockwork of cells, effectively turning the cellular clocks 'off' by turning 'on' a genetic regulator called MYC. Human patients with a type of cancer called a 'neuroblastoma' were then studied, as this type of tumour expresses low or high levels of MYC. Very interestingly, individuals expressing high levels of MYC in their neuroblastoma (stopped clocks) died much earlier than individuals with low levels of MYC (clocks still ticking). These findings provide strong evidence that the possession of a functional circadian clock in tumour cells would greatly slow tumour progression and increase patient survival.[689]

Radiation therapy is commonly used to treat solid tumours of the breast, lung and oesophagus. Sadly, the heart is often unintentionally affected, leading to heart problems and ultimately heart failure. In studies on mice, where their circadian rhythms were intentionally disrupted, radiation exposure caused greater levels of DNA damage and increased heart problems, suggesting that normally the circadian clock protects the body from ionizing radiation.[690]

Similar impacts of SCRD have been shown in us. As discussed in chapter 4, individuals who have experienced many years of night shift work or rapidly changing shift work schedules have significantly higher rates of cancer[193], including breast[691] and prostate cancer[692-4]. Nurses who work night shifts are reported to have higher rates of breast, endometrial, and colorectal cancers[691, 695-8], and this risk increases with the time spent doing shift work[699] (chapter 4). And it is not just nurses. For example, women who mainly work at night have a higher risk of cancer[700], and, significantly, pre-menopausal women, but not post-menopausal women, had a higher breast cancer risk if they were current or recent night shift workers[701]. This, and a host of other evidence, drove the International Agency for Research on Cancer (IARC) to classify night shift work as a probable carcinogen.[702] So a significant percentage of the working population is being knowingly exposed to a 'Category 2A human carcinogen'. I bet that wasn't in the job description.

The same findings emerge in other occupations where circadian rhythm disruption is routine. Female flight attendants show an increased risk of breast cancer and malignant melanoma[703], and a study on Canadian and Norwegian pilots showed they had higher rates of prostate cancer[704-5]. A recent study showed that chronic jet lag drives hepatocellular carcinoma (HCC). HCC is the most common type of primary liver cancer in adults, and is the most frequent cause of death in people with non-alcoholic fatty liver disease, or NAFLD. The circadian rhythm disruption induced by jet lag seems to alter the regulation of multiple genes. This then changes metabolic pathways, which lead to insulin resistance (cannot easily take up glucose from the blood in response to insulin), NAFLD (build-up of fat in the liver) and liver steatohepatitis (liver inflammation due to the build-up of fat).[706] Some caution about cause and effect needs to be considered in these cases as aeroplane travel also increases exposure to ionizing radiation, which is a carcinogenic agent. The bottom line is that multiple studies have linked SCRD to an increased susceptibility to cancer development in all key organ systems in humans, including breast, ovarian, lung, pancreatic, prostate, colorectal and endometrial cancers, non-Hodgkin's lymphoma

(NHL), osteosarcoma, acute myeloid leukaemia (AML), head and neck squamous cell carcinoma and hepatocellular carcinoma. The importance of DNA repair mechanisms in cancer development comes from a recent study showing that night shift work schedules disrupt DNA repair pathways and increase the chances of cancer development.[707] So the clear pattern emerging is that the loss of robust circadian control of physiology represents an independent risk factor for the development of cancer.[708]

Like the mouse studies, defective clocks in the cells of the body have been linked to higher rates of cancer. For example, some ovarian tumours have lower levels of key clock genes, including PER1 and PER2.[708] Decreased clock gene levels have also been found in chronic myeloid leukaemia[709] and breast cancer tumours[710]. Indeed, disruption of the molecular clock appears to be a common feature of cancer cells. This has led to some interesting new treatment approaches designed to restore the circadian rhythms of cancer cells in an attempt to stop the cancer. In one study on mice, drugs were used that act as key circadian drivers of the molecular clockwork. Remarkably, these drugs restored the circadian rhythms of the tumour and reduced cancer growth. In addition, and very significantly, these drugs were lethal to cancer cells, but had no impact on normal cells.[711] The same strategy has been shown to work with other drugs that increase the 'strength' of the clock leading to the inhibition of cancer cell spread.[712] Conversely, a study using a drug to suppress the circadian clock increased tumour development.[713] These exciting findings represent a new pathway for cancer treatments – by restoring a robust circadian rhythm within a cancer cell, tumour progression seems to be inhibited, or at least slowed.

In addition to the development of new drugs designed to reset the circadian rhythms of cancer cells, a parallel path is taking a more holistic approach by trying to stabilize the circadian rhythms of the whole person – as much as possible. As might be predicted from the cellular findings, SCRD is frequent in individuals with cancer. For example, measures of sleep/wake cycles in patients with advanced lung cancer showed significantly disturbed 24-hour sleep/ wake cycles and poorer-quality sleep compared to healthy controls.[714] This has also been shown in young individuals with acute lymphoblastic leukaemia. SCRD was associated with more cancer-related fatigue, which is manifest as a persistent sense of physical, emotional tiredness and exhaustion.[715] Disrupted circadian rhythms have also been reported in colorectal cancer,[716] and the more severe the disruption the lower the chances of survival.[717-18] Such findings have given rise to the idea that approaches which stabilize the circadian timing of cancer patients will not only improve their quality of life, but also their chances of survival. These methods are detailed in chapter 6, but include: 1. the timing of meals; 2. appropriate light exposure around dawn and dusk; 3. the reduction of light in the evenings; 4. consistent sleep/wake schedules; 5. appropriate sleeping space, including night-time darkness, appropriate temperature, good mattress and pillows, etc.; 6. minimal use of sleeping medications; 7. increasing daytime alertness and discouraging napping; 8. avoiding stimulants such as caffeine close to bedtime, etc. All this certainly makes sense but has yet to be tested.

The point is that the disruption of our circadian rhythms, as in night shift work, disrupts our physiology and especially the immune (chapter 11) and metabolic systems (chapter 12). This disruption prevents us from getting the correct materials in the right place, in the right amount, at the right time of day. This weakens our ability to fight off tumours at an early stage. A cell with a weak or no clock, as in a cancer cell, loses the protective action of the circadian system, which normally acts as a 'break' to prevent unrestricted cell division and tumour growth.

*Should we take circadian timing into account when using the current anti-cancer drugs?*

As discussed above, although drugs *are* being developed that restore circadian rhythms to cells to decrease tumour growth, the current non-surgical approaches to attack cancer deploy a range of anti-cancer drugs or use some form of radiation. The overwhelming challenge in using these approaches is to kill the rogue cancer cells without killing the patient. Anti-cancer drugs used for chemotherapy are highly poisonous and can cause damage to the main organs of the body, including the kidneys and heart. Radiation treatment can also produce very damaging side effects upon the body. The frustration is that it is very difficult to destroy every cancer cell, and only a few cancer cells need to survive to multiply and seed a new batch of tumours. So treatments have to be aggressive and the side effects are generally hideous, with nausea, vomiting, diarrhoea, loss of feeling in the hands and feet, and hair loss being common.

But let's step back and consider the biology. Under normal circumstances, cells multiply through cell division, which involves growing bigger, producing a duplicate set of DNA wrapped up in chromosomes, and the separation of a single cell into two new 'daughter' cells. In 2001 the Nobel Prize was awarded to Paul Nurse, Leland Hartwell and Tim Hunt for their research on how cells multiply by going through the set stages in a 'cell cycle'. The cell cycle consists of key steps. In the first phase (G1) the cell grows. When the cell has reached its appropriate size, it enters a phase of DNA-synthesis (S), where the DNA and chromosomes are duplicated. During the next phase (G2) the cell prepares for division. With cell division, called mitosis (M), the chromosomes separate, and the cell divides into two new 'daughter' cells with identical sets of chromosomes. After division, the cell is back in G1 and the cycle kicks off again.[719] Cells divide repeatedly to build tissues and organs and replace damaged cells. We started life as a single cell and most of us have around 37.2 trillion cells (a trillion is a million million or 1,000,000,000,000) and many of these cells, like red blood cells, need to be replaced frequently. It takes an immense amount of cell division to make us and then keep us going. And it is remarkable that there are so few mistakes in this complicated process. Normally, once body parts have been built and the necessary on-going repairs made, cell division stops. But cancer cells keep dividing. The normal systems that stop unrestricted cell division are damaged, and the damage is usually due to a small change or mutation in some of the key regulatory proteins associated with the cell cycle. One family of proteins, called the 'RAS' proteins, are mutated and defective in over one third of all human cancers.[720] Significantly, RAS proteins are regulated by the circadian clock[721], and RAS proteins in turn act to help regulate the circadian machinery[722]. There is an intimate relationship between key cell cycle proteins and the molecular clock. What has emerged recently is that cell cycle proteins such as RAS are embedded within the core of the circadian 'architecture' of cells.

Before I continue, I need to stress that the development of cancer involves more than a single mutation in just one cell cycle protein such as RAS. Of the approximately 21,000 genes in each of our cells, there are at least 140 genes that, when mutated, can promote or 'drive' tumour growth, and a typical tumour contains 2–8 of these 'driver gene' mutations. This is an important point and illustrates why mutations in either one of the two breast cancer susceptibility genes BRCA1 and BRCA2 means you have a much higher risk of developing breast cancer or ovarian cancer compared with someone who doesn't have the mutation. *But* a positive result doesn't mean you will certainly develop cancer – it will depend upon the other mutations you carry, and environmental factors such as smoking[723], and circadian rhythm disruption such as night shift work[724]. On average, a woman with a BRCA1 or BRCA2 gene mutation has up to a 7 in 10 chance of getting breast cancer by the age of 80[725] – a very high chance but not an automatic guarantee. The discovery that cancers arise from mutations in cell cycle genes and their regulatory systems such as the clock genes is one of the triumphs of twentieth-century biology, and not only explains the origin and development of many cancers, but also provides the basis for potential new treatments in the coming years.

## *The timed use of current anti-cancer drugs*

Non-surgical cancer treatments such as chemotherapy and radiotherapy are designed to kill cancer cells by preventing them from growing, dividing and making more cancer cells. And because cancer cells usually grow and divide faster than normal cells, chemotherapy and radiotherapy have more of a damaging effect upon cancer cells. But these treatments also affect non-cancerous rapidly dividing cells such as those in bone marrow (where red blood cells are made), hair follicles and the stomach lining. Which is why chemotherapy is associated with anaemia, hair loss and feeling sick. Under normal circumstances, there is a circadian rhythm of cell division in many human tissues,[726-7] and the key point is that the circadian timing of the cell cycle in healthy cells is often different from cancerous cells. This being the case, if most of the daily chemotherapy or radiotherapy is confined to times of lowest DNA synthesis in the non-cancerous cells, then the toxicity is reduced and consequently higher doses can be given.

William (Bill) Hrushesky is an oncologist in Columbia, South Carolina, and somebody whose work I have followed since my time at the University of Virginia in the late 1980s. Bill has argued for chronopharmacological approaches to treat cancer for decades, and in a pioneering experiment in the 1980s compared the time of day of chemotherapy in women with ovarian cancer. He divided the women into two groups. Each group received two standard cancer drugs, adriamycin and cisplatin, but one group received adriamycin at 6 a.m. and cisplatin at 6 p.m., while this daily schedule was inverted in the other group. He found that the women on the adriamycin 6 a.m., cisplatin 6 p.m. schedule developed roughly half the side effects. There was less hair loss, less nerve damage, less kidney damage and less bleeding, and fewer blood transfusions were needed. As Bill commented: 'Every toxicity was markedly diminished several-fold simply depending on what time of day the drugs were given'.[728] And what about survival? In the same year, children with acute lymphoblastic leukaemia were given the anti-cancer drug 6-mercaptopurine in the morning or evening as part of their treatment therapy. Disease-free survival was far better for children on evening chemotherapy. The risk of relapse was 4.6 times greater for the morning schedule than for the evening schedule.[729] Similar findings have been shown for colorectal cancer.[730] Such results, of reduced toxicity and improved survival with timed chemotherapy, have been shown with different cancers in multiple studies.[731-2] In addition to timed chemotherapy, timed radiotherapy also seems to provide treatment options for aggressive brain tumours.[733]

In general, patients receive anti-cancer drugs at times that are convenient for the staff administering the drugs. Clinical capacity and cost are key issues, and there are important logistical issues in the delivery of toxic drugs in busy hospitals. However, recently developed ambulatory medical pumps could deliver anti-cancer drugs to patients at an appropriate time at low cost and potentially in the home.[732] Practicalities aside, when I discuss this issue with clinical colleagues, some doctors are just not convinced of the value of chronotherapy. Many acknowledge that there could be some benefits, but these are sometimes dismissed as too small for routine use. Another major hurdle is simple lack of knowledge. I will say it again, in a five-year training, circadian rhythms and sleep represent a mere footnote in most medical school training programmes. It is also worth mentioning that many of our doctors with little or no knowledge of circadian rhythms advise drug companies about the development of new drugs. Until circadian rhythms become a serious topic for study in our medical schools, there will always be a barrier between exciting laboratory findings, medical application and new drug discovery. This has to change.

## Questions and Answers

### 1. Will my chronotype alter when I should take my medications?

Unless you are an extreme morning or evening type this is unlikely to be much of a problem because most drugs have a relatively long half-life and so will still be effective over a time window of several hours at least. So, if advised to 'take a particular drug at bedtime' you will probably be fine. However, if you are an extreme morning or evening type it may be wise to discuss drug timing with your medical practitioner. More of a problem is taking drugs or medications after chronic jet lag, as your circadian system could be at any time/phase. Also, because of the nature of radiotherapy, where delivery and effect occur at the same time with no extended half-life as in chemotherapy, the precise timing of radiotherapy may be more important than timing in chemotherapy.

### 2. I have read that cluster headaches are seasonal – is this true?

In many individuals who experience cluster headaches, they not only occur at a particular time of day, but also at a particular time of the year over consecutive years. How these annual rhythms are generated remains unknown, but highlights the rhythmic nature of cluster headaches.[669]

### 3. Are strokes linked to dementia?

Both transient ischemic attacks (TIA) and minor ischemic stroke (IS) are associated with an increased risk of cognitive impairment and dementia later in life. And an explanation by Philip Barber and colleagues has been proposed very recently relating to 'hippocampal atrophy' following a TIA or IS. The hippocampus (Figure 2) plays a major role in learning and memory, but as we age the hippocampus begins to atrophy (shrink) slowly. However, patients who experienced TIA or IS showed a significantly higher hippocampal atrophy compared to healthy controls over a three-year study period. The higher rate of hippocampal atrophy was correlated with reduced episodic memory (conscious recollection of previous experiences) and executive function (the mental processes that enable us to plan, focus attention, remember instructions and multi-task successfully) over the same three-year period. These data provide a direct link between stroke, dementia and cognitive decline.

### 4. Should I worry when I have surgery?

The first point to make is that surgery is not without risk at any time, but there has been growing concern for some time that surgery undertaken by surgeons who have been working on extended shifts and have not had the opportunity for sleep for many hours are more likely to make mistakes and commit medical errors. There have been calls to stop surgeons from operating if they have had too little sleep. In a recent study, patient survival was assessed following high-risk cardiovascular surgery that was performed either in the afternoon or in the morning. Death was significantly reduced with afternoon versus morning surgery.[734] This whole issue of surgeon fatigue and the time of surgery is still being debated and has not been resolved with formal guidelines.[735] See also chapter 14.

### 5. Can you automate drug delivery to a particular time of day?

The short answer is yes. Hospitalized and non-hospitalized patients can receive infusions of chemotherapy using a 'chrono-programmable' electronic pump. These pumps can deliver up to four drugs at specific times over several days. In addition to pumps, other timed-release systems are being developed.[736] Such systems will reduce one of the major barriers for chrono-pharmacology and hopefully embed this approach across many areas of medicine.

# II

# A Circadian Arms Race

## The immune system and enemy attack

The struggle for existence never gets easier. However well a
species may adapt to its environment, it can never relax,
because its competitors and its enemies are also adapting to
their niches. Survival is a zero-sum game.

Matt Ridley

During the Spanish Flu pandemic of 1918–19, it is estimated that
about 500 million people, or one third of the world's population,
became infected with this virus. The number of deaths was estimated to be at least 50 million worldwide. It is thought that 25 per
cent of the British population were affected, resulting in 228,000
deaths. Young adults between 20 and 30 years old were particularly
vulnerable and onset was remarkably rapid. You could be fine and
healthy at breakfast and dead by teatime. The first symptoms of
fatigue, fever and headache would progress rapidly into pneumonia
and individuals would turn blue due to a shortage of oxygen. Death
by suffocation would follow. I write this towards what I hope is the
last phase of the COVID-19 pandemic in the UK in January 2022, and
to date over 150,000 people have died in the UK and more than 5.5
million worldwide. But numbers are still rising. As things stand,
there have been relatively fewer deaths than in the pandemic of
1918–19, although the crisis is not over yet. The widespread and rapid
introduction of vaccines, not available 100 years ago, has made an
immense difference, and although applauded and greeted with a

sense of relief, I am not entirely convinced that many of us fully appreciate how close we came to a much greater disaster – if it were not for the vaccines. Science has saved the world, and there now seems to be some level of control over this appalling infection. The 'collateral damage', in terms of social isolation, has also been immense across all levels of society; not least the reduced medical care for the vulnerable. Although I will add a note of caution: only vaccination of all people around the world, and the ability to deal with new variants, will deliver some approximation of victory. Given that many of us are thinking about infection, it seems like a good time to reflect upon the relationships between the circadian system, sleep and our ability to fight off infection. The links are both fascinating and important. New research has shown that our individual responses to infection change over the day, and, even more importantly, if we experience sleep and circadian rhythm disruption (SCRD) our immune system is impaired. Such information is important to all of us, but especially to our front-line staff.

The immune system is the body's defence against infections and provides us with multiple layers of protection. It's horribly complicated. Indeed, I am reminded of the joke:

> An immunologist and a cardiologist are kidnapped. The kidnappers threaten to shoot one of them, but promise to spare whoever has made the greater contribution to humanity. The cardiologist says, 'Well, I've identified drugs that have saved the lives of millions of people.' Impressed, the kidnappers turn to the immunologist. 'What have you done?' they ask. The immunologist says, 'The thing is, the immune system is very complicated . . .' And the cardiologist says, 'Just shoot me now.'

It's an old joke, but it certainly rings true with me! The different parts of the immune response are truly fascinating but they really are immensely complicated, and the situation is made worse because immunologists keep changing the story and the names of the characters. It's like the different convoluted versions of the old Norse sagas, told differently by the Vikings, Anglo-Saxons and Icelanders. When the rest of us begin to understand the immune response, immunologists discover something new, change the narrative, and leave the rest of us baffled. I'm beginning to think this might be deliberate, providing 'job security' for immunologists. Anyway, the bones of the immune story, with some of the leading characters, are outlined in Appendix II. This will provide additional background to the discussion that follows.

## Immune and Circadian System Interactions

We now appreciate that every aspect of the immune response is being regulated by the circadian system.[737-8] The skin is one of the most important, but most overlooked, parts of our immune defence, providing an incredibly effective barrier preventing disease-causing microbes such as viruses, bacteria or other pathogens from entering the body. The circadian system plays an important role in skin porousness (permeability). Permeability is increased in the evening and at night, with lowered permeability in the morning and during the day.[739] This means that there is more water loss from the skin in the evening, which is part of the reason why we experience increased skin itchiness towards the evening and at night as our skin dries out, made worse by conditions like eczema and psoriasis (Figure 8, p. 226). This also means we are more at risk of bacteria and viruses getting in through the skin in the evening and at night. The increased permeability of the skin, combined with scratching itchy skin, increases the chance of pathogens entering the body. Interestingly, blood flow to the skin increases at night[739] (remember the discussions on heat loss), allowing immune defences in the blood a better chance of attacking invaders as soon as they get in. These are not the only changes seen in the skin across the day. The topmost layer of the skin consists of dead cells that form a dense physical layer that resists invasion. Skin proliferation has also been found to have a daily rhythm, with the highest rate of proliferation and shedding of old skin occurring around midnight,[739] which will act to shed bacteria attached to old skin. If you cut or burn your skin, it will heal more than twice as fast if the injury was inflicted during the day compared to the night.[738] This all makes sense. It's more likely that we will damage our skin or encounter an invading pathogen when we are moving around our environment and meeting other people, or animals, with pathogens. In the middle of the night we are mostly immobile and less likely to encounter new individuals carrying disease. I appreciate that this may not be the case for many university undergraduates.

If pathogens do enter the body, then cells and protective molecules are waiting to defend us (Appendix II). Our white blood cells, or 'leukocytes', account for only about 1 per cent of our blood, but these are the cells of the immune response, and every aspect of their 'behaviour' is regulated by the circadian system. For example, one type of leukocyte is the 'macrophages', which are amoeba-like cells that rush to the site of infection, recognize an invader, either directly or because of an attached antibody, and have the ability to ingest (phagocytose) and kill the pathogen. Macrophage sensitivity to attack varies across the day, driven by a circadian clock, with sensitivity increased during the day when we are usually awake.[740]

A study published in 2016 exposed mice to the herpes virus at different times of the day and night and showed that if the virus was given to mice at the start of their sleep, the virus multiplied 10 times faster than 10 hours later, when the mice were ready for activity.[741] This shows that the immune system is heightened at a time when mice would normally be active. This was confirmed in another study. Mice were infected with flu virus delivered to the lungs either just before sleep or at the beginning of activity. The immune system triggered a much greater response, with higher levels of protective inflammation, at the beginning of activity.[742] This increased immune and inflammatory response makes sense – it anticipates the need for increased protection from virus attack when the mouse is active and more likely to encounter another infected animal. Similar time-of-day differences have been shown in us. Elderly individuals were vaccinated against the N1H1 flu virus either in the morning (9–11 a.m.) or in the afternoon (3–5 p.m.). Those individuals who were vaccinated in the morning showed an antibody response three times higher than those vaccinated in the afternoon.[743] Data for night-time vaccination has not been collected. Whether there will be similar time-of-day effects with the various COVID-19 vaccines is an interesting question, and is currently being studied. In the future, such optimally timed vaccination may be an important additional weapon to prevent infection and disease spread, especially in the elderly.

As the circadian regulation of the immune system helps protect us when we are most vulnerable (during the day), it is perhaps no surprise that some pathogens actually try to disrupt the circadian system to weaken our immune responses. There is evidence that the human immunodeficiency virus (HIV) does this, although the precise mechanisms remain unclear. We know more about the hepatitis B and C viruses, which infect the liver and are a major cause of liver disease. Both hepatitis B and C viruses attack the circadian-clock-regulated pathways that protect liver cells from infection.[744] For example, the hepatitis C virus directly interferes with the molecular clockwork of liver cells and this seems to reduce the capacity of liver cells to resist viral attack.[745] More recent studies have shown that the replication (production of more viruses) of the influenza viruses in mouse cells is much higher in mice with defective circadian clocks, compared to mice capable of producing normal circadian rhythms.[741] These examples suggest that if the host circadian clocks are weak or defective, more virus can be made. However, and amazingly, the herpes simplex virus, an infection that causes oral herpes and genital herpes, does the opposite – it exploits our circadian system. This virus hijacks the host's molecular clock, and the cellular clock is actually needed for viral replication.[746] One possible explanation for this 'hijacking' is that the host clock is used to make and release millions of new viruses all at the same time. Such a coordinated release of new virus might effectively 'swamp' the host defences. I will also come back to this idea later.

## Why bother to regulate immune responses by the circadian system?

Our immune system is generally turned up in advance of when we are active and are more likely to encounter pathogens in the environment or from other people, while at night we are not as good at resisting infection, arguably coinciding with the time when the chances of encountering new pathogens is much reduced. A key question is why is the immune system not on full throttle all the time? Part of the reason is that this would be 'expensive' and not cost-effective. Maybe it is better to 'tune' immune responses to when they are most likely to be needed. However, perhaps the much more important reason is that although an immune response and inflammation are necessary for fighting infection, a balance must be achieved between defence against bacteria and viruses, and damage to ourselves from an overly enthusiastic immune response like a 'cytokine storm' (Appendix II). An overactive immune system can lead to auto-immune disorders whereby elements of the immune system can't tell the difference between invaders and self. So, perhaps, clock regulation changes the aggressiveness of the immune system to a time when it is most likely to be useful, and this helps reduce the chances of our immune system attacking us accidentally, and driving an autoimmune disease such as rheumatoid arthritis, inflammatory bowel disease, multiple sclerosis (MS), psoriasis or Hashimoto's thyroid disease.

## The impact of SCRD on the immune responses

As discussed earlier, mice exposed to a virus when awake showed reduced levels of infection compared to when they were exposed to the same virus when asleep. Very interestingly, when the experiment was repeated using mice with disrupted body clocks, the immune response was poor, with high levels of infection, whenever the mice encountered the virus.[741] The finding that disrupted circadian rhythms are associated with weaker immune responses has been shown repeatedly. In another study, mice were immunized against the flu virus. In one group of mice, animals were deprived of sleep for seven hours immediately after immunization. In the non-sleep-deprived mice, immunization prevented infection, but in the sleep-deprived group viral infection spread to high levels.[742] This is also true for humans. Two groups of people were immunized against the flu virus while keeping two different sleep schedules; the group who were allowed to sleep only four hours per night when given the jab had less than half the level of protective antibodies to the flu virus than the group who slept their usual 7.5–8.5 hours per night after the jab.[747] Another study has shown that insomnia is a risk factor for decreased influenza vaccine responses.[748] Similar results have been shown for the antibody response to a hepatitis B and hepatitis A vaccination, which was less effective in those who were sleep deprived.[749–50] So along with the timing of vaccination, ensuring that individuals have had sufficient sleep can enhance vaccine effectiveness. Not easy, I know, especially during a pandemic.

## SCRD and stress

Sleep and circadian rhythm disruption lower our ability to resist infection – but why? As discussed above, SCRD disrupts our ability to coordinate an effective immune response to a pathogen. The beautifully orchestrated immune defence network breaks down (Appendix II). However, there is an additional reason why SCRD is associated with reduced immune function (chapter 4). Individuals experiencing SCRD release more of the stress hormones cortisol and adrenaline. As mentioned, stress is a bit like the first gear of a car engine – it provides rapid acceleration – which can be very useful short term. But if you keep the engine in first gear for a long journey you will destroy the engine. Sadly, this analogy is less meaningful if you only know how to drive an automatic. Anyway, the stress response prepares us for 'fight or flight', getting the body ready for rapid and vigorous action. SCRD keeps the stress response in first gear and one of the consequences is a suppressed immune system,[751] and the results can be devastating. As mentioned previously, a recent study has shown that night shift workers are more likely to be admitted to hospital if they contract COVID-19.[192] SCRD can clearly increase our chances of infection, but it can also lead to the reawakening of dormant viruses lurking in the body and to abnormal inflammatory responses, which can lead to both weakened immunity and overall poor health.[109, 752] In a remarkable study, the effect of late-night partial sleep deprivation on the immune system was studied in healthy males who were sleep deprived between 3 and 7 a.m. The activity of natural killer cells (Appendix II) was reduced the next day by around 28 per cent, showing that even modest disturbances, in this case the loss of four hours' sleep in one night, can lead to an impairment of the immune response.[753]

There is a key link between sleep loss, stress and immune suppression in the form of the stress hormone cortisol. Increased levels of cortisol prevent the release of a range of substances in the body that cause inflammation and trigger the immune response.[109] Inflammation is really helpful as it allows the 'weapons' of the immune system to mobilize and move to where they are needed – the site of infection. Cortisol-based drugs are used to treat conditions resulting from overactivity of the immune system. For example, in rheumatoid arthritis the early-morning joint pain (Figure 8) is triggered by inflammatory signals.[624] Cortisol suppresses this inflammatory response, and, interestingly, naturally low levels of morning cortisol (Figure 1) are a typical feature of rheumatoid arthritis.[754]

## So what does all this mean?

The circadian regulation of the immune system prepares us to attack bugs when we are most likely to encounter them, and reduces the aggressiveness of the immune response when it is less likely to be needed, lowering the chances of our immune system attacking ourselves. The circadian system also acts to coordinate an immensely complex set of responses which are 'primed' to get the correct materials in the right place, in the right amount, at the right time of day. SCRD not only disrupts the regulation of the immune system by disrupting its circadian timing, but also results in the release of stress hormones such as cortisol which act to reduce the effectiveness of the immune system to resist infection. These are important observations and raise the question of how could we use this information for vulnerable individuals and front-line medical staff? Looking ahead, the following actions should be seriously considered:

### PROTECTIVE CLOTHING

As we are more vulnerable to infection at night, then the use of protective clothing by front-line night shift staff is likely to be even more important at this time.

### VIGOROUS WASHING

Our personal first line of defence is the skin, and so a final shower or vigorous hand and face wash to remove bugs from the skin before we sleep is probably sensible.

### VACCINATE AT AN OPTIMAL TIME

As there is evidence that vaccination against some viruses is most effective during the first part of the day, given the opportunity, we may want to time vaccinations to the optimal time for a particular vaccine. This might be important for the elderly, who often show an overall weakened immunity.

### MINIMIZE SCRD

SCRD reduces the effectiveness of our immune responses. So reducing SCRD, and the stress that results from SCRD, prior to and immediately after vaccination will enhance the immune response. As a result, it is essential that we recognize this and prioritize sleep where possible in front-line workers. Obviously, minimizing SCRD and stress in front-line staff is easier said than done, but should be attempted by limiting continuous night shift work where possible.

## 'Bigger' Bugs

I have focused much of the discussion so far on bacteria and particularly viruses. But we also have to fend off bigger pathogens, often just called 'parasites', such as parasitic protozoa, worms and flukes, and ectoparasites such as ticks and fleas. Estimates vary, but parasitic infections are thought to cause over a million deaths worldwide annually, with many more individuals suffering from the complications of infection. Different parasites contribute to this shocking statistic of a million deaths, but malaria leads the field by far.[755] Importantly there is increasing evidence that parasites use a circadian clock to try to get past our immune defence systems. There is an 'arms race' between us and our parasite attackers to get the upper hand. This is an exciting and new area of research, and we know most about this complex circadian dance in malaria.

Malaria, today, is a disease of the tropics, with Africa experiencing 94 per cent of malaria cases and deaths. The World Health Organization estimated that in 2019 there were 229 million cases of malaria worldwide, and approximately 409,000 deaths. Children under the age of five years are the most vulnerable, accounting for 67 per cent of all malaria deaths worldwide. Although currently rare in non-tropical regions, malaria is predicted to increase with global warming.[756] Historically, malaria ('bad air') was even more widespread. The Campagna region of Italy, an area which surrounds Rome, was a notorious place to catch malaria. Many Popes died of malaria, as did the cardinals who gathered at the Vatican to elect a new Pope. Extract of cinchona bark, also known as 'Jesuit's bark', was recognized as an effective treatment for malaria and was introduced into Europe by Spanish Jesuit missionaries based in Peru who were taught the healing power of the bark by native Peruvians, sometime between 1620 and 1630. The marshlands of coastal southern and eastern England had high levels of malaria from the sixteenth century to the nineteenth,[757] and although 'Jesuit's powder' was available in London in the 1650s, anti-Catholic prejudice was so strong that many Puritans refused to use any remedy endorsed by the Pope, including Oliver Cromwell, who died of malaria in 1658. Ten years earlier he had cancelled the celebration of Christmas, so his death was not considered a tragedy by all. Despite the initial opposition, cinchona extracts, consisting of finely ground bark mixed with wine, gradually became the treatment of choice, and finally, in 1820, the French chemist Pierre Joseph Pelletier and French pharmacist Joseph Bienaimé Caventou isolated quinine from powdered cinchona bark, allowing the standardization of treatments. But it was expensive, which is why people investigated willow bark, and in the process discovered not quinine but another fever-reducing drug, salicylic acid and ultimately Aspirin (chapter 10).

Malaria is caused by a single-celled protozoan parasite called a *Plasmodium*. Female *Anopheles* mosquitoes pick up the parasite from infected people when they bite an individual to obtain the blood nutrients needed to develop their eggs. This biting behaviour varies between different species of mosquito, but usually occurs at night and is characterized by looking for human victims after dusk, increasing to a peak bite rate around midnight, with 60–80 per cent of bites estimated to occur between 9 p.m. and 3 a.m. Activity in the mosquito, and hence biting behaviour, seems to be largely driven by a circadian clock. Mosquitoes attempt to locate us by sensing our heat, odours and the carbon dioxide from our breath at a particular time (9 p.m. to 3 a.m.)[615] (Figure 8). Inside the mosquito, the parasites reproduce and develop. When an infected female mosquito bites a human, the multiplied plasmodium parasites are injected into the bloodstream. Whether the night-time biting evolved because our immune system is less effective at night is an intriguing question. The parasites travel to and infect liver cells and can remain dormant or divide multiple times to produce thousands and thousands of parasites, which then burst from the liver cells and infect red blood cells (erythrocytes). More parasite cell division occurs in the erythrocytes to produce yet more parasites. The erythrocytes then all rupture at the same time releasing billions of parasites simultaneously into the bloodstream. These then invade more erythrocytes. A very nasty business.

Some plasmodia enter a different stage (gametocyte) and are programmed to develop into a form that is picked up by the female *Anopheles* mosquito during its blood meal. The gametocytes migrate to the capillaries just below the skin, where a mosquito will inadvertently suck them up when feeding. This is all aided by increased blood flow to the skin at night. By migrating into these capillaries at night the gametocytes increase the chances of being picked up by a mosquito. This behaviour is thought to be driven by a circadian clock in the gametocyte. This is an old idea, and makes good sense, but there are surprisingly few data to support this hypothesis.[758-9] The gametocytes then develop in the mosquito gut, producing many more parasites ready to be injected into a new victim.

Eruption of the parasites from the erythrocytes triggers fever as a result of the activation of most elements of the immune system, especially due to the release of tumour necrosis factor (TNF) by macrophages and other immune cells[760] (Appendix II). A key action of TNF is to promote inflammation and turn up the body's temperature thermostat, giving the infected individual the sensation of extreme coldness, which induces shivering that acts to raise the body temperature to 39–40 °C, and this is often accompanied by drenching sweats. This inflammation and increase in temperature is often associated with joint pain, headaches, frequent vomiting and delirium. If severe, convulsions, coma and death result. When the fever subsides, the whole cycle is called a 'paroxysm'.

Infected individuals show oscillations of fever with periods that are multiples of 24 hours. The fever matches precisely the development of the plasmodium parasite within the erythrocytes, which is either 24, 48 or 72 hours, depending on the species of malarial parasite. Fever follows the circadian rhythm of body temperature, peaking when body temperature normally peaks in the early evening.[761] It is remarkable that all the malarial parasites progress through their developmental cycle at the same time.[762] Such synchronization immediately suggests the involvement of a circadian clock – but whose clock, a clock in the parasite or a circadian rhythm in us? Experiments on malarial parasites in mice show that circadian rhythms in the parasites continue even when there are no circadian rhythms in mice possessing defective clocks. So each of the plasmodia must have its own molecular clock. However, the parasites do require signals (entrainment signals) from us to allow all the parasites to synchronize to each other and generate a coordinated eruption from the red blood cells.[763] Currently it remains unclear what circadian entrainment cue(s) are being used by the plasmodium clock. Body temperature cycles, melatonin rhythms and rhythms in nutrients following eating may all play a role under natural conditions, but individually none of these bodily cycles is required for synchronizing red blood cell rupture and plasmodium release.[764] Significantly, if the host circadian rhythms are disrupted, parasite development and synchrony are also disrupted.[765]

As discussed earlier, some infectious diseases such as viruses benefit from disrupting the circadian rhythms of the host, blunting the immune response. For example, the influenza virus seems to interfere with the circadian rhythms of its host to enhance viral replication.[741] But malarial attack does not disrupt our circadian rhythms.[761] This is thought to benefit the parasite by allowing its synchronous development.[766-7] So for the malarial parasite and for some viruses,[746] the synchronous production of new pathogens using a circadian clock seems to be important, and the question is: why? Why produce millions of pathogens all at the same time? Synchronous breeding, or, more accurately, the synchronous production of offspring during a narrow window in time, is widespread in many and diverse forms of life. We are all familiar with this phenomenon as it is often discussed in wildlife documentaries that show bird colonies gathering to produce their young at a specific time of year. There are several advantages to synchronous breeding. One such advantage is that when members of the population all breed at the same time, the huge numbers of offspring produced will swamp or 'glut' the predator population.[768] This 'glutting effect' means that while some offspring will be eaten by predators, offspring are produced in such huge numbers, and all at the same time, that many will survive. If offspring were produced over an extended period of time, the predators could pick them off more easily. And, in my view, this explains why parasites such as the malaria plasmodium and some viruses use a circadian clock to produce offspring at the same time – to simply overwhelm the immune system of the host. If this is indeed the case, one future treatment strategy would be to target parasite circadian timing systems, which would prevent the host immune system from being overwhelmed.

## Questions and Answers

### 1. Are there links between multiple sclerosis (MS), the immune system and circadian rhythms?

Multiple sclerosis is a disease whereby the insulating covering of nerve cells (myelin sheath) in the brain and spinal cord becomes damaged. This damage disrupts the ability of the nervous system to transmit signals, resulting in physical, mental and sometimes psychiatric problems. Common symptoms include double vision, blindness (usually in one eye), weakness in muscles and problems with coordination. The disease usually begins between the ages of 20 and 50 and is twice as common in women as it is in men. The disease is caused as a result of a chronic autoimmune condition in which the body's immune system attacks the insulating covering of our nerve cells. SCRD is much more frequent in people with MS, and is especially common in those suffering from fatigue. Interestingly, genetic changes in some of the genes of the molecular clock seem to be associated with an increased risk of multiple sclerosis[769], and, remarkably, night shift work at a young age is associated with an increased risk for developing MS[770]. And finally, the management of SCRD in MS patients has been linked to improved health and wellbeing in these individuals.

### 2. Are there annual changes in our immune responses?

Seasonal disease outbreaks are a common feature of human societies. For example, most respiratory viruses cause winter-time infection and polio was/is primarily a summer-time disease.[771] However, the reasons for these annual patterns are unclear. A recent study using data from the UK Biobank examined the seasonal variability of multiple immune markers, including inflammatory proteins in the blood, lymphocytes and antibodies. Key seasonal changes were identified in most of the immune markers examined. This could certainly explain why our chances of infection change over the year, but what drives these annual immune changes remains unclear. Whether this is due to the action of an endogenous annual (circannual) clock or whether these changes are driven by some environmental set of signals is an intriguing mystery.

## 3. What are the links between the circadian and immune systems and asthma?

Asthma is a condition whereby the airways narrow and swell and may produce extra mucus. This makes breathing difficult, which can trigger coughing, wheezing when you breathe out and shortness of breath. Triggers for asthma include infections like colds and flu or allergens such as pollen, dust mites, animal fur or feathers, smoke, fumes and pollution. A characteristic feature of asthma is that these symptoms get significantly much worse overnight and peak around 4 a.m., which is when sudden death from asthma is most likely (Figure 8). Such asthma attacks at night can severely disrupt sleep.[772] Interestingly, lung function shows a 24-hour rhythm in healthy individuals with peak respiratory flow around 4 p.m. and with the lowest flow at 4 a.m. In asthma, the normally lower respiratory flow at 4 a.m. is very much worse. Some asthma has been linked to an abnormal level of eosinophil activity. When activated, eosinophils trigger inflammatory responses, attracting cells like macrophages to the site of infection (Appendix II). They also release cytotoxic proteins which normally attack pathogens, but under conditions of overactivity can damage our own cells. Eosinophils usually show a circadian rhythm of activity in the lung, but in individuals with asthma, eosinophil and macrophage levels are significantly higher at 4 a.m.

It is not precisely clear what is going on, but circadian disruption and circadian-driven changes of immune activity within the lung have been suggested. One possibility is that the molecular clocks in the lung are driving up (over-activating) the sensitivity of the immune response in the lung, which causes increased inflammation and a narrowing of the airways along with extra mucus.[620] In addition, allergens present in the bedding or bedroom at night could interact with the circadian regulation of the immune system to drive up allergic responses. Work is currently on-going to understand these links in greater detail.

### 4. Could circadian rhythms be important in the defence against COVID-19?

It is still early days, but at the time of writing this chapter there are at least two studies about to be published showing that SCRD in night shift workers increases the chances of infection and hospitalization from COVID-19. So, in this respect, COVID-19 is like other infections, and this needs to be considered when managing the risk of infection in different groups, especially in front-line medical staff. Whether vaccination against COVID-19 at different times of the day will alter vaccine efficacy remains unknown. However, as circadian regulation of the immune system impacts upon other respiratory viruses, such as influenza, there is an urgent need to study circadian/immune and COVID-19 interactions.[773]

### 5. What about vitamin D, light exposure and COVID-19?

I discussed in the 'Questions and Answers' section of chapter 4 the potential use of vitamin D supplements for those individuals, such as night shift workers, who are exposed to little natural light. And that supplementation would seem to be a good idea for improved health in these individuals. However, whether vitamin D supplements might help prevent COVID-19 infection remains unproven. Observational studies show that certain groups are both more likely to have vitamin D deficiencies and to catch

COVID-19, including the elderly, individuals with obesity and people with darker skin (including black and South Asian people). However, Dr Aurora Baluja makes the very important point that 'although vitamin D deficiency was a well-established risk factor among people who die in intensive care, vitamin D supplementation alone has always failed to reduce the risk of those patients'. Indeed, a recent paper claiming that vitamin D supplementation would improve COVID-19 survival has been withdrawn due to weaknesses in the methodologies. The short answer is that vitamin D is important for health, but there are no scientific data showing that you can reduce the impact of COVID-19 infection by taking vitamin D supplements alone. Randomized controlled trials are currently being undertaken in several locations around the world to assess any effects.

# 12.

# Eating Time

## Circadian rhythms and metabolism

By 'life', we mean a thing that can nourish itself and grow and decay.

Aristotle

In the rich countries of the world there is an apparent surplus of food, and dieting is an overwhelming preoccupation for many, fuelled by an avalanche of advice from social and every other sort of media. At any given time, about half of all Americans are trying to lose weight. In the rich nations, and across all age groups, obesity is rising. Obesity is also a feature of countries that are transitioning from poor to rich, especially in their children. A recent survey suggested that China has the largest number of obese children in the world at 15 million, with India a close second at 14 million.[774] Widespread obesity is a modern phenomenon, less than a century old. The scarcity of food throughout most of human history led to the association that corpulence was desirable. This was certainly reflected in the arts. Think of the 'Venus of Willendorf', the very rounded limestone female figure made about 25,000 years ago, or the paintings of the Flemish artist Peter Paul Rubens (1577–1640) and the depiction of fleshy 'Rubenesque' men and women. It was only in the second half of the nineteenth century, with the passion for classical art, that corpulence began to be stigmatized by the ruling classes for aesthetic reasons. And although the health consequences of obesity were linked to fatigue, gout and breathing difficulties in the rich, it was only after the 1950s that obesity began to be recognized and linked to poor health in the broader population.[775]

In case you needed any reminding, obesity greatly increases the risks of high blood pressure (hypertension), Type 2 diabetes, coronary heart disease, stroke, obstructive sleep apnoea and osteoarthritis. A very recent study showed that individuals who were obese in their twenties and thirties, or have high blood pressure and/or glucose, will suffer a sharper cognitive decline in later years.[776] And we can now add to this sad list of problems a greater chance of dying from COVID-19.[777] The condition of obesity (excess body fat, especially stored around the waist), along with its associated risks of heart disease, stroke, Type 2 diabetes, increased blood pressure, and high blood sugar have been grouped together and are referred to using the collective term '**metabolic syndrome**'. It has been estimated that the direct cost to the UK's National Health Service (NHS) attributable to metabolic syndrome is projected to reach £9.7 billion by 2050, with the wider costs to society estimated at £50 billion per year. The World Health Organization estimates that treating the ill-health caused by metabolic syndrome around the world will top $1.2 trillion every year from 2025. There's an old Spanish Catalan proverb that says: 'The table kills more people than war does.' I'm not sure that was true during the first half of the twentieth century, but there is no question that this proverb is true for much of the world today.

In this chapter, and the next, I want to consider the causes and consequences of metabolic syndrome and how this is influenced by our circadian biology. And my message is simple. By gaining a better understanding of our metabolism, and how our metabolic pathways are regulated by the circadian and sleep systems, we will be better armed to navigate the difficult path between healthy eating and metabolic syndrome with greater confidence. The relationship between the circadian system and metabolism is an emerging area of science, but already it is transforming our understanding of what makes us healthy and what can make us sick. In preparation for this discussion, and before we jump into the circadian story, let's kick off with some essential key facts about metabolism that are relevant for this and the next chapter.

## *Metabolism – key facts*

The first point to make is that food provides the energy to drive our metabolism, which in turn drives the processes of life. In an absolute sense, death can be defined as 'the absence of metabolism'. But how is food converted into energy? The remarkable molecule adenosine triphosphate (ATP) functions as the energy currency for all cells. When one phosphate group is removed from ATP and it is converted to adenosine diphosphate (ADP), energy is released to drive metabolic processes. The supply of more ATP requires the conversion of ADP back into ATP in the mitochondria of cells. This rebuilding of ADP into ATP needs energy, and this comes from the process of 'cellular respiration'. Cellular respiration is where glucose is broken down within cells using oxygen to release energy to re-make ATP, along with water and carbon dioxide. This reaction will be familiar to many of you and can be summarized as: glucose + oxygen + ADP > carbon dioxide + water + ATP. When we breathe in, we take in oxygen from the air to drive cellular respiration to make ATP. When we breathe out, we are getting rid of carbon dioxide and excess water vapour, which are the 'waste products' of cellular respiration.

So glucose, like oxygen, is essential for almost all animal life. Plants, by the way, make their own glucose through photosynthesis. When we are awake and active, glucose comes mainly from the food we eat. But during sleep, and effective 'starvation', glucose has to be mobilized from stores. The circadian system anticipates these different metabolic states of sleep and wake, and adjusts our metabolism accordingly.[778] The key elements of

(A)

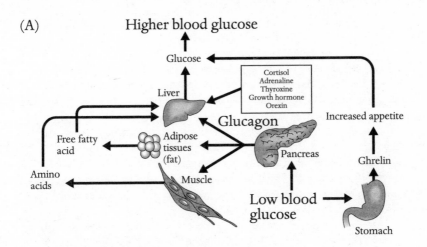

(B)

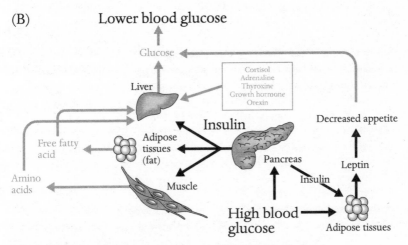

**Figure 9. The mechanisms that increase and decrease blood glucose.**
**(A) Increasing blood glucose.** In response to the circadian anticipation of sleep or as
a result of low blood glucose, the *alpha* cells of the pancreas are stimulated to release
glucagon. When glucagon reaches the liver, it stimulates the conversion of stored
glucose (glycogen) into glucose, a process called glycogenolysis. Glucose is then
released into the blood. Glucagon also acts on adipose tissue (body fat) beneath the
skin, around internal organs (visceral fat) and other places to break down stored fat
(triglycerides) into free fatty acids, which travel in the blood to the liver and are
converted into glucose. Under starvation conditions, glucagon can act to break down
muscle to liberate amino acids, which can then be converted to glucose in the liver.
The result is to increase blood glucose.[779] In addition to low blood glucose, other
processes act to increase circulating glucose. Under conditions of stress, cortisol and

adrenaline promote the liver to make glucose; in addition, cortisol and adrenaline are also strongly regulated by the circadian system. Cortisol and adrenaline levels rise from the middle of the night and peak just before waking in the morning, then gradually decline throughout the day, with lowest levels reached during the first part of the night and our deepest sleep. As cortisol and adrenaline promote glucose production in the liver, the pre-wake increases in these hormones prepare the body for activity, driving glucose production from stores before glucose can be obtained more directly from food; thyroxine from the thyroid gland is another important metabolic regulator and allows us to mobilize and utilize glucose from stores when we are asleep. Normally, thyroxine levels increase sharply at the beginning of sleep, then decrease after waking;[780] growth hormone (GH) is involved in tissue repair and cell growth, and helps to produce glucose when we are asleep, with levels peaking between 2 and 4 a.m. (Figure 1, p. 20). This ensures that glucose is released from stores during sleep and energy is available to drive tissue repair and growth. Interestingly, if you don't sleep and remain active, levels of GH release are markedly decreased,[781] diverting our biology away from growth and preparing us for 'fight or flight'. Orexin is produced in the lateral hypothalamus (Figure 2, p. 24). It is not only involved in driving the wake state during the day (chapter 2), but also plays a key role in food intake and energy metabolism, increasing glucose production in the liver, which in turn allows increased heart rate, body temperature and muscle activity. Ghrelin is a hormone produced mainly by the stomach. It is often called the 'hunger hormone', because when ghrelin reaches the brain it increases the sensation of hunger and promotes eating to increase levels of glucose in the blood. Ghrelin levels are high during the day to promote feeding, but low at night, when feeding is not compatible with sleep. Let me stress, the activity of the pancreas, liver, muscles, adipose tissue and their regulatory centres in the brain has been shown to be either directly or indirectly regulated by the master circadian clock in the SCN, mostly via the autonomic nervous system (chapter 1).

(B) Decreasing blood glucose. In response to circadian drivers that anticipate food intake during the day, and hence the need to keep blood glucose under control, and as a direct result of high levels of blood glucose, such as after a meal, excess blood glucose is lowered and then stored for later use. Glucose storage is mainly in the form of both glycogen and triglycerides in the liver and triglycerides in adipose tissues. This critical activity is achieved by the release of **insulin**. In response to circadian signals and high blood glucose, the *beta* cells of the pancreas release insulin. Insulin acts to reduce blood glucose. In the liver, insulin inhibits glucose production, stimulating the liver to convert glucose into glycogen for storage. In metabolically active cells such as muscle, insulin stimulates glucose uptake from the blood; in adipose tissue, insulin suppresses the breakdown of triglycerides into free fatty acids and drives the synthesis of stored fat (triglycerides) from free fatty acids. The net effect is to lower blood glucose and to increase stored glucose as either glycogen or triglycerides. **Leptin** is a hormone made mainly by adipose cells, and when released into the blood acts on the brain to reduce hunger. This in turn reduces food intake and glucose consumption. Leptin levels peak during sleep to suppress appetite at a time when feeding is not compatible with sleep. Leptin levels are low during the day, when ghrelin levels are high, so that feeding is encouraged. The release of leptin is partly regulated by insulin but also involves other metabolic and circadian signals.[782] Diabetes Type 1 is caused by insufficient or non-existent production of insulin from the beta cells of the pancreas, while Type 2 is mainly due to a failure of the muscles, adipose tissue and liver to fully respond to insulin, called **insulin resistance**. Both types of diabetes result in too much glucose remaining in the blood (hyperglycemia).

glucose metabolism are summarized in Figure 9 (p. 275), and are truly amazing. Our understanding of metabolic physiology is a scientific success story, but sadly it is overshadowed by more glamorous disciplines. Scientists working on the brain always attract a horde of acolytes at a party, while those working on the liver, gut or stomach often get left to drink alone. Sad but true . . . Take a quick look at Figure 9 to help guide you through the next part of our discussion.

## The Circadian Regulation of Feeding and Glucose Metabolism

The circadian system influences every aspect of metabolism, from hunger and digestion to the regulation of the metabolic hormones. For example, under normal circumstances we eat during the day. So, no surprise, there is a circadian rhythm in saliva production that rises over the day and falls at night. Saliva allows normal speech, taste, chewing and swallowing.[783] The human stomach empties faster, after identical meals, in the morning than in the evening. Colon contraction has a circadian rhythm, with movement during the day and much less during the night.[784] Normally we defecate during the day, with over 60 per cent of individuals reporting a 'bowel movement' in the morning, and less than 3 per cent late at night. Gastric acid secretion varies depending on when we eat, but there is an underlying daily rhythm with increased production towards the afternoon and early evening.[785] Interestingly this correlates with the highest incidence of a perforated stomach ulcer, which is reported to be between 4 and 5 p.m., when gastric acid secretion is getting going (Figure 8, p. 226).[617] Eating before bed results in increased gastric acid secretion, which is the reason why the pain from untreated stomach ulcers or acid reflux is often worse at night, and can prevent sleep. Taking proton-pump inhibitors (PPIs) can greatly reduce stomach acid production, as discussed in the 'Question and Answer' section in this chapter.

### Circadian control of metabolic hormones

Metabolism is regulated by multiple hormones and enzymes, and the circadian system adjusts the levels of these signalling and regulatory molecules across the 24-hour day (Figure 9). The first evidence that the 'master biological clock', the SCN, was involved in glucose metabolism came from studies in rats in which the SCN had been lesioned. The result was that daily rhythms in glucose, insulin, glucagon and feeding behaviour were completely abolished.[786-7] The liver is a key source of stored glucose in the body in the form of glycogen. In addition to the rhythmic action of insulin on the liver, predicting activity and sleep, the SCN partly regulates the liver and the production of glucose via the autonomic nervous system (ANS) to produce a daily rhythm in blood glucose. But if the neural connections to the liver are cut, the daily rhythm in blood glucose is compromised, showing a direct role of the SCN in glucose metabolism – not just via insulin. Although the SCN acts as the 'master' biological clock, it is not the only clock. Most, if not all, cells of the body have the capacity to generate a circadian rhythm. Interestingly, in mice with a damaged SCN, the individual liver clock cells continue to show circadian rhythms. But in the absence of an input from the SCN, and without the rhythmic production of insulin, the circadian rhythms of individual liver cells drift apart and coordinated glucose metabolism is lost.[778]

Abnormalities in clock genes have also been linked to altered glucose metabolism, Type 2 diabetes and obesity in both mice and us.[778] Mice with defective circadian rhythms, due to mutations in their clock genes, fail to show a clear night/day feeding rhythm. The mice eat excessively, are obese and develop metabolic abnormalities, including fatty liver disease and insulin resistance.[788] The key point is that in the absence of an internal circadian regulator, the hormones of metabolism, as illustrated in Figure 9, become chaotic, and the anticipation of the different metabolic states of sleep and wake is lost. Metabolism fails.

## Contradictory circadian signals

The complete loss of circadian signalling leads to metabolic failure, but what about 'mixed' or contradictory circadian signals? The light/dark cycle acts as the key entraining agent in setting the SCN to the external world, but for the synchronization of the clock cells in the rest of the body – the 'peripheral clocks', such as liver cells – metabolic signals can also play a key entraining function. This has been shown in mice. In a series of experiments, mice were exposed to 12 hours of light and 12 hours of darkness, and were only given access to food for a few hours during the day. The mice were essentially forced to feed during the day, when normally they would be inactive and asleep. Remarkably, the peripheral clocks in the liver, muscles, gut and other organs 'moved' their circadian rhythms to coincide with the new feeding time. However, the SCN clock remained locked on to the light/dark cycle and continued to drive most of the animal's behavioural activity at night. The SCN and peripheral clocks were no longer aligned to each other, and the normal metabolic pathways that underpin the different demands of sleep and activity were severely disrupted.[789] So, under certain circumstances, the liver and other organs can uncouple themselves from the SCN and respond to metabolic (feeding) signals. While an uncoupling of the SCN and liver might be useful short term, to cope with an acute and immediate need, such a misalignment in the long term ultimately leads to major metabolic problems, including obesity and insulin resistance.[790]

Based on the mouse studies, an obvious question is: what are the consequences of mixed circadian signals in us? And we return to sleep and circadian rhythm disruption (SCRD) in night shift workers, who are forced to work when their biology is in the sleep state. I introduced this topic in chapter 4 but now want to consider metabolism in a bit more detail. Multiple studies have shown a clear link between night shift work metabolic syndrome and notably Type 2 diabetes. And the longer the time spent doing night shift work, the greater the risks.[791-4] But such problems are not confined to night shift workers. Sleep loss and sleep disruption in other groups of workers are also associated with a higher risk of metabolic syndrome.[795-7] Once again, we see that glucose metabolism and the differing demands of sleep and activity are misaligned. This in turn can activate the stress axis, which, if sustained, helps promote metabolic syndrome (see chapter 4). However, the link between SCRD and metabolic disruption involves more than the release of the stress hormones of cortisol and adrenaline.

### Leptin and ghrelin

Recent studies have shown a robust SCN-driven rhythm in leptin release from adipose tissues, with a peak around 2 a.m. (when sleeping) and a low point of release around noon (when active). Leptin acts on cells within the hypothalamus to suppress appetite, hence its nickname of the 'satiation hormone' (Figure 9B). Normally, high levels of leptin at night suppress appetite so that hunger does not disrupt sleep. The action of leptin is opposed by ghrelin, which is primarily secreted by the stomach. Ghrelin stimulates hunger by activation of other pathways in the brain, hence its name of the 'hunger hormone' (Figure 9A). Ghrelin levels actually increase before regular meal times and this 'food anticipation' is driven by the circadian system, acting to increase our appetite before we start eating.[798] A whole industry has been built up around this phenomenon in the form of pre-dinner snacks and cocktails – often more enjoyable than the meal that follows!

Individuals who are forced to endure reduced sleep have lower levels of leptin (satiation hormone) and increased levels of ghrelin (hunger hormone), and experience increased hunger and eat more food.[799, 800] A 'classic' study examined the impact of only four hours of sleep on two consecutive nights on leptin and ghrelin levels, along with hunger and appetite. In these healthy young men, blood levels of leptin decreased by 18 per cent, and there was a 24 per cent increase in ghrelin, a 24 per cent increase in hunger and a 23 per cent increase in appetite. Notably, the appetite for high-carbohydrate food increased by 32 per cent following sleep deprivation.[801] These findings strongly suggest that we are metabolically programmed to consume more calories when we are sleep-deprived, driven by decreased leptin and increased ghrelin. Such findings explain our personal experiences of feeling hungry when we get less sleep,[108] sometimes called 'the munchies'.

SCRD in night shift workers leads to internal desynchrony, mixed circadian signals and a shift into an 'obese metabolic state'.[802] You might expect that in obese individuals leptin (satiation hormone) will be low and ghrelin (hunger hormone) will be high. But it's more complicated than that. Obese individuals still show a circadian rhythm in leptin release, but the rhythm is greatly flattened, losing the clear 'high at night' and 'low during the day' change.[803-4] Along with a flattened rhythm, leptin release from adipose tissue in obese individuals is high. So with high leptin this should mean that obese individuals should *not* feel hungry. But this is not the case. Obese individuals show what is called 'leptin resistance'.[805] The trigger of high blood glucose following food intake (Figure 9B) stimulates the adipose tissues to pump out loads of leptin. However, the brain has become insensitive to this constant leptin bombardment. The result is that although leptin levels are high, the leptin-driven satiation signal in the brain of obese individuals is greatly blunted. The brain is not responding to the leptin. Although the leptin satiation signal is largely lost, the hunger signals from ghrelin remain, and may even increase. The result is that obese individuals tend to feel hungry more of the time.

## The Circadian Regulation of Fat Metabolism

Adipose tissue ('fat') is the main site for long-term energy storage and is found throughout the body, but notably around the internal organs (visceral fat) – which is particularly obvious in scientists of a certain age! Normally, when we eat, our liver converts any surplus glucose we don't need immediately into triglycerides, but the liver has a limited storage capacity for triglycerides. High levels of triglycerides stored in the liver cause non-alcoholic fatty liver disease, or NAFLD, the extreme form of which is called NASH, standing for non-alcoholic steatohepatitis. When the liver is 'full', triglycerides are transported in the blood and stored within adipose tissues, which makes us 'fat'. The liver converts excess sugar into triglycerides, which explains why eating excessive sugar-based food makes us fat. You don't have to eat fat to get fat! When the body requires fatty acids to make glucose, the hormone glucagon from the alpha cells of the pancreas drives the breakdown of the triglycerides into free fatty acids in both the liver and adipose tissue.[806] Stored triglycerides represent unused calories, and for many of us this bank account of stored energy remains deposited around our waist, driving up our body mass index (BMI).

### SCRD and fat metabolism

The mobilization of triglycerides, their transport by proteins in the form of 'lipoproteins', and their breakdown for energy utilization or for physiology are all regulated by the circadian system. Even the absorption of fats by the digestive system is under the control of both the SCN and local circadian clocks in the small intestine. SCRD, arising from night shift work, jet lag or 'social jet lag' is associated with major metabolic abnormalities, not least in fat metabolism, along with a greater risk of obesity. [807-8] For example, reduced sleep (six hours or less) is associated with an increased BMI, along with an increased risk of Type 2 diabetes. [809] There is a clear relationship between obesity/BMI and total sleep time. [810] The greater the sleep loss, the higher the BMI. In addition to disrupted metabolic control, it seems that overweight individuals find it more difficult to get to sleep and stay asleep, and are at a higher risk of obstructive sleep apnoea (OSA) [811] (see chapter 5). We also need to factor in that being awake, along with leptin resistance in obese individuals, increases both the opportunity and desire to eat more, particularly later in the day. Finally, eating a high-energy meal before bedtime can lead to increased body temperature. [812] As a slightly lowered core body temperature seems to be important for getting to sleep, a heavy meal before bedtime may therefore act to delay sleep. Although the evidence to support this idea is not strong. But as we shall see in the next chapter, a major meal in the evening is a very bad idea for a whole host of other reasons too . . .

## Questions and Answers

### 1. When should I take my proton-pump inhibitors (PPIs)?

I am often asked about the use of drugs to control stomach acid production and it is an interesting topic. As discussed in chapter 10, drug action depends upon the half-life of the drug, its length of action on the target cells and the time of day – its pharmacokinetics. The family of drugs called proton-pump inhibitors, such as Omeprazole, sold under the brand names Prilosec and Losec, stop the cells that line the stomach from producing too much stomach acid by turning off their 'proton pumps'. A key point is that once the proton pumps producing gastric acid are turned off by PPIs, they remain inhibited. More stomach acid can't be produced until new pumps are made by the stomach cells, which takes around 36 hours.[813] These drugs can help to prevent ulcers from forming, or aid the healing process if you have a stomach ulcer. In addition, PPIs can help prevent gastro-oesophageal (acid) reflux, especially at night when lying down, and in this regard can help with sleep. After you take PPIs, they are cleared from the blood very rapidly, within just a few hours. The advice is to take PPIs in the morning, but why if stomach acid production peaks in the late afternoon/evening (Figure 8)? A critical point is that the PPIs are only able to effectively turn off the proton pumps if the pumps have been activated by the presence of new food in the stomach. Without food the PPIs are much less effective at lowering stomach acidity.[814] Taking PPIs in the morning, 30 minutes before food, shuts off the proton pumps effectively until new pumps can be made. So taking a PPI in the morning will effectively reduce the circadian-driven stomach acid production for the next 36 hours. Taking a PPI before bed, when eating has finished for the day, means that PPI action would be much less effective. Many individuals still take PPIs just before bed, in the mistaken belief that night-time acid reflux will be controlled by night-time PPIs. It won't. Some individuals, who take their PPIs in the morning, still experience gastro-oesophageal reflux at night.[815] The explanation for the reflux is often related to not eating breakfast after taking morning PPIs, and the failure to activate the proton pumps. Additionally, too few pumps might have been turned off by the single morning dose of the PPI. When this occurs, a twice-daily dose of PPI is recommended, one before breakfast and another before the evening meal, but not before bed. This approach of 'hitting the pumps twice' has been shown to help reduce severe gastro-oesophageal reflux before bed.[816]

## 2. Why is high glucose in the blood such a bad thing?

We know that high levels of glucose in the blood cause damage to blood vessels. Damaged blood vessels increase the risk of heart disease and stroke, kidney disease and problems with vision. But what is actually causing this damage? It seems that high blood glucose causes the activation of an enzyme called protein kinase C (PKC), which in turn triggers a series of pathways that result in blood vessel constriction.[817] Constricted blood vessels raise blood pressure, which can damage blood vessels, and LDL (bad) cholesterol (see below) accumulates in the damaged artery walls. The net effect is to increase the workload of the circulatory system while decreasing its efficiency, ultimately causing organ failure. I should also mention that, over time, high blood glucose can damage the small blood vessels that supply the nerves of the body, stopping the delivery of essential nutrients to the nerves. As a result, the nerve fibres can become damaged or die, which is called neuropathy. Neuropathy can damage or destroy sensations such as touch, temperature, pain and other messages from the skin, bones and muscles to the brain. It usually affects the nerves in the feet and the legs, but can occur in the arms and hands. In addition to these problems, the ability of cells of the immune response (chapter 11 and Appendix II), such as neutrophils, to engulf bacteria is reduced when glucose and fructose levels are increased in the blood. Fasting strengthens the neutrophils' phagocytic capacity to engulf bacteria.[818] Perhaps this explains why individuals who have higher levels of blood glucose, as occurs in Type 2 diabetes, are more vulnerable to infection. See also the next question.

## 3. Why does infection alter levels of blood glucose?

There are important relationships between levels of blood glucose and infection. First, infection triggers the stress response and the body produces more cortisol and adrenaline. These hormones work against the action of insulin and promote the generation of glucose (gluconeogenesis) (Figure 9A). As a result, the body's production of glucose increases, which results in high levels of blood glucose. This is why long-term infection can increase the chances of metabolic syndrome. Significantly, raised levels of blood glucose (hyperglycemia), as in Type 2 diabetes, seem to impair immune responses, and especially the spread of invading pathogens. This is the reason why Type 2 diabetic individuals are more at risk from infection. Indeed, an early sign of Type 2 diabetes can be boils and spots that won't go away.

## 4. Is cholesterol involved in energy metabolism?

As discussed in this chapter, triglycerides are stored in adipose tissues and represent 'stored calories' and can be converted to glucose for metabolism. The other form of stored fat in the adipose tissues is cholesterol, which is *not* directly involved in energy metabolism but is essential for building cells and key hormones, including cortisol, oestrogen, progesterone and testosterone. And as a brief reminder, cholesterol is transported in the blood in the form of a protein complex (lipoprotein), and two types of lipoproteins carry cholesterol to and from cells. One is low-density lipoprotein, or LDL. The other is high-density lipoprotein, or HDL. LDL cholesterol is considered the 'bad' cholesterol, because it contributes to fatty build-ups in arteries (atherosclerosis) by delivering cholesterol to these sites. This narrows the arteries and increases the risk for heart attack and stroke. HDL cholesterol is thought of as the 'good' cholesterol because a healthy level may protect against heart attack and stroke. HDL carries cholesterol away from the fatty deposits in the arteries and back to the liver, where the cholesterol is broken down and excreted from the body. Sadly, HDL doesn't completely remove cholesterol from the atherosclerosis-causing fatty deposits in the arteries. Note that, as well as lowering LDL cholesterol, statins (chapter 10) can also lower triglycerides, which can be useful as high triglycerides have been linked to NAFLD, heart disease and diabetes, and the whole spectrum of conditions seen with metabolic syndrome.[806]

# 13.

# Finding Your Natural Rhythm

## Circadian rhythms, diet and health

If we shouldn't eat at night, why do they put a light in the fridge?

Anonymous question to me

We evolved with very little sugar in our diet. Chemically refined sugar was probably first produced in India around 3,500 years ago, and from there spread east to China and west through Persia and the early Islamic world, reaching the Mediterranean in the thirteenth century. Until the end of the Middle Ages (around 1500) sugar was rare and expensive. Refined sugar was produced in Cyprus, in Sicily and on the Atlantic island of Madeira. The Portuguese then established sugar plantations in Brazil, which were sustained by a slave-based plantation economy. The introduction of sugar cane to the Caribbean from Brazil in the 1640s led to an explosion in sugar production, along with the slave trade. Other European nations were keen to get involved in sugar production, and also used slave labour to grow and harvest the sugar cane. It is estimated that more than 12 million human beings were shipped from Africa to the Americas until the early 1800s when the transatlantic slave trade was abolished,[819] although slavery itself did not end, more or less, until after the American Civil War in 1865. This appalling trade in human misery, degradation and death fuelled the 'sugar craze'. Such unspeakable inhumanity, all for a substance that nobody actually needed but everyone craved. The trade in sugar has stamped itself upon the modern world, providing both a shameful legacy of cruelty and an inheritance of monumental poor health.

Tooth decay is seen in about 20 per cent of skeletons unearthed from the Middle Ages, but by the twentieth century more than 90 per cent of skeletons show tooth decay. My colleague and old friend Ben Canny told me that the leading cause of preventable hospital admissions for children in Tasmania is general anaesthesia for teeth extraction due to tooth decay. Until the 1700s sugar was a luxury and a symbol of wealth. Elizabeth I of England (r.1558–1603) was apparently addicted to sugar and, as a result, by the time she was in her early fifties her teeth had turned black or fallen out. As the queen's teeth turned black, the nobles and higher gentry at court decided that black teeth were a symbol of beauty and wealth and so women began colouring their teeth black with soot. Those who could not afford sugar also coloured their teeth black to fool others into thinking they were rich. It remains unclear if the foul stench of bad breath was also considered fashionable. I should not be too scathing; perhaps someday blackened teeth will return as a fashion statement. After all, who would have predicted that flared trousers and platform shoes from the 1970s would crawl from the pit of oblivion as high fashion? But the role of sugar in tooth decay is just the tip of the poor health iceberg.

Individuals living in Europe and North America get about 15 per cent of their daily calories from sugar, but this figure is an average, and many individuals consume much more. Refined sugar is crystallized sucrose, which is made from a molecule of glucose and a molecule of fructose, and obtained by extracting sugar from foods like sugar cane, sugar beet or corn. The problem is that it is frequently added to nutrient-poor, processed foods ranging from sugary drinks, breakfast cereals and sauces to even many types of bread. The World Health Organization suggests that refined sugar should form less than 10 per cent of our daily calories. However, most of us consume much more than this. A 15-year study published in 2014 found that individuals who got 17–21 per cent of their calories from added sugar had a 38 per cent higher risk of dying from cardiovascular disease compared with those who consumed 8 per cent of their calories as added sugar.[820] High sugar is also linked to metabolic syndrome – higher blood pressure, inflammation, weight gain, diabetes and fatty liver disease. How sugar actually causes these problems is not well understood (see chapter 12, 'Questions and Answers', question 2, page 285), but it may be due to the way it is metabolized in the liver. Excess sugar, and by that I mean more than is needed to satisfy the body's immediate energy demand for glucose, overloads the liver, which converts it into fat. This can result in fatty liver disease, which, in turn, contributes to diabetes and the risk of heart disease.[821] So perhaps a moment of reflection is prudent before deciding to have that 'sticky toffee pudding' for dessert! Incidentally, the practice of having a sugar-rich sweet dessert at the end of a large meal originates from Tudor times and the belief that sugar aided digestion and was good for the stomach.

In the last chapter I outlined the role that the circadian system plays in orchestrating our metabolism – basically how we regulate food intake and convert food into glucose as an energy source, and the role of the circadian system and impact of sleep and circadian rhythm disruption (SCRD) on these processes. Next, I want to address what we can do to maximize our chances of maintaining a healthy level of blood glucose and stored fat. At the core of this discussion will be how circadian knowledge and biological timing can be used to help us achieve a more robust and healthier metabolism.

## How Do I Know If I Have a Healthy Metabolism?

Just a quick recap from chapter 12. Metabolic syndrome is a term which describes a cluster of linked conditions, including obesity and excess body fat stored around the waist, an increased risk of heart disease, stroke, Type 2 diabetes, increased blood pressure and high blood sugar. Insulin acts to lower blood glucose (Figure 9B, p. 275), and insulin resistance is a pre-diabetic state in which there is a failure of the muscles, adipose tissue and liver to respond to insulin fully and reduce glucose in the blood. This results in glucose intolerance (also called impaired glucose tolerance), whereby your blood glucose is raised beyond the normal range but it is not so high that you have Type 2 diabetes. The point is, a good measure of metabolic health is your measure of blood glucose. One approach to measuring your blood glucose is to visit your medical practitioner. Alternatively, accurate measures can also be collected from a home-use glucose-monitoring device. Using such a device, blood glucose values for a healthy individual are under 6mmol/l before eating and below 7.8mmol/l two hours after a meal; an individual who has glucose intolerance will have glucose values between 6.0 and 7.0 mmol/l before a meal and between 7.9 and 11.0mmol/l two hours after it; and an individual with Type 2 diabetes will have glucose values over 7.0mmol/l before a meal and over 11.0mmol/l two hours after a meal. Fasting glucose values are collected from individuals who have not eaten for at least eight hours before the test. As you progress from glucose intolerance and acquire Type 2 diabetes, you have high blood sugar, advanced insulin resistance (your body does not respond to insulin), and lower levels of insulin production from the beta cells of the pancreas (Figure 9B). Symptoms may include feeling thirsty, the need to pee a lot, feeling hungry, feeling tired and infections that take a long time to heal. If left untreated high blood sugar can lead to heart disease, stroke, blindness, kidney failure, gout, and poor blood flow in the feet and legs, which may lead to amputation.

Another measure of blood glucose is called the glycated haemoglobin, or HbA1c, measure. Glycated haemoglobin is made when the glucose in your body sticks to your red blood cells. HbA1c provides an average blood glucose level for the last 2–3 months. Diabetes

UK advise that the normal range for people without diabetes is an HbA1c level below 6.0 per cent (below 42mmol/mol). Prediabetes HbA1c levels are in the 6.0–6.4 per cent (42–7mmol/mol) range, and HbA1c levels over 6.5 per cent (48mmol/mol) indicate diabetes. HbA1c is an important indicator of long-term glucose control, providing a measure of chronic elevated glucose and the risk of long-term diabetes complications.[822] Such measures are likely to replace the current 'snapshot' before- and after-meal measures.

It isn't exactly clear what triggers insulin resistance, and why some people are affected and others escape. A friend of mine expressed this as 'dodging the bullets'. However, a family history of Type 2 diabetes, being overweight (especially around the waist), inactivity and SCRD all raise the risk. So, as usual with disease, a mixture of genetic risk factors and environmental interactions are involved. In the early stages of insulin resistance, the numbers of beta cells in the pancreas (Figure 9B) increase, producing more insulin to compensate for the decreased insulin sensitivity and raised blood glucose. But as the disease continues the beta cells die. An individual with full-blown Type 2 diabetes will have lost about half of their beta cells. The point is, if you have glucose intolerance or Type 2 diabetes, you need to act urgently to reverse these conditions.

In an attempt to prevent metabolic syndrome, many people diet to lose weight. But this can be frustratingly difficult. For about 98 per cent of us, weight loss from dieting alone is followed eventually by weight gain.[823] And the problem relates to a fundamental part of our physiology called 'homeostatic control'. Homeostasis is the process whereby the body maintains a more or less stable environment such as temperature, hormone levels, blood pressure, heart rate, blood glucose or calorie intake. Our bodies continuously monitor these essential processes so that they are maintained at a particular 'set point'. A significant change from the set point will normally be corrected (increased or decreased) as homeostatic mechanisms are triggered in the form of a 'negative-feedback loop' – where the response to a change reverses the direction of the change. For example, an increase in body temperature above a set point triggers a change in our physiology that then lowers temperature. Conversely, a significant drop in temperature feeds back to increase temperature.

For clarity, the term 'set point' implies that it is fixed. But this is deeply misleading, and dates back to a time when the set point was considered fixed, and changes in the set point were thought to indicate disease. This idea originates with Claude Bernard (1813–78), the 'father of physiology'. He wrote: 'All the vital mechanisms, varied as they are, have only one object, that of preserving constant the conditions of life in the internal environment.' However, as we have discussed throughout this book, the circadian system is the agent for change across the 24-hour day, and homeostatic set points are finely controlled by the circadian system. Changed set points anticipate the altered demands of activity and rest. Body temperature may average 37°C, but at 4 a.m. it is closer to 36.5°C or lower, and at 6 p.m. it is closer to 37.5°C or higher. Resting heart rate is around 64 beats per minute at 5 a.m. but around 72 beats in the early afternoon.[824] Leptin, which suppresses appetite, is low during the day, when we are active and need to eat, but high at night (suppressing hunger), when we are normally asleep and unable to eat.[803] Until recently, the failure to inform medical students during their training that homeostatic set points are under circadian control has resulted in both misdiagnosis and inappropriate medication.[825] Claude Bernard also wrote shockingly about his animal experimentation (vivisection):

> The physiologist is no ordinary man. He is a learned man, a man possessed and absorbed by a scientific idea. He does not hear the animals' cries of pain. He is blind to the blood that flows. He sees nothing but his idea, and organisms which conceal from him the secrets he is resolved to discover.

Deeply awful now, and also to many in the nineteenth century too. To the horror of his wife and daughters, Bernard apparently dissected the family dog. His wife left him in 1869 and went on to campaign against the practice of vivisection. With such experiences you can understand why!

Returning to homeostasis: homeostatic 'correcting' mechanisms arise from a negative-feedback loop, where a drop is followed by a corrective increase, and an increase is followed by a loss; the change in physiology alters the direction of the change and things stay where they are. And this is the problem when we diet. When we try to 'lose weight' we lose stored fat. But our brain then detects this depletion of stored calories and then starts to correct for this loss. As mentioned previously, leptin is produced in adipose tissue (fat cells), and as we reduce the amount of stored fat through dieting, we produce less leptin (Figure 9). Less leptin means that after a meal you still feel hungry. Along with less 'I'm full' leptin, the body releases more ghrelin from the stomach, which increases hunger, so you eat more (Figure 9). In addition, as the body detects we have less fat, this stimulates the thyroid gland to produce less thyroxine to lower metabolic rate and so burn fewer calories when we sleep (remember, thyroxine is under circadian control and normally shows a higher release at night). This allows the body to save calories – which leads to an increase in stored fat. So to lose more weight, you have to cut back the calories even further just to maintain the weight loss you have achieved. You feel hungry, and especially for sugars, and you have a slower metabolism, especially at night. The set point for your levels of body fat has been 'defended' and not changed. The faithful brain 'thinks' that its owner is being starved. So, with this somewhat depressing information in mind, what can be done *in addition* to dieting to achieve a healthier metabolism?

Working with our circadian system can help metabolic health in four important areas: the role of activity and the timing of exercise; the prevention of SCRD; appropriate eating times; and working with the circadian rhythms of our gut bacteria.

## The Role of Activity and the Timing of Exercise

In the UK in the 1950s, the average woman was a size 12 and had a 27-inch waist. Today, the average woman is a size 16 and has a 34-inch waist. This is a striking difference and is partly related to levels of activity. A study published in 2012 suggested that the 1950s woman burnt around 1,300 calories a day compared to 670 calories today. Most of this calorie burn is related to running a household. And, to be clear, I am not suggesting that we should return to that singular pattern of domestication, but it makes the point that you can help control how many calories you burn by your level of physical activity. The more active you are, the more glucose and stored glucose (fat) you turn into energy. Aerobic exercise such as walking, running, cycling, rowing or using an elliptical machine is very effective at burning calories, and a minimum of 30 minutes, five days a week, is recommended. In parallel, strength training such as lifting weights, working with resistance bands, climbing stairs, push-ups, sit-ups and squats two times a week builds muscle. And muscle burns more calories than fat, so increasing your muscle mass will help you liberate stored calories from fat stores. Also, let's not forget gardening. Three hours of serious gardening can have the same calorie burn as an hour in the gym. The typical gardener spends more than five hours a week tending the garden, burning around 700 calories. Gardening will also help you find a place for that bottle of urine you may have collected overnight (chapter 8).

Exercise is clearly important, and two questions need to be addressed: is there an optimal time of day to exercise?; and could there be additional benefits from exercise in helping to prevent SCRD?

## *The timing of exercise for a maximum calorie burn*

Our capacity for exercise, along with peak performance, changes across the day. Studies in both humans and mice show that muscle strength and the ability of muscle cells to take up oxygen and glucose for respiration varies across the day.[826-7] In general, our muscle strength and muscle respiration peak in the late afternoon and early evening.[828] This helps explain why peak athletic performance occurs in the late afternoon and early evening.[829] On average, muscular strength coincides with our peak in core body temperature, which is on average 4–6 p.m. (Figure 1, p. 20). Increased body temperature will act to increase metabolic rate and muscle power. Even when resting, we burn about 10 per cent more calories in the late afternoon and early evening, compared to the early morning.[830] And, in general, we achieve better athletic performance in the afternoon/evening compared to morning. But is this the best time to do exercise to burn calories? There are two complications here. The first relates to your **chronotype** (Appendix I). An interesting study on athletes showed that chronotype has an important impact upon peak athletic performance. Morning, intermediate and evening chronotypes performed better as the day progressed, but the late types showed a *much* better performance later in the day, with as much as a 26 per cent difference between 7 a.m. and 10 p.m.[831] The second issue relates to the **metabolic status** of your body. Afternoon/early-evening exercise makes intuitive sense for most of us, but there is good evidence that exercising first thing in the morning, on an empty stomach (you can drink water) may be better for some individuals. During morning exercise, before breakfast, the body is still using stored fat as fuel, and so exercise at this time of day will burn more fat.[832-3] So here is the dilemma – you are more likely to burn stored fat first thing in the morning. However, chronotype-dependent exercise, when your body is better adapted for physical activity,[830] will allow more vigorous exercise. The bottom line is that if your chronotype pushes you towards being a morning type and exercise is more appealing/easier for you in the morning, then you should exercise at this time – before breakfast. However, if you are an evening type, and morning exercise is difficult, develop an afternoon or early evening routine. An added advantage of afternoon/evening exercise is that it may help prevent injuries as muscles are already 'warmed up', preventing muscular strain. Aware of these effects, some of my colleagues tell me that they try to do a daily short morning exercise of around 20 minutes, followed by a 30- to 40-minute block of exercise later in the day. As discussed in chapter 6, try not to exercise too close to bedtime as this will increase core body temperature and delay sleep onset. In addition, very vigorous exercise close to bedtime can cause a spike in cortisol and an activation of the stress axis. If cortisol is still high immediately before bed, then this will act to delay sleep (chapter 4). One final tip is that if you walk for 30–45 minutes after an evening meal, rather than before, this can assist in the control of blood glucose and hence weight loss.[834-5]

## The effects of exercise on the clock and reducing SCRD

As discussed in chapter 3, light at dusk delays the circadian clock (you get up and go to bed later), while light at dawn advances the clock (you get up earlier and go to bed earlier), and light in the middle of the day has little impact. But, in addition to light, exercise can also play a role in helping entrain circadian rhythms. This has been known for a long time in rodents such as hamsters and mice,[836] but the evidence has been less clear in humans. However, in a recent report about 100 people were studied to see whether exercising at different times affected their sleep/wake timing. The exercise, which consisted of one hour of walking or running, was undertaken at different times during the day (1 a.m., 4 a.m., 7 a.m., 10 a.m., 1 p.m., 4 p.m., 7 p.m. or 10 p.m.) over three days. Those individuals who exercised between morning and mid-afternoon (7 a.m. to around 3 p.m.) woke earlier, whereas those exercising later

(7–10 p.m.) woke later, with little impact on sleep timing when exercise was undertaken between 4 p.m. and 2 a.m.[837] So routine morning/early-afternoon exercise will help you get up earlier in the morning – which could be useful for adolescents (chapter 9). All this means that appropriately timed light exposure, combined with appropriately timed exercise, can help reinforce entrainment and act to stabilize the circadian system, preventing SCRD. And as discussed previously and again below, reduced SCRD leads to improved metabolic health.[838] With this in mind, the practice of keeping individuals confined to bed during the day in the care home or hospital setting, when bed rest is not actually required, may be easier for staff but unhelpful for the patient. Activity during the day (morning/early evening) should be encouraged where possible to promote circadian health. Which leads me to the next topic.

## *Preventing SCRD Promotes Good Metabolic Health*

As discussed in chapter 12, sleep loss is associated with increased release of the 'hunger hormone' ghrelin from the stomach and reduced release of the 'satiation hormone' leptin from adipose tissue.[801] The net result is an increase in appetite and the consumption of sugar-rich foods, and an increased risk of metabolic syndrome.[839] Poor sleep is also associated with higher levels of cortisol in the evening, which will increase levels of blood glucose (Figure 9A, p. 275), which, if not used, will be converted into stored fat,[840-41] predisposing individuals to weight gain and obesity. And, just another reminder, these conditions can lead to obstructive sleep apnoea (OSA), causing further poor sleep (chapter 5). This all emphasizes the point that sleep loss is much more than feeling tired at an inconvenient time, but associated with major health problems (Table 1, p. 77), not least metabolic syndrome. The links between SCRD and metabolic syndrome have been demonstrated repeatedly. For example, mice without a circadian clock rapidly develop insulin resistance, glucose intolerance and obesity.[842] It also seems likely that the loss of synchrony between central and peripheral clocks (internal desynchrony), and a loss of amplitude or robustness of circadian clocks, can lead to insulin resistance,[843] which helps explain why the elderly are more vulnerable to Type 2 diabetes, as a reduction in the amplitude and internal desynchrony are both common features of the aged circadian system. And, as discussed above, exercise can help consolidate circadian timing, and chapter 6 provides general advice about how to reduce SCRD. However, there is an important additional link between SCRD and metabolism relating to alcohol consumption.

## SCRD and alcohol

SCRD can turn moderate drinkers into heavy drinkers. This has been shown in long-term night shift workers[844] and the chronically tired, who use alcohol to induce sedation in the mistaken belief that it promotes normal sleep[481]. In addition, a disrupted circadian system exaggerates the toxic impact of alcohol on metabolism, as shown in mice with mutant clocks. Circadian-defective mice were provided with alcohol in their drinking water and had a greater level of fatty liver disease than mice with normal clocks. They also showed increased leakiness of the gut, which allowed endotoxins (fragments of broken-down bacteria) to pass into the blood, causing endotoxemia, which causes multiple illnesses, including liver damage.[845] This probably means that individuals with SCRD, such as night shift workers, long-haul aircrew and members of the business community are more vulnerable to metabolic damage from alcohol. They drink more and the alcohol causes more liver damage. Alcohol also has direct effects on the molecular clock. This has been shown in mice, in which alcohol consumption advanced the timing of liver clocks, but left the SCN unchanged. The result was to uncouple the liver clocks from the SCN. Also, alcohol flattens the amplitude of the liver clocks.[846] Such alcohol-induced internal desynchrony between the SCN and liver, combined with weaker circadian regulation of liver metabolism, acts to disrupt glucose metabolism and promote fatty liver disease and the other metabolic abnormalities associated with insulin resistance.[846] This will also increase vulnerability to infection, as discussed in chapter 11. Alcohol is also likely to alter the circadian rhythms of other organs. For example, the circadian-driven rhythm of core body temperature is advanced if alcohol is consumed in the evening[847], and the amplitude of the rhythm in core body temperature, like liver clocks, is almost halved[848]. Because circadian temperature amplitude has been shown to decrease in mood disorders[849], and that sleep is linked to the rhythm in core body temperature[850], this has led to the suggestion that a flattening of the temperature rhythm by alcohol may contribute to sleep disruption, and, by extension, mood disruption[848]. I think that this is an interesting idea, but it needs to be explored further.

The impact of alcohol on sleep and depression goes way beyond changes in core body temperature. Alcohol also disrupts neurotransmitter and hormone release within the brain, which acts to alter sleep structure and mood directly. Alcohol causes brain activity to slow down and induces feelings of relaxation and sleepiness, but excess alcohol can lead to poorer sleep and marked insomnia. Alcohol reduces REM sleep during the first part of the night,[851] alters slow-wave sleep, decreases sleep quality, results in a shorter sleep duration and produces more fragmented sleep. Alcohol can also make the symptoms of OSA worse by relaxing the muscles at the back of the throat.[852] Because alcohol can cause insomnia, daytime sleepiness is often a problem. This leads to a cycle of consuming caffeine-rich drinks during the day to stay awake, and then using alcohol as a sedative to offset the effects of these stimulants at night – the so called 'sedative–stimulant feedback loop'.[481] The point is, the increased vulnerability to metabolic syndrome that results from SCRD is made worse by alcohol consumption, and the tendency to drink more alcohol exacerbates SCRD.

## Optimal Eating Times

### Meal timing and 'chrononutrition'

The Sephardic Jewish philosopher, astronomer and physician Moses ben Maimon (1138–1204), usually known as Maimonides, remains a somewhat controversial figure, with a lasting legacy for Jewish philosophy and faith. But for our discussion he is remembered for the quotation 'Eat like a king in the morning, a prince at noon, and a peasant at dinner.' This philosophy makes Maimonides the founding father of the branch of circadian research called 'chrononutrition'. Chrononutrition incorporates the idea that the time of food intake, along with the amount and type of food you eat, is critical for our overall metabolic and physical health.

We now appreciate that the same meal consumed at different times of day can produce very different levels of blood glucose due to the circadian-driven changes in glucose uptake and metabolism.[843, 853] This has very important consequences, not least for those of us who eat most of our daily calories in the evening. For clarity, I define evening loosely as between 6 p.m. and bedtime. Evening eaters are at a greatly increased risk of impaired glucose tolerance, Type 2 diabetes, weight gain and obesity.[854-6] A detailed study compared individuals who were given the same 20-week reduced-calorie diet, but consuming most of their calories either early in the day or late in the day (between 6 p.m. and bedtime). Those who ate late in the day lost less weight and showed slower weight loss compared to individuals who ate early in the day. Both groups were equally active and had the same length of sleep.[857] A similar study showed that more weight was lost by eating calories in the morning compared to evening, and this was also associated with lower blood glucose, reduced glucose intolerance and reduced levels of Type 2 diabetes.[858-9] In addition, high-energy food intake in the evening and fasting in the morning (as in night shift workers or the business sector) has been associated with the development of obesity, while just skipping breakfast is linked to worsened glucose intolerance.[860]

Importantly, the same meal in the evening results in higher levels of blood glucose (hyperglycaemic response) compared to identical meals that are eaten in the morning.[861] This increased glucose intolerance in the evening is the result of a lower circadian-driven insulin release from the pancreas, as well as circadian-driven changes in insulin resistance by the liver.[856] In tightly controlled laboratory studies, glucose *intolerance* has been shown to increase from the morning to the evening in healthy individuals, leading to raised levels of blood glucose. In a recent study from Harvard, young individuals were given the same meal in the morning at 8 a.m. and then again 12 hours later at 8 p.m. Blood glucose levels were significantly higher (17 per cent) after the evening meal, demonstrating that the healthy subjects had higher glucose intolerance in the evening. The researchers then simulated a night shift work pattern, where participants were only allowed to sleep during the day. After only three days of this circadian rhythm disruption, glucose intolerance was even worse in the evening. Clearly, circadian misalignment exaggerates glucose intolerance and increases the risk of Type 2 diabetes and obesity.[862] What could be responsible? It seems likely that some of the health problems seen in night shift workers could arise from an uncoupling of the SCN and the peripheral clocks due to conflicting signals – internal desynchrony. The SCN is 'set' to the light/dark cycle and promotes sleep at night. Eating at a time when the SCN 'thinks' it's sleep time means that the metabolic regulation of physiology becomes misaligned with the peripheral clocks. The clocks in the liver, adipose tissue, pancreas and muscles are shifted by feeding signals and adopt an entirely different timing from the SCN. The circadian network comprising the SCN and peripheral clocks has evolved to work together and instruct the metabolic axis to get the correct materials in the right place, in the right amount, at the right time of day. A breakdown in the circadian network results in metabolic meltdown.

Returning to the wisdom of Maimonides and when to eat: interestingly, there has been a gradual change in our eating habits over the centuries. In England and Europe in the Middle Ages (around 1100–1500) the main meal of the day, or 'dinner' (from Old French *disner*, meaning 'to dine'), had moved from early in the day to around noon, and this is true for both aristocrats and peasants. But as the use of artificial lighting in the form of candles, oil lamps and then electricity was adopted, first by the very wealthy and finally by the poor, dinner (main meal of the day) shifted later. This was reinforced by industrialization and changes in working practices. The main meal of the day occurred after the 'breadwinner' had commuted home from work. In the north of England, 'dinner' is still at lunchtime and 'teatime' is the last meal of the day, also called 'supper'. Today, the breakdown of the nuclear family, long commutes, irregular working times, increased night shift work, the pressures of school work and the availability of easy to prepare (e.g. microwaveable) highly processed food have all pushed the major sugar-rich meal of the day to an irregular mid- to late-evening time slot. If you were designing a schedule to be particularly bad for our circadian-regulated metabolism, this would be it.

## Gut Microbiota – Working with Our 'Bugs'

Our bodies are not our own. They are home to a vast collection of micro-organisms (microbiota) including bacteria, fungi, viruses and protozoa. In fact, our own cells make up only about 43 per cent of the body's total cell count. If you think that's a lot, earlier estimates suggested that for every human cell there were 10 microscopic colonists – this now seems like an overestimate, although you will still see this number frequently quoted.[863] These micro-organisms are mostly made up of bacteria located in the gut. And these bacteria contribute to how the lining of the gut functions, keeping unwanted substances out of the bloodstream but allowing nutrients through. A key realization in recent years has been that SCRD can change our microbiome. Remarkably, this altered microbiome can then change our metabolism, energy balance and even immune pathways. Such disruption can lead to the problems associated with metabolic syndrome. As discussed in chapter 12, there is a global rise in metabolic syndrome, and it has been argued that this is driven in large part by our so-called 'Western lifestyle', characterized by the consumption of a diet high in fat and sugar (including alcohol) but low in plant fibre, along with a lack of exercise and, critically, increasing levels of SCRD.[864] This lifestyle may have originated in the West but is now endemic across the world. So what is the link between gut bacteria, circadian rhythms and a disrupted metabolism?

Although most bacteria were originally thought not to be able to generate circadian rhythms, we now know that this is wrong. Bacteria do show circadian changes in their biology[59, 865], and the bacteria that live in our gut can synchronize their circadian rhythms to the circadian rhythms of *our* gut cells[866-7]. If that were not amazing enough, even more remarkable is that some of our gut bacteria also 'talk back' to us.[868] We know this because the loss of gut bacteria disrupts the circadian biology of the cells that line our gut (intestinal epithelium).[866] It seems this communication results from a variety of signals from the bacteria. Physical contact of our intestinal epithelial cells with the proteins in the cell walls of the bacteria may provide an important cue[869], along with chemical signals produced by bacteria themselves[870]. This cross-talk between us and some of our gut bacteria seems to be very important, not least because obesity and metabolic syndrome are associated with changes in gut bacteria.[871] An increase in 'bad bacteria' such as *Streptococcus* and *Clostridium* can promote metabolic syndrome. By contrast, an improvement in metabolic syndrome, along with a reduction in obesity, is associated with fewer 'bad bacteria' and an increase in 'good bacteria' such as *Akkermansia*. Many of us will be familiar with *Akkermansia muciniphila* as these bacteria are used in commercially available probiotic supplements. And, unlike many supplements, there is now good evidence that they are genuinely useful and have a positive impact upon our metabolism.[872] As is usual, the most detailed studies have been in mice. For example, obesity in mice can be reversed by delivering gut bacteria from non-obese mice into the gut of obese mice.[873] The link between metabolic syndrome and gut bacteria seems really important, but how is the circadian system involved?

In both mouse and human studies, SCRD changes the gut bacteria and promotes metabolic dysfunction[874], and the transfer of gut bacteria from mice with SCRD into non-SCRD mice will cause metabolic abnormalities[875]. Remarkably, gut bacteria have been shown to programme the circadian rhythms of metabolic activity in gut cells. Gut epithelial cells have receptors called 'pattern recognition receptors', or PRRs, which recognize the surface of 'friendly bacteria'. Activation of these PRRs can synchronize the molecular clock in the gut epithelial cells and regulate genes linked to the regulation of metabolism.[876] Mice with defective PRRs show abnormal rhythms in their metabolic pathways.[870] Very recently, it has been found that chemical signals from bacteria can have a direct action on gut epithelial PRRs. For example, several 'friendly bacteria' have been shown to release metabolites that increase the amplitude and lengthen the period of gut cell circadian clocks.[877]

There is now strong evidence that SCRD can alter gut bacteria, which in turn can lead to disrupted clocks within our gut epithelia. And this circadian disruption can give rise to metabolic disorders. Fully understanding these links will have an important role in the development of new therapeutics that decrease the personal and economic burden of metabolic syndrome. However, the importance of our gut bacteria may extend well beyond just the gut. There is some evidence that signals from gut bacteria can influence the circadian machinery of the liver and circadian-linked liver metabolism.[878] There are even links between gut bacteria and the circadian regulation of the immune system. Mice who have no gut bacteria show markedly abnormal immune responses[879], including abnormal populations of T-cells and B-cells (chapter 11 and Appendix II)[877]. There are hints that such immune impairment, and the loss of circadian precision, affects our ability to fight infection and in the development of auto-immune diseases such as multiple sclerosis.[880] Finally, there also seems to be a 'microbiome–gut–brain axis' that helps regulate our sleep and mental states. We know that if we experience SCRD this will disrupt our gut microbiome, but the recent suggestion is that the disrupted microbiome leads to disrupted sleep, with an increased risk of depression.[881]

It's still early days, and it is difficult in some cases to tease apart cause and effect, particularly when talking about the links between bacteria, sleep and depression. However, what is clear is that our circadian rhythms influence the circadian rhythms of our gut bacteria, and that the circadian activity of the bacteria we accommodate in our gut impacts upon our metabolism. In view of the fact that around 50 per cent of the cells of our body are bacterial cells, it seems very likely that in the coming years we will discover more and more links between human and bacterial rhythms, and that these links will be recognized as very important for our health. We are on the verge of another exciting branch of medicine – the 'circa-microbiome'.

## Questions and Answers

### 1. Is it true that our metabolic rates can't be changed?

No. While it's true that genetics help determine our metabolic rates, we can increase metabolism by increasing lean muscle mass. Muscle is metabolically active, which means that people with lean, muscular bodies need more energy to function than people with a higher percentage of body fat. So exercising regularly at a time that is best for your chronotype will help you burn fat and increase muscle mass. This will act to increase your metabolic rate.

### 2. Why do you put on weight as you age – is this anything to do with the clock?

Metabolism changes markedly as we age, making it easier to gain weight. Part of the reason is that the amplitude (robustness) of the circadian rhythms that regulate metabolism decreases, and the alignment of all the multiple circadian rhythms involved in metabolism is less well synchronized. Collectively, metabolism is less tightly controlled, and this dysregulation paves the way for weight gain and obesity. The situation is similar to the problems experienced by night shift workers. In addition, my clinical colleagues suggest that as we get older we become much more rigid regarding routines, never missing a meal. So we tend to eat because it is a 'mealtime' and not because we are hungry.

### 3. What can we do to encourage our 'friendly' gut bacteria?

The first point to make is that SCRD encourages the growth of 'unfriendly' gut bacteria such as *Salmonella*,[874] and circadian health encourages 'friendly' bacteria in the digestive tract and robust metabolic health. An established population of friendly bacteria will outcompete pathogenic 'unfriendly' bacteria for food and space and in some cases will change the local gut environment, making it more difficult for pathogens to survive. It is also worth noting that many antibiotics usually only target bacteria (both good and bad indiscriminately), and do not kill fungi, which can lead to an 'overgrowth' of fungi and cause yeast infections.[882] As a result, following antibiotic treatment or SCRD, the re-introduction of friendly bacteria, such as *lactobacilli* found in unpasteurized yoghurt, can help restore a normal balance and aid metabolic health.

### 4. What is the impact of Ramadan, and eating after dusk, on circadian rhythms and health?

During the month of Ramadan, Muslims are not permitted to eat or drink during the hours of daylight. Modern-era Ramadan practices in Saudi Arabia have been associated with disturbed feeding and sleep patterns, increased daytime sleep, and staying awake to eat food and drink water until dawn. Ramadan was associated with high levels of cortisol in the evening, when cortisol is normally low (Figure 1, p. 20), and increased insulin resistance, whereby muscles, fat and liver cells don't respond well to insulin and can't easily take up glucose from the blood. This study suggested that such changes might contribute to the high levels of obesity, hypertension, metabolic syndrome, Type 2 diabetes and cardiovascular problems observed in the Kingdom of Saudi Arabia.[883] However, a recent review suggested that when meal timings are confined to the early-evening and predawn periods, combined with an adequate night of sleep, cardiovascular and metabolic problems may be less obvious.[884] This is clearly a complex issue and requires a large-scale study that considers the multiple factors such as age, health status, job pressures, early versus late eating times, and the impact of SCRD.

# 14.

# The Circadian Future
## What happens next?

Somewhere, something incredible is waiting to be known.

Carl Sagan

The overwhelming message in this book has been that circadian rhythms are embedded within every aspect of our biology, and that we ignore this rhythmic biology at our peril. I have discussed the actions, and the reasons for these actions, needed to enhance our circadian and sleep health, the point being that such 'corrective measures' will act to improve our cognition, overall wellbeing, metabolism, fitness and life expectancy. These actions are not particularly demanding, especially in view of the benefits. Putting the individual personal costs aside for one moment, in pure economic terms society needs to take the impact of sleep and circadian rhythm disruption (SCRD) seriously. A detailed study ('Asleep on the Job') by the Sleep Health Foundation of Australia estimated that inadequate sleep cost the Australian economy in 2016–17 around 26 billion Australian dollars. The gross domestic product of Australia that year was around 1,500 billion Australian dollars. So SCRD created an enormous financial burden for the Australian economy, and it is likely that other nations will take a similar percentage 'hit'.

Mahatma Gandhi said: 'The future depends on what you do today', and I want to use this final chapter to discuss what we could, should and have started to do to improve our circadian health. The first part of this chapter is a 'call to arms' to use education to change societal attitudes towards circadian rhythms and sleep. Education, across multiple levels of society, could allow a journey of personal responsibility that will improve the health of future generations. While this is critical, education is not always enough. The second part of this chapter considers how the new science of circadian rhythms is being used to develop novel therapeutics to correct SCRD. Today, there are multiple diseases and conditions where SCRD cannot be corrected, resulting in appalling health for the sufferer, and misery for their carers and family. New 'circadian drugs' are currently under development that will hopefully transform key areas of health.

## Changing Behaviour

In view of the personal and economic costs, why has circadian and sleep health not been embraced more enthusiastically by society as a whole? The answer to this question must be strongly related to education. And perhaps there are some parallels to be learnt from the campaign against smoking cigarettes. Education regarding the harmful effects of smoking cultivated a marked change in societal attitudes. Smoking moved from a fashionable and 'cool' practice to an activity that is now regarded, at least by most, as socially unacceptable and deeply irresponsible. Smokers are banished from the workplace and passive exposure to cigarette smoke is no longer tolerated. Analogous to smoking, the short- and long-term impact of SCRD on our individual health can be profound (Table 1, p. 77), while the impact of 'passive' SCRD on family, friends, colleagues and broader society can be devastating. In chapter 10 we considered some examples of major SCRD-related accidents including Three Mile Island nuclear plant, Chernobyl nuclear plant and the *Exxon Valdez* oil spill. Education has changed attitudes to smoking, and a similar educational strategy is now needed to address SCRD. If implemented, and successful, then those individuals who turn up for work boasting they did another 'all-nighter' will be regarded with the same contempt as smokers, and hopefully the machismo culture of long hours and little sleep will go the way of the ashtray.

A key place to start this change in attitudes will be in our schools. Across all levels of education, from schools and colleges to universities, there is only scant information about why our sleep and circadian rhythms are important; how sleep changes as we age; and in what ways our sleep biology can be affected by different societal and biological circumstances or events. Such information, appropriately packaged, should be taught and embedded within the school curriculum from an early stage, and I am reminded of the variously attributed quote: 'Give me just one generation of youth, and I'll transform the whole world.' The short- and long-term impact on health and wellbeing could be astounding.

Currently, if the topic of sleep is taught at all, it is because of dedicated and motivated teachers who try to carve out a few lessons in the gridlocked curriculum. And this is not easy. Appropriate and standardized teaching materials are lacking, and support from head teachers is often lukewarm, chilled by the constraints of fulfilling the burdensome demands of the National Curriculum, which does not include sleep education. Yet, as I have hoped to outline in this book, good sleep and circadian health will not only enhance cognition and educational performance, but also lead to major health advantages across the lifespan of the individual. This is fully recognized by many teachers. One teacher that our team worked with said: 'Sleep is the foundation on which all else we do at school is built.' Interestingly, child and student welfare is often trumpeted as a priority by decision makers, but sleep is seldom, if ever, discussed in this context.

In view of the obvious need, and support from many teachers, it has been deeply frustrating that our repeated attempts to develop a standardized set of teaching tools for the National Curriculum, addressing the consequences and correction of SCRD, have consistently hit a brick wall. Such proposals, developed in partnership with teachers, have not been considered a high enough priority by funders. This is all the more puzzling as the stated aims of one of the funders we approached is to 'raise the attainment and wider outcomes of 3 to 18-year-olds, particularly those from disadvantaged backgrounds'. Let us hope in the coming years that such laudable aims will embrace the importance of sleep and circadian health for our young people.

However, the need for SCRD education extends well beyond the classroom. Our front-line staff and 'key workers' in health and social care, including doctors, nurses, midwives, paramedics, and those in public safety and national security, including police, the armed forces and fire service employees, all have to combine profoundly demanding duties with the added burden of night shifts and extended work schedules. The impact of SCRD can be huge. The account below is from a serving police officer in the UK who described his experiences to me:

Earlier in my career, I was fortunate to survive an attack by a mentally ill individual armed with a knife. Although not physically injured, I couldn't sleep. The insomnia got worse and worse. Then doing shift work just made my sleep fall apart. Before work I would try and look after the kids but I would be desperate for a sleep before starting my shift. It would never happen; I'd be too anxious and angry. I fell into the habit of using far too much caffeine and gym pre-workouts to get through the night shifts. This ultimately led to poor physical health as I was so dehydrated I damaged my kidneys and I developed gout. I was really struggling at this point. I lost count of the times I nearly nodded off on the way home.

Eventually I asked for help from my GP [medical practitioner], who prescribed me Zopliclone [a sleeping tablet]. But this was not the solution for me. I was snappy at home. Lethargic. My moods would be up and down. I put weight on. I couldn't relax. Following the advice from a psychiatrist, I stopped working at night. For the first time in years, I remembered what it was like to feel normal: not to go crazy at the first sign of trouble. To stop going totally over the top. Not everything was a conspiracy against me, and as my sleep improved my mental health recovered. I became a better father, husband and professional. I had more capacity to do the things I loved. I took up hobbies and completed qualifications I wanted to do. I just had more about me.

I was fortunate, but some of my colleagues have been less lucky. I lost a young colleague who crashed his car on the way home after the night shift. I made a promise to this young officer, to his family and to his colleagues that I would use some of my extra capacity to highlight the impact of night shift work and alert colleagues to the dangers. I now tell people, don't deprive yourself of sleep as in the long term it is counterproductive. You may feel like you're getting stuff done but you are reducing your effectiveness. You are putting your life at risk.

The police officer who gave this account, now a friend, eventually resolved this difficult situation, recognizing that lack of sleep was a key contributing factor that needed to be addressed. He is now a senior member of the police force, as well as being a wonderful husband and father. But his colleague was not so lucky. This young policeman was not alerted to the dangers of sleep loss, and no provision was made by his employers to help him deal with his accumulated sleep debt. He fell asleep at the wheel and crashed his car on the way home after the night shift and died. Tragically, this is not uncommon. According to the Department for Transport in the UK, around 300 people each year are killed as a result of falling asleep at the wheel. As I have discussed in this book, SCRD is too often a bedfellow of death.

What about our health care workers? An Institute of Medicine report from the USA estimated that as many as 98,000 deaths annually are the direct result of medical errors,[885] and that night shift work and long working hours are a major contributing factor to this problem. The death of an 18-year-old woman while under the care of medical residents in an emergency department in New York in 1984 stimulated the drive for reform of resident hours.[886] But it has taken a long time to achieve a significant change for medical residents – those individuals who have finished medical school and are receiving training in a specialized area, such as surgery. In their first year, US residents were frequently expected to work a 24-hour shift every third night, which amounts to 96 hours per week. Two studies examined the impact of these schedules and found that sleep-deprived surgical residents showed twice the number of errors in a simulated surgical procedure.[887-8] Another study showed that residents who worked more than 80 hours a week were 50 per cent more likely to make a significant medical error that harmed the patient, compared to those working less than 80 hours.[889] Currently, regulations drawn up by the Accreditation Council for Graduate Medical Education in the USA have capped the working week for residents at 80 hours. But there is good evidence that actual working hours are frequently under-reported.[890] In the rest of the world, junior doctors are expected to work much shorter shifts. For example, the European Working Time Directive set a maximum working week of 48 hours for all workers, including junior doctors.[891] But again, in the UK at least, many junior doctors exceed this cap.[892] Certainly, reducing working hours and allowing more sleep will improve cognition,[567] but the problems of night shift work and extended working hours, as discussed in chapter 4, won't go away. Even with reduced working hours, a fairly recent survey of junior doctors reported that 60.5 per cent stated that they had made an error which 'played on their mind' and that a 'high workload' was the most commonly identified contributory factor.[893] And, worryingly, a recent study from the UK reported that 57 per cent of junior doctors had experienced either a motor vehicle crash or a near miss after working on the nightshift.[894]

While the working hours of junior doctors are better than they were 20 years ago, those in the banking sector, which means those key individuals looking after our pensions and financial futures, may be much worse. In March 2021, the BBC reported on a survey of first-year bankers at the investment bank Goldman Sachs who warned that they might resign unless their working conditions improved. These first-year bankers averaged 95 hours of work a week and slept around five hours a night. One respondent in the survey said: 'The sleep deprivation, the treatment by senior bankers, the mental and physical stress . . . I've been through foster care and this is arguably worse.' This is not unique to Goldman Sachs. A recent review reported that mental health problems had increased markedly in the banking sector, and were likely to be linked to the increased stress and lack of sleep due to the pressures of work.[895] Problems began with too little sleep, then anxiety and depression, followed by maladaptive behaviours such as excessive alcohol consumption, and ended in job burnout – all symptoms illustrated in Table 1. The Banking Standards Board (BSB) was established by the UK government to improve behaviour in the banking sector after the financial crisis and a subsequent interest rate rigging scandal. A survey of bankers published in 2020 by the BSB showed that nearly 40 per cent of respondents said they slept for six hours or fewer each night, with almost 30 per cent saying they felt tired at work every or almost every day. In response to such findings the BSB said: 'Given the importance of sufficient sleep not only for physical and mental health, but also for the ability to exercise professional and ethical judgement, this may be something that the industry wishes to explore further.' I think the penny has dropped at the BSB, but wonder about those in government. I have long wanted to assess SCRD in our politicians.

It is perplexing that while there is raised awareness about the importance of sleep, and the consequences of sleep disruption, in the media, decision makers across all sectors of society have not done much about it. It is noteworthy that, at the time of writing this book, Australia has yet to act on the 'Asleep on the Job' report and the 26 billion Australian dollars lost every year due to sleep loss by the workforce. And the COVID-19 crisis is not a real excuse. The past two years could have been spent getting this sorted out. So what *should* be done? In my view the first action is to develop evidence-based and sector-specific advice and educational tools that address SCRD. This should be established in schools as part of the curriculum, providing long-term knowledge to explain why SCRD occurs, why it is likely to be a life-long risk, and what to do to mitigate SCRD as we change and grow older, and as our circumstances alter over time. In parallel, employers have a duty of care to act in three important ways: alert their workforce to the dangers of SCRD; not to promote or encourage SCRD in the workplace; and where possible to mitigate the effects of work-related SCRD. In the short term, there is no 'magic bullet' to defend against the impact of night shift work or work-related SCRD. Employers and employees have to accept that there will always be significant health consequences, especially with night shift work. Currently, and as outlined in chapter 6, the best we can hope to achieve is a reduction in the severity of symptoms – but this is still overwhelmingly a critical and important thing to do – and do now: 'The future depends on what you do today.'

One last point. Because, currently, we can only mitigate some of the problems of SCRD, society needs to consider very carefully the circumstances in which the consequences of SCRD are balanced against the gains. Just because we can run a 24/7 economy, encompassing all sectors of work, should we? And ignoring the moral dimension, in view of the economic costs discussed at the beginning of this chapter, is it even cost-effective for society long term in view of the lost productivity due to poor health? Such decisions must emerge from evidence-based discussions involving scientists, government, industry and above all the workforce, and, hopefully, these discussions will take place and a consensus will emerge, before litigation polarizes and derails any constructive debate.

## When Behavioural Change is Not Enough

Much of this book has considered how our actions and behaviours can inflict, minimize and sometimes resolve the problems of SCRD. However, there are circumstances in which severe SCRD occurs, and there is very little we can do about it. Such is the case in profound blindness and neurodevelopmental diseases such as attention deficit hyperactivity disorder (ADHD) (Figure 4, p. 103). This is also true for severe dementia, as discussed in chapter 8. Such conditions produce a level of SCRD that devastates the lives of both the individual and the people they live with. Some personal testaments will be used below to illustrate how very difficult these conditions can be. But there is a real and exciting flame of hope. I will finish this chapter, and this book, with a look to the near future and the research being undertaken to develop new drugs that are being designed to correct SCRD in multiple health conditions. Let's start with neurodevelopmental disorders.

### Neurodevelopmental disorders (NDDs)

These are a group of disorders caused by abnormalities in early brain development that result in marked behavioural and cognitive changes. They are often, although not always, caused by genetic conditions. NDDs occur in approximately 1–2 per cent of the general population, and common types include: intellectual disability; other learning disabilities; cerebral palsy; autism spectrum disorder; and attention deficit hyperactivity disorder (ADHD). Specific conditions include Smith-Magenis syndrome, Angelman and

Prader–Willi syndrome, Rett syndrome and a huge range of other genetic conditions. Typical problems for such children are speech and language difficulties, problems with movement, memory and learning difficulties, and behavioural problems.[896] A striking feature of NDDs is that up to 80 per cent of children with these conditions have some form of severe SCRD.[896] Disrupted nighttime sleep is accompanied by poor daytime performance, including increased disruptive behaviour, and poorer cognition, growth and overall development. The nature and development of SCRD across different NDDs are very variable, and managing SCRD in children with NDDs is difficult for both the child and for the family. My colleague in Oxford, Andrea Nemeth, who is Professor of Neurogenetics and is based at the Oxford Centre for Genomic Medicine, very kindly provided the following account of how difficult SCRD can be in NDDs for both the child and the wider family:

In children with neurodevelopmental disorders (and ongoing into adult life), the normal difficulties in developing routine sleep habits can be compounded and magnified. The effects on the child and family may be catastrophic. Some will never settle into a sleep routine despite all efforts. On one day they may fall asleep at 7 p.m., on another day, not until 3 a.m. There are no patterns and no obvious reasons for the differences. The children find it difficult to get to sleep and stay asleep, effects that may be disease or mechanism specific. During the night, they wake their parents or carers on a regular basis. In the words of one parent, 'sleep deprivation is a way of life', and managing the problem takes over family life. The children can have little appreciation of their environment, requiring constant supervision, day or night. Some parents have special 'sleeping tents' to prevent the child from wandering unsupervised. By morning, the child and their carers are exhausted. The carer may face an entire day's work and may need to look after other children. Many parents, often the mother, end up leaving their jobs. The affected child will be tired during the day and doze off unless monitored, leading to a vicious cycle of going to sleep later and waking later. The exhaustion contributes to challenging behaviours and may reduce already limited abilities to engage in educational or social activities. If sleep hygiene programmes fail, then medications may be tried such as melatonin, sedating or other CNS active drugs, but these can cause side effects such as drowsiness, additional sleep problems, behavioural problems and cognitive impairment. Furthermore, the evidence base for efficacy is not strong. For some families with severely affected children, they have no option but to place the child or young adult in an institution which can provide 24-hour care, leaving the affected individuals isolated from their families as well as having major financial implications for both families and society.

In this account Professor Nemeth mentioned that medications have been used as treatments to address SCRD, and so it is worth considering a few of these in a bit more detail here. **Iron supplementation** is sometimes used in general for sleep-related movement disorders such as periodic limb movements (chapter 5), and there is some evidence that low iron (serum ferritin) levels may occur in children with ADHD and autism spectrum disorder (ASD). However, there is no clear evidence that iron supplementation improves periodic limb movement during sleep in these conditions, and may be counterproductive as it may cause gastrointestinal (gut) problems.[896] **Melatonin**, the major neurohormone released by the pineal gland (Figure 2, p. 24) has often been used for the management of SCRD in children with NDDs. One study showed that 5–15 mg of melatonin 20–30 minutes before bedtime can lead to a small improvement in total sleep time at night of approximately 30 minutes, primarily as a result of a reduction in the time it took to get the child to sleep (shortened sleep onset).[897] However, overall, melatonin is reported to have a variable impact upon the improvement of sleep in NDDs, and parents and caregivers often describe melatonin as being beneficial only for initiating sleep, not for maintaining it. These findings are consistent with our previous discussion, and the discussion below, that melatonin acts as a mild modulator of sleep rather than a sleep hormone (see chapter 2). Comparable findings have been reported with Ramelteon, a drug designed to have similar properties to melatonin.[896] **Benzodiazepine sleeping tablets** can shorten the time to get to sleep, increase the total time asleep and improve the maintenance of sleep, but they are associated with daytime sleepiness and addiction, and so their recommended use is for only limited periods.[898] **Non-benzodiazepine (Z drug) sleeping tablets**, including zolpidem, zaleplon and eszopiclone, do not seem to be particularly effective in helping children with NDDs to sleep. Zolpidem provided little improvement over a placebo and produced adverse events such as dizziness, headaches and hallucinations in many of the children.[899] In short, there are currently no robust drug options available for the normalization of SCRD in children with NDDs. Part of the problem is that NDDs represent such a varied group of conditions, with multiple genetic causes and environmental modifiers. Indeed, the situation is probably analogous to mental illness, where the circuits in the brain affected by any neurodevelopmental disorder will overlap to some degree with the brain circuits driving circadian rhythms and sleep. In addition, the NDD will exacerbate SCRD, and the SCRD will exacerbate the NDD (see Figure 7, p. 218). And again, like mental illness, new drugs to improve SCRD may provide an important therapeutic approach in future years.

### Profound blindness

Profound blindness can result from complete eye loss, severe eye diseases that destroy the retina, or major damage to the optic nerve or the ganglion cells that give rise to the optic nerve. As discussed in chapter 3, such a condition renders an individual unable to reset or entrain their SCN to the light/dark cycle, and in the absence of this daily resetting the person will drift through time following the rhythm of their body clock. As discussed earlier, this is called freerunning. For most of us the circadian sleep/wake cycle is naturally a bit longer than 24 hours, and so we would experience daily delays in our sleep/wake cycle, getting up later and later each subsequent day. There will be a few days when the sleep/wake cycle is more or less correctly timed, before drifting off again, and the drive to eat and be active will be at the wrong time of day. To give you some sense of how disorienting this can be, the personal account of a veteran blinded during military service over 24 years ago is recorded below. He has no light perception. I thank Professor Renata Gomes, Chief Scientific Officer of Blind Veterans UK, for allowing me to publish his account here:

> I am an optimist, and most people won't notice I am completely blind until I say so, or I get my cane out, but honestly, sometimes I don't know what to do! My body tricks me. I used to arrive at work at 4 a.m. thinking it was 9 a.m.; luckily my workplace is open 24 hours. I started living by the sounds of the talking clock; still after all these years, my body keeps trying to trick me. I have very nice neighbours, they don't complain to me, only to my wife . . . I can get disorientated, sometimes I go to the shed in the garden, do some work, but then again, the time is wrong. It wakes up the neighbours. My body tricks me a lot. It's not that I think I am crazy, sometimes it is as if I am a child, I have no notion of time. I used to take medication which my doctor prescribed. I took it for seven years, it never helped, and I was worried about side effects. So I decided to stop taking it, I don't take any medication. I force myself to live by my talking clock. Absolutely no naps in the afternoon. My wife also helps me keep track of time.

Another account by a blind veteran who lost his sight over 22 years ago reinforces the disorientation experienced by the profoundly blind:

At the beginning I didn't understand what was going on, when I left the hospital, I went straight to the [Blind Veterans] rehabilitation centre, I used to wake up, shave, put on clothes and go down to the cafeteria for breakfast. At the beginning, it was very quiet . . . because there was no one there . . . The nursing staff told me it was midnight! I went to bed three hours before! How was that possible? I slept for three hours and I believed it was morning.

No matter how hard they try, these blinded individuals cannot fully entrain their circadian rhythms. They suffer the dual tragedy of being both visually blind and also 'time blind'. I have also provided an account below of a mother who recorded the impact of SCRD on her child and family. The child has a condition known as aniridia. Aniridia is a rare birth defect affecting 1 in 40,000–100,000 people. It is caused by genetic changes (mutations) in a gene called PAX6 or mutations in how PAX6 is regulated. They result in an under-developed or absent iris, and usually other very severe problems within the eye, and possibly the brain, including the loss of the pineal gland (Figure 2).[900] PAX6 mutations can occur with mutations in other genes which combine to make the condition even worse.[901] Aniridia is also associated with severe SCRD.[902] The reasons for SCRD probably combine damage to the eye and, like NDDs, damage to the brain. We don't know, but the short 'Question and Answer' account below by the mother of a child with aniridia illustrates the impact of SCRD on both the child and the family. The young child's name has been changed here.

Could you describe the pattern of sleep/wake disturbance shown by your child?

*If I am honest there doesn't seem to be a pattern. What I do notice is being overtired, anxious or too much stimulation before bed can often trigger poor sleep with multiple wakings every couple of hours.*

What impact does this disruption have on your sleep and other family members, and how does this affect your ability to cope?

*At times it has affected us greatly. If Johnny is having an episode of poor sleep it normally means frequent wakings or not sleeping at all! This means we are all kept awake. I would say this has caused myself to nearly become completely burnt out at times.*

How important would it be for you if there was some way to correct the sleep/wake problems of your child?

*Massively important. It has such an impact on all of the household. Also, for Johnny the lack of sleep on top of his sensory fatigue causes huge issues for him. I think if we could correct his sleep he would find his school life much easier.*

I would like to thank Professor Mariya Moosajee, Consultant Ophthalmologist at Moorfields Eye Hospital and Great Ormond Street Hospital for Children, for connecting me with Johnny's mother, who kindly took the time to answer the questions I sent her. What I find particularly distressing about this account, and the reflections provided by Professor Nemeth, is that Johnny and children like him, along with their immediate families, are all suffering, and there is not much that can be done to help. Melatonin, and drugs that mimic the action of melatonin, are currently the only therapeutic option, but sadly they are not the solution.

### Melatonin in the blind

Melatonin has been used as a treatment for non-24-hour sleep/wake disorders or freerunning circadian rhythms in the profoundly blind (Figure 4, p. 103). In general, if melatonin is taken at the same time over a period of very many weeks or months, the freerunning rhythm of *some* blind individuals will eventually 'lock on' and become entrained to the daily administration of melatonin. For example, in the most successful published study ever undertaken, entrainment was achieved in 12 of 18 patients (67 per cent).[903] I stress that this is the best study reported – most have either shown a small effect or no significant impact. Acknowledging the limitations of using melatonin in the blind, the current recommendations suggest that the most effective approach is to give low doses of melatonin (0.5–5mg), taken daily, approximately six hours before the desired bedtime. But this is no guarantee of success.[902] **Tasimelteon** is sold under the brand name Hetlioz, as a medication for non-24-hour sleep/wake disorders in the blind. The drug resembles melatonin in that it is designed to activate melatonin receptors, and has been used to see if it will entrain the freerunning circadian rhythms of blind people. In one study tasimelteon entrained 8/40 or 20 per cent of the patients after four weeks of treatment. In a second study tasimelteon entrained 24/48 or 50 per cent of subjects after 12–18 weeks.[904] So, in this regard, the 'best' results for entrainment with tasimelteon (~50 per cent of subjects) are worse than for melatonin (~67 per cent of subjects), although a full back-to-back comparison between melatonin and tasimelteon needs to be undertaken. The bottom line is that melatonin and 'melatonin-like' drugs such as tasimelteon can achieve entrainment, in some but not all people, after multiple weeks or months of treatment. This is consistent with melatonin's role as a 'biological marker of the dark' to augment the dawn/dusk light signal detected by the eyes.[905] It is also possible that as melatonin has a slight sleep-inductive effect in some people, the induction of sleep behaviour might act to feed back upon the circadian system and help entrain the clock.[906] So melatonin's action on circadian timing may be via its action on sleep rather than its direct effects upon the clock.

As mentioned, some forms of aniridia are associated with a reduced or absent pineal gland (Figure 2; the major source of melatonin in the body), and because melatonin has been used in the severely blind to address non-24-hour rest/activity cycles, and because it is often misleadingly called a 'sleep hormone', melatonin treatment has been used clinically in attempts to correct SCRD in aniridia. In the USA the recommendation is that individuals who have been found to have a small or no pineal gland should take melatonin supplements to improve their quality of sleep and regulate their sleep patterns. A reduced or absent pineal is indeed correlated with low levels of serum melatonin.[902] However, to my knowledge and the knowledge of Professor Mariya Moosajee, no detailed studies have been undertaken to assess whether melatonin treatment has any effect upon sleep at all in this group.

Perhaps this is a good point to consider why the effects of melatonin are not more robust for non-24-hour sleep/wake disorders. We have touched on this topic in various chapters of this book (e.g. chapter 2), and the first point to make is that it has been quite difficult to show any effect of removing the pineal gland on rest/activity rhythms in animal models. The first studies were on rats, which showed essentially normal rest/activity cycles after the removal of the pineal,[907] and these findings have been confirmed more recently.[908] Rather surprisingly, if animals were treated to a simulated jet lag, by abruptly shifting the light/dark cycle, they would adapt faster if they had *no* pineal,[909] suggesting that pineal melatonin acts as a brake on rapid shifts. This idea is supported by studies in humans, whereby melatonin production from the pineal was largely eliminated using a beta-blocker, which led to a faster adaptation to simulated jet lag.[910] So ironically while melatonin has been used to replace light and hasten the adaptation to a new light/dark cycle, it is possible that an important function of melatonin is to do the exact opposite.[249]

### *What can we do when behavioural change is not enough?*

Ideally, the first line of approach for the improvement of SCRD is to implement behavioural changes, and this has been discussed in detail in chapter 6, with additional advice in other chapters.

However, and as illustrated by profound blindness and neuro-developmental disorders in this chapter, and dementia in earlier chapters, there are trauma, genetic and age-related conditions where behavioural approaches to improve SCRD will have little or no impact. Melatonin has been used to correct non-24-hour circadian rhythm disorders in the blind. But this is neither a fast nor a robust treatment even when it does work, and for many individuals it fails completely. In addition, beyond non-24-hour conditions, melatonin is largely ineffective in alleviating SCRD across different domains of health. If we were to judge melatonin using the criteria that we evaluate a vaccine with, then it would be considered a notable failure. The problem is that melatonin and drugs that act on melatonin receptors have been our only weapons to try to correct circadian rhythm abnormalities, and the question many circadian biologists are now asking, including our team in Oxford, is can we find something better than mela-tonin? The good news is that currently research is being undertaken in laboratories around the world to develop new drugs to address SCRD that are based upon our recent and emerging knowledge of how circadian rhythms are generated and regulated at a molecular level.

I have not gone into any detail in this book regarding the molecular pathways of the circadian system. I talked about a 'molecular feedback loop' (Figure 2), but did not populate this discussion with any detail. This a profoundly exciting area, and a major focus of the work we are doing in Oxford, but getting to grips with this detail requires more than a general understanding of biology, and my aims for this book were to excite interest, provide a knowledge platform, and encourage the reader to dig deeper using the references provided. In truth, some of my colleagues even find the molecular stuff a bit scary! If you would like to learn more, a good first step would be these references.[14, 9-11-12] What I would like to stress is that this fundamental molecular analysis has moved forward faster than I think any of us would have predicted. We can now link directly multiple genes with specific actions involved in circadian rhythm generation and regulation. And, critically, we are developing a real understanding of how mutations in these genes, and how they are turned on and off, can influence individual susceptibility to different sorts of health risks and conditions. This research has been driven forward largely by curiosity. But we are now in the exciting position of being able to use this information to develop evidence-based and condition-specific therapeutics to correct circadian rhythm disruption of the types illustrated in Figure 4 (p. 103). This figure illustrates the different circadian sleep/wake patterns along with an example of a disease condition or state associated with these altered patterns.

Let me stress, understanding the molecular mechanisms that generate and regulate circadian rhythms, and appreciating how circadian rhythm abnormalities are associated with different health states, provides the basis for developing drugs to correct these defects. For example, our group in Oxford has several drugs in development. For the purposes of transparency, I should say that this effort is part of an Oxford University commercial 'spin out' called Circadian Therapeutics. One drug 'mimics' the effects of light on the clock, activating the same pathway that light uses to bring about entrainment. In a sense we are trying to 'fool the clock' that it has seen light, and we hope that this drug will be used to address freerunning, non-24-hour sleep/wake disorders in the profoundly blind (Figure 4). Such work will involve a close collaboration with our research team and Blind Veterans UK. This light-mimicking drug is a 'repurposed drug', meaning it was originally developed for another purpose but was found not to be effective. Although it was not effective in these early-stage clinical trials, the drug was found to be safe.[913] Our circadian drug-screening programme identified this drug as having a big effect on the circadian system, and so we have been able to move quickly and safely into the human trials because the groundwork has already been done. The most famous repurposed drug to date is Viagra. The company Pfizer was working on a drug to treat angina (chest pain caused by reduced blood flow to the heart muscles), but early clinical trials showed it wasn't working, and the project was just about to be shut down. Then the lead researcher heard from some of the male participants in the trial that they were experiencing a lot more erections than usual. Pfizer switched its clinical-trials focus to erectile dysfunction, and the drug went from being almost cancelled to a lead drug, with sales in the USA of more than $400 million in the first three months after its launch.[914]

Other, non-repurposed, drugs we are developing act by increasing the amplitude of the clock, and in mice, at least, this has been shown to eliminate aspects of metabolic syndrome. Ultimately, we envisage that this drug could address the problems associated with fragmented sleep and insomnia, and their associated poor health states in conditions like dementia (Figure 4). Yet another drug increases the sensitivity of the clock to light, and should be useful in conditions such as mental illness and advanced old age, where there is evidence that the circadian system is less sensitive to light and entrainment is a problem. We are not alone in this quest to develop new evidence-based drugs to correct circadian rhythm disorders for the improvement of health. Other researchers like us around the world are working with financial backers and charities to develop drugs that will correct conditions such as advanced or delayed sleep phase syndrome (Figure 4), and drugs that target the circadian regulation of cell division and cancer progression. All of us are a few years away from delivering a treatment to the clinic, and some of the drugs may not work in the end, but success is getting close and I am immensely optimistic. It is my earnest hope, and primary motivation, that those voices you have heard in this chapter, describing the impact of SCRD resulting from blindness or neurodevelopmental disease, will represent a historical account, and not a lived experience in the coming years.

## *Finally . . .*

The circadian research community around the world has spent much of the last sixty years trying to understand the 24-hour rhythms exhibited by most life on our planet. Progress in understanding the fundamental nature of circadian biology has been astonishing, and certainly this knowledge has added to our wonder at, and appreciation of, the biological world. In parallel with this appreciation there has been an emerging realization of the fundamental importance of circadian rhythms to our health and wellbeing. What we do *when* really matters. The time of day will influence to a greater or lesser extent our decision-making skills, our vulnerability to infection, stroke or making a mistake, how our food will be processed, the efficacy of our medications and treatments, and even the impact of exercise. This is life-changing information that, as individuals and as a society, we have largely ignored. Circadian and sleep health is not taught in our schools or to our medical students, and is in many cases absent in the working environment. SCRD damages educational performance and adolescent health. Our key workers have to combine immensely demanding jobs with the additional insults inflicted by SCRD. The economy is in the hands of chronically tired and stressed individuals, which is not a great start for delivering solutions to the mess caused by the COVID-19 pandemic. The fact that society is not embracing the science of circadian rhythms represents an immense squandering of resources, and a major missed opportunity to improve health at every level.

Some of the most challenging diseases of our time are associated with, and made worse by, SCRD. A reduction in SCRD can potentially ameliorate and even eliminate these disease states. By understanding the mechanisms that generate and regulate circadian rhythms, new therapeutic drug targets have been identified that will address the damaging impacts of SCRD. This new class of 'circadian drugs' will drive forward a revolution in medicine, providing evidence-based and condition-specific treatments for disease states that until now have been largely untreatable.

Life is so often all about gaining an opportunity in the briefest of moments, either avoiding harm such as an infection, or achieving an advantage by making a wise decision. And it is our circadian rhythms that help us increase our chances of success in a dynamic world. Circadian rhythms are all about timing and not about time itself. They regulate actions to produce the best effect. Our bodies need the correct materials in the right place, in the right amount, at the right time of day, and a clock anticipates and delivers these different needs. The lives of both wise and foolish people all end in death, but in the context of this book the circadian wise will, on balance, live longer, be happier and lead more fulfilled lives.

# Appendix I
## *Studying Your Own Biological Rhythms*

### *Part I. Developing a Sleep Diary*

Keeping a record of your sleep/wake patterns can be useful if you think there might be a problem with your sleep or are just interested. You can design your own **Sleep Diary**, but the sort of information that you should consider collecting is listed below. Collect information over several weeks, and you should make a note of any significant events in your life that you think may affect your sleep.

Fill in all questions after you wake up each morning, based on the sleep you have just had. The sorts of question you may want to include in your questionnaire could be:

1. *What time did you get into bed?*
2. *What time did you start to try to sleep?*
3. *How long did it take you to fall asleep?*
4. *How many times did you wake up before your final awakening?*
5. *In total, how long did these awakenings last?*
6. *What time was your final awakening?*
7. *What time did you get out of bed after waking up for the final time?*
8. *How would you rate the quality of your sleep?*
    a. *Very poor*
    b. *Poor*
    c. *Fair*
    d. *Good*
    e. *Very good*

9. *Did you dream and what was the nature of your dream?*
10. *Make a note of any other observations about your sleep.*
11. *Make a note of any events during the day that you think may have affected your sleep that night such as difficult situations at work or in the home.*

# Part II. Chronotype Questionnaire

Do you have a Morning (lark) Evening (Owl) or Intermediate Chronotype?

*Morningness/eveningness questionnaire*

For each question, please select the answer that describes you by circling the value that best indicates how you have felt in recent weeks. At the end of the questionnaire add up your score and find out your chronotype.

**1. *Approximately* what time would you get up if you were entirely free to plan your day?**

| | |
|---|---|
| 5.00–6.30 a.m. | 5 |
| 6.30–7.45 a.m. | 4 |
| 7.45–9.45 a.m. | 3 |
| 9.45–11.00 a.m. | 2 |
| 11.00 a.m.–12 noon | 1 |

**2. *Approximately* what time would you go to bed if you were entirely free to plan your evening?**

| | |
|---|---|
| 8.00–9.00 p.m. | 5 |
| 9.00–10.15 p.m. | 4 |
| 10.15 p.m.–12.30 a.m. | 3 |
| 12.30–1.45 a.m. | 2 |
| 1.45–3.00 a.m. | 1 |

**3. If you usually have to get up at a specific time in the morning, how much do you depend on an alarm clock?**

Not at all              4
Slightly                3
Somewhat                2
Very much               1

**4. How easy do you find it to get up in the morning (when you are not awakened unexpectedly)?**

Very easy               4
Fairly easy             3
Somewhat difficult      2
Very difficult          1

**5. How alert do you feel during the first half-hour after you wake up in the morning?**

Not at all alert        1
Slightly alert          2
Fairly alert            3
Very alert              4

**6. How hungry do you feel during the first half-hour after you wake up?**

Not at all hungry       1
Slightly hungry         2
Fairly hungry           3
Very hungry             4

**7. During the first half-hour after you wake up in the morning, how do you feel?**

Very tired              1
Fairly tired            2
Fairly refreshed        3
Very refreshed          4

**8. If you had no commitments the next day, what time would you go to bed compared to your usual bedtime?**

| | |
|---|---|
| Slightly or no later | 4 |
| Less than one hour later | 3 |
| 1–2 hours later | 2 |
| More than two hours later | 1 |

**9. You have decided to do physical exercise. A friend suggests that you do this for one hour twice a week, and the best time for him is between 7 and 8 a.m. Bearing in mind nothing but your own internal 'clock', how do you think you would perform?**

| | |
|---|---|
| Would be in good form | 4 |
| Would be in reasonable form | 3 |
| Would find it difficult | 2 |
| Would find it very difficult | 1 |

**10. At *approximately* what time in the evening do you feel tired, and, as a result, in need of sleep?**

| | |
|---|---|
| 8.00–9.00 p.m. | 5 |
| 9.00–10.15 p.m. | 4 |
| 10.15 p.m.–12.45 a.m. | 3 |
| 12.45–2.00 a.m. | 2 |
| 2.00–3.00 a.m. | 1 |

**11. You want to be at your peak performance for a test that you know is going to be mentally exhausting and will last two hours. You are entirely free to plan your day. Considering only your internal 'clock', which one of the four testing times would you choose?**

| | |
|---|---|
| 8.00–10.00 a.m. | 6 |
| 11.00 a.m.–1.00 p.m. | 4 |
| 3.00–5.00 p.m. | 2 |
| 7.00–9.00 p.m. | 0 |

**12. If you got into bed at 11 p.m., how tired would you be?**

| | |
|---|---|
| Not at all tired | 0 |
| A little tired | 2 |
| Fairly tired | 3 |
| Very tired | 5 |

**13. For some reason you have gone to bed several hours later than usual, but there is no need to get up at any particular time the next morning. Which one of the following are you most likely to do?**

| | |
|---|---|
| Will wake up at usual time, but will not fall back asleep | 4 |
| Will wake up at usual time and will doze thereafter | 3 |
| Will wake up at usual time, but will fall asleep again | 2 |
| Will not wake up until later than usual | 1 |

**14. One night you have to remain awake between 4 and 6 a.m. in order to carry out a night watch. You have no time commitments the next day. Which one of the alternatives would suit you best?**

| | |
|---|---|
| Would not go to bed until the watch is over | 1 |
| Would take a nap before and sleep after | 2 |
| Would take a good sleep before and nap after | 3 |
| Would sleep only before the watch | 4 |

**15. You have two hours of hard physical work. You are entirely free to plan your day. Considering only your internal 'clock', which of the following times would you choose?**

| | |
|---|---|
| 8.00–10.00 a.m. | 4 |
| 11.00 a.m.–1.00 p.m. | 3 |
| 3.00–5.00 p.m. | 2 |
| 7.00–9.00 p.m. | 1 |

**16. You have decided to do physical exercise. A friend suggests that you do this for one hour twice a week. The best time for**

her is between 10 and 11 p.m. Bearing in mind only your internal 'clock', how well do you think you would perform?

| | |
|---|---|
| Would be in good form | 1 |
| Would be in reasonable form | 2 |
| Would find it difficult | 3 |
| Would find it very difficult | 4 |

17. Suppose you can choose your own work hours. Assume that you work a five-hour day (including breaks), your job is interesting and you are paid based on your performance. At *approximately* what time would you choose to begin?

| | |
|---|---|
| 5 hours starting between 4.00 and 8.00 a.m. | 5 |
| 5 hours starting between 8.00 and 9.00 a.m. | 4 |
| 5 hours starting between 9.00 a.m. and 2.00 p.m. | 3 |
| 5 hours starting between 2.00 and 5.00 p.m. | 2 |
| 5 hours starting between 5.00 p.m. and 4.00 a.m. | 1 |

18. At *approximately* what time of day do you usually feel your best?

| | |
|---|---|
| 5.00–8.00 a.m. | 5 |
| 8.00–10.00am | 4 |
| 10.00 am–5.00 p.m. | 3 |
| 5.00–10 p.m. | 2 |
| 10 p.m.–5 a.m. | 1 |

19. One hears about 'morning types' and 'evening types.' Which one of these types do you consider yourself to be?

| | |
|---|---|
| Definitely a morning type | 6 |
| Rather more a morning type than an evening type | 4 |
| Rather more an evening type than a morning type | 2 |
| Definitely an evening type | 1 |

_____ **Total points for all 19 questions**

## Appendix I

### Interpreting and Using Your Morningness/Eveningness Score

This questionnaire has 19 questions, each with a number of points. First, add up the points you circled and enter your total morningness/eveningness score here:

Scores can range from 16 to 86. Scores of 41 and below indicate 'evening types.' Scores of 59 and above indicate 'morning types.' Scores between 42 and 58 indicate 'intermediate types'.

| 16–30 | 31–41 | 42–58 | 59–69 | 70–86 |
|-------|-------|-------|-------|-------|
| Definite evening | Moderate evening | Intermediate | Moderate morning | Definite morning |

This questionnaire is based upon the original paper by J. A. Horne and O. Ostberg (1976) A self-assessment questionnaire to determine morningness-eveningness in human circadian rhythms. *International Journal of Chronobiology*, vol. 4, pp. 97–110.

# Appendix II

## *The Key Elements and Overview of the Immune System*

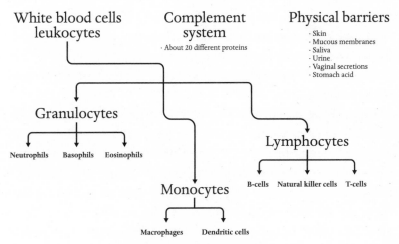

White blood cells leukocytes

Complement system
· About 20 different proteins

Physical barriers
· Skin
· Mucous membranes
· Saliva
· Urine
· Vaginal secretions
· Stomach acid

Granulocytes

Neutrophils   Basophils   Eosinophils

Lymphocytes

B-cells   Natural killer cells   T-cells

Monocytes

Macrophages   Dendritic cells

**Figure 10.** Immune system.

As mentioned in chapter 11, the first physical barrier preventing infection is the skin, which is composed of tightly packed cells that normally prevent infectious viruses, bacteria or parasites entering the body. The topmost layer of the skin consists of dead cells that form a dense physical layer which resist invasion. In addition, the skin's surface is coated in secretions that inhibit the growth of some pathogens. Some bacteria and viruses, however, can remain alive on the skin for some time and can be transferred to less well-protected areas such as the eyes and nose and be taken into the lungs, where they can enter and cause havoc. To help prevent this, the nose and lungs are lined with mucous membranes which produce mucus that can trap invaders. In the lungs, tiny hair-like structures (cilia) move the mucus away from the lungs to the upper airway, where the mucus/bug cocktail is swallowed and destroyed by stomach acids. Alternatively, the mucus can be coughed up or sneezed out with the trapped bugs. Robust hand washing will remove invaders from the skin surface[915], and trapping your sneezes in a handkerchief prevents the spread of pathogens that are temporarily captured in the mucus[916]. In addition, the flushing action and antiseptic properties of urine from the bladder[917] and tears in the eyes[918] also help wash away bacteria and viruses from these vulnerable sites of infection. And, finally, vaginal secretions contain antibacterial compounds[919], as does semen[920], providing an additional element of protection (Figure 10).

Because the skin is such an effective barrier, infection is usually through another route, usually the lungs. But if pathogens do enter the body, then the cells and protective molecules of the immune systems are waiting to defend us (Figure 10). Our white blood cells account for only about 1 per cent of our blood, but these are the cells of the immune response. When our body is attacked, white blood cells rush in to help destroy the invader. White blood cells are also called **leukocytes** and there are three main groups: lymphocytes, monocytes and granulocytes:

- **Lymphocytes** are key cells of the immune response. Two types of lymphocyte are the B-cells (B-lymphocytes) and T-cells (T-lymphocytes). **B-cells** recognize an invader by detecting a specific antigen, usually some type of protein, on the surface of the bug. Each B-cell has its own receptor that binds to a specific antigen, and, remarkably, these receptors are made in advance, just in case that particular antigen should make an appearance. It is a bit like carrying tens of thousands of keys in your pocket just in case you encounter a door with that particular lock. Following activation, a number of different things can then happen. The B-cells can form different types of cell, two of which I will mention here. Plasma B-cells are triggered to make antibodies which act to lock on to specific parts (antigens) of the invader. But antibodies can't kill bacteria and viruses without help; the complement system (see below) is triggered to do this job. In addition, antibodies attached to pathogens make it easier for macrophages (see below) to recognize, attack, engulf and kill the pathogen. Antibodies usually remain in the blood, and more antibodies can be made quickly by the memory B-cells, if the immune system is triggered again by the same pathogen. This is partly how some vaccines prevent diseases. An immunization takes a non-infectious protein from the pathogen (or in the old days a dead pathogen), which is then injected into the body. **T-cells**, like B-cells, also have their own receptor that will bind a specific antigen. These receptors are made in advance, and it has been estimated that there are maybe up to a billion different T-cells in the body, each with its own T-cell receptor that will bind to only one kind of antigen, just in case that antigen should make an appearance. Because they are made in advance, they are ready for action waiting for an attack. Once a T-cell has been activated by binding to its specific antigen, the T-cell will proliferate massively and differentiate into many more T-cells. There are two major types of T-cell, the helper T-cells and the cytotoxic T-cells. Helper T-cells play a key role in the immune responses. Once activated, helper T-cells stimulate the B-cells to make antibodies to their specific antigen – allowing an attack by complement and phagocytes of the innate immune system. In addition, helper T-cells produce factors (e.g. cytokines) that activate virtually all the other immune system responses.

They rally a generalized attack upon one pathogen recruiting phagocytes and the complement cascade. Helper T-cells do not have to meet a pathogen antigen directly to be activated. After digesting a pathogen, a macrophage or dendritic cell (see below) will present an antigen of the pathogen on its cell surface, which is then detected by the corresponding helper T-cell, and this triggers an attack. Cytotoxic T-cells when activated by their antigen punch holes in the cell, triggering death. Cytotoxic T-cells can also recognize a cell that has been infected by a virus and kill it, preventing more virus from being made. There are also regulatory T-cells (also known as suppressor T-cells), which modulate the immune system, driving tolerance to self-antigens (those made by the body). Regulatory T-cells are very important in preventing autoimmune disease such as rheumatoid arthritis, inflammatory bowel disease or multiple sclerosis (MS).[921] **Natural killer cells** are a type of lymphocyte in the same family as T-cells and B-cells, and so are able to recognize an invader by detecting a specific antigen. These cells respond quickly to a wide variety of pathogens and are best known for killing cells infected with a virus, and detecting and controlling early signs of cancer. They can also trigger inflammation.

- **Monocytes** represent another whole class of white blood cells (leukocytes). One type is the **macrophages**. These are amoeba-like cells that rush to the site of infection, recognize an invader, either directly or because of an attached antibody, and have the ability to ingest (phagocytose) and kill the pathogen. They can also be stimulated to attack by cytokines (from T-cells), and, interestingly, the sensitivity of macrophages to these signals changes across the day due to a circadian clock in the macrophages. Sensitivity is increased during the day, when we are usually awake.[740] Another type of monocyte is the **dendritic cells**. These are less well understood and after detecting and phagocytosing a pathogen, they present pathogen antigens on their cell surface. This in turn activates T-cells and other defence mechanisms of the immune system. **Tumour necrosis factor** (TNF) comprises multiple proteins produced mainly by activated macrophages, T-cells and natural killer cells. TNF acts as a type of cell-signalling protein (cytokine) promoting immune responses and inflammation.

- **Granulocytes** are the third major type of leukocyte, and surprise, surprise, they also come in multiple forms. Some are called **neutrophils**, which are the most abundant type of granulocyte and make up 40–70 per cent of all our white blood cells. Neutrophils detect, phagocytose and digest bacteria and fungi. Interestingly, the ability of neutrophils to engulf bacteria is reduced when simple sugars like glucose and fructose are increased in the blood. Fasting strengthens the neutrophils' phagocytic capacity to engulf bacteria.[818] Perhaps this explains why individuals with Type 2 diabetes are more prone to infection. **Eosinophils** are capable of responding to a broad range of antigens, and when activated release a variety of cytokines which attract B-cells and T-cells. They also release cytotoxic proteins which attack cells and are an important defence against parasitic infections, but may also cause massive tissue damage in allergic conditions like asthma (see 'Questions and Answers' in chapter 11). **Basophils** promote inflammatory reactions. They can release the anticoagulant heparin, which prevents blood from clotting too quickly to allow the cells and proteins of the immune defence to get access to the site of infection, and for the same reason they also contain the vasodilator histamine, which promotes blood flow to tissues. Like eosinophils, basophils are also associated with defence against parasites, and their over-activation has been implicated in allergic reactions.

- **Complement system**, which consists of about 20 different proteins that are activated by an invader. These amazing proteins can: 1. detect and puncture the cell walls of bacteria and kill them directly; 2. bind to antibodies produced by the B-cells, which have already recognized pathogens; complement proteins then kill the pathogens by puncturing the cell walls or by attracting macrophages; 3. bind directly to pathogens and attract macrophages; the macrophages then kill the pathogens; 4. trigger further inflammatory responses, recruiting more help from the other elements of the immune system.[921]

**Note:** The immune response is usually classified into **innate immunity**, which is what we are born with, and **adaptive immunity**, which is what we acquire following disease exposure. Innate immunity would involve physical barriers, the complement system, granulocytes and monocytes. Adaptive immunity involves the lymphocytes and the B-cells, T-cells and natural killer cells.

# References

1 Ho Mien, I. et al. Effects of exposure to intermittent versus continuous red light on human circadian rhythms, melatonin suppression, and pupillary constriction. *PLoS One* **9**, e96532, doi:10.1371/journal.pone.0096532 (2014).

2 Trenell, M. I., Marshall, N. S. and Rogers, N. L. Sleep and metabolic control: waking to a problem? *Clin Exp Pharmacol Physiol* **34**, 1–9, doi:10.1111/j.1440–1681.2007.04541.x (2007).

3 Blatter, K. and Cajochen, C. Circadian rhythms in cognitive performance: methodological constraints, protocols, theoretical underpinnings. *Physiol Behav* **90**, 196–208, doi:10.1016/j.physbeh.2006.09.009 (2007).

4 Bagatell, C. J., Heiman, J. R., Rivier, J. E. and Bremner, W. J. Effects of endogenous testosterone and estradiol on sexual behavior in normal young men. *J Clin Endocrinol Metab* **78**, 711–16, doi:10.1210/jcem.78.3.8126146 (1994).

5 Kleitman, N. Studies on the physiology of sleep: VIII. Diurnal variation in performance. *Am J Physiol* **104**, 449–56 (1933).

6 Herculano-Houzel, S. The human brain in numbers: a linearly scaled-up primate brain. *Front Hum Neurosci* **3**, 31, doi:10.3389/neuro.09.031.2009 (2009).

7 Swaab, D. F., Fliers, E. and Partiman, T. S. The suprachiasmatic nucleus of the human brain in relation to sex, age and senile dementia. *Brain Res* **342**, 37–44, doi:10.1016/0006–8993(85)91350–2 (1985).

8 Schulkin, J. In honor of a great inquirer: Curt Richter. *Psychobiology* **17**, 113–14 (1989).

9 Moore, R. Y. and Lenn, N. J. A retinohypothalamic projection in the rat. *J Comp Neurol* **146**, 1–14, doi:10.1002/cne.901460102 (1972).

10 Stephan, F. K. and Zucker, I. Circadian rhythms in drinking behavior and locomotor activity of rats are eliminated by hypothalamic

lesions. *Proc Natl Acad Sci USA* **69**, 1583–6, doi:10.1073/pnas.69.6.1583 (1972).

11  Ralph, M. R., Foster, R. G., Davis, F. C. and Menaker, M. Transplanted suprachiasmatic nucleus determines circadian period. *Science* **247**, 975–8 (1990).

12  Welsh, D. K., Logothetis, D. E., Meister, M. and Reppert, S. M. Individual neurons dissociated from rat suprachiasmatic nucleus express independently phased circadian firing rhythms. *Neuron* **14**, 697–706 (1995).

13  Tolwinski, N. S. Introduction: Drosophila – A model system for developmental biology. *J Dev Biol* **5**, doi:10.3390/jdb5030009 (2017).

14  Takahashi, J. S. Transcriptional architecture of the mammalian circadian clock. *Nat Rev Genet* **18**, 164–79, doi:10.1038/nrg.2016.150 (2017).

15  Lowrey, P. L. et al. Positional syntenic cloning and functional characterization of the mammalian circadian mutation tau. *Science* **288**, 483–92, doi:10.1126/science.288.5465.483 (2000).

16  Jones, S. E. et al. Genome-wide association analyses of chronotype in 697,828 individuals provides insights into circadian rhythms. *Nat Commun* **10**, 343, doi:10.1038/s41467-018-08259-7 (2019).

17  Nagoshi, E. et al. Circadian gene expression in individual fibroblasts: cell-autonomous and self-sustained oscillators pass time to daughter cells. *Cell* **119**, 693–705, doi:10.1016/j.cell.2004.11.015 (2004).

18  Richards, J. and Gumz, M. L. Advances in understanding the peripheral circadian clocks. *FASEB J* **26**, 3602–13, doi:10.1096/fj.12–203554 (2012).

19  Balsalobre, A., Damiola, F. and Schibler, U. A serum shock induces circadian gene expression in mammalian tissue culture cells. *Cell* **93**, 929–37, doi:10.1016/s0092–8674(00)81199-x (1998).

20  Albrecht, U. Timing to perfection: the biology of central and peripheral circadian clocks. *Neuron* **74**, 246–60, doi:10.1016/j.neuron.2012.04.006 (2012).

21  Jagannath, A. et al. Adenosine integrates light and sleep signalling for the regulation of circadian timing in mice. *Nat Commun* **12**, 2113, doi:10.1038/s41467–021–22179–z (2021).

22  Rijo-Ferreira, F. and Takahashi, J. S. Genomics of circadian rhythms in health and disease. *Genome Med* **11**, 82, doi:10.1186/s13073–019–0704–0 (2019).

23  Lewczuk, B. et al. Influence of electric, magnetic, and electromagnetic fields on the circadian system: current stage of knowledge. *Biomed Res Int* **2014**, 169459, doi:10.1155/2014/169459 (2014).

24  Postolache, T. T. et al. Seasonal spring peaks of suicide in victims with and without prior history of hospitalization for mood disorders. *J Affect Disord* **121**, 88–93, doi:10.1016/j.jad.2009.05.015 (2010).

25  Foster, R. G. and Roenneberg, T. Human responses to the geophysical daily, annual and lunar cycles. *Curr Biol* **18**, R784–R794, doi:10.1016/j.cub.2008.07.003 (2008).

26  Underwood, H., Steele, C. T. and Zivkovic, B. Circadian organization and the role of the pineal in birds. *Microsc Res Tech* **53**, 48–62, doi:10.1002/jemt.1068 (2001).

27  Kovanen, L. et al. Circadian clock gene polymorphisms in alcohol use disorders and alcohol consumption. *Alcohol Alcohol* **45**, 303–11, doi:10.1093/alcalc/agq035 (2010).

28  Levi, F. and Halberg, F. Circaseptan (about-7-day) bioperiodicity – spontaneous and reactive – and the search for pacemakers. *Ric Clin Lab* **12**, 323–70, doi:10.1007/BF02909422 (1982).

29  Walker, M. P. The role of slow wave sleep in memory processing. *J Clin Sleep Med* **5**, S20–26 (2009).

30  Clemens, Z., Fabo, D. and Halasz, P. Overnight verbal memory retention correlates with the number of sleep spindles. *Neuroscience* **132**, 529–35, doi:10.1016/j.neuroscience.2005.01.011 (2005).

31  Mednick, S. C. et al. The critical role of sleep spindles in hippocampal-dependent memory: a pharmacology study. *J Neurosci* **33**, 4494–504, doi:10.1523/JNEUROSCI.3127–12.2013 (2013).

32  Forget, D., Morin, C. M. and Bastien, C. H. The role of the spontaneous and evoked k-complex in good-sleeper controls and in individuals with insomnia. *Sleep* **34**, 1251–60, doi:10.5665/SLEEP.1250 (2011).

33  Ben Simon, E., Rossi, A., Harvey, A. G. and Walker, M. P. Overanxious and underslept. *Nat Hum Behav* **4**, 100–110, doi:10.1038/s41562–019–0754–8 (2020).

34  Meaidi, A., Jennum, P., Ptito, M. and Kupers, R. The sensory construction of dreams and nightmare frequency in congenitally blind and late blind individuals. *Sleep Med* **15**, 586–95, doi:10.1016/j.sleep.2013.12.008 (2014).

35  Lerner, I., Lupkin, S. M., Sinha, N., Tsai, A. and Gluck, M. A. Baseline levels of rapid eye movement sleep may protect against excessive activity in fear-related neural circuitry. *J Neurosci* **37**, 1123–44, doi:10.1523/JNEUROSCI.0578–17.2017 (2017).

36  Giedke, H. and Schwarzler, F. Therapeutic use of sleep deprivation in depression. *Sleep Med Rev* **6**, 361–77 (2002).

37  Mann, K., Pankok, J., Connemann, B. and Roschke, J. Temporal relationship between nocturnal erections and rapid eye movement episodes in healthy men. *Neuropsychobiology* **47**, 109–14, doi:10.1159/000070019 (2003).

38  Schmidt, M. H. and Schmidt, H. S. Sleep-related erections: neural mechanisms and clinical significance. *Curr Neurol Neurosci Rep* **4**, 170–78, doi:10.1007/s11910–004–0033–5 (2004).

39  Oliveira, I., Deps, P. D. and Antunes, J. Armadillos and leprosy: from infection to biological model. *Rev Inst Med Trop Sao Paulo* **61**, e44, doi:10.1590/S1678–9946201961044 (2019).

40  Schenck, C. H. The spectrum of disorders causing violence during sleep. *Sleep Science and Practice* **3** (**2**), 1–14 (2019).

41  Cramer Bornemann, M. A., Schenck, C. H. and Mahowald, M. W. A review of sleep-related violence: the demographics of sleep forensics referrals to a single center. *Chest* **155**, 1059–66, doi:10.1016/j.chest.2018.11.010 (2019).

42  Mistlberger, R. E. Circadian regulation of sleep in mammals: role of the suprachiasmatic nucleus. *Brain Res Rev* **49**, 429–54, doi:10.1016/j.brainresrev.2005.01.005 (2005).

43  Greene, R. W., Bjorness, T. E. and Suzuki, A. The adenosine-mediated, neuronal-glial, homeostatic sleep response. *Curr Opin Neurobiol* **44**, 236–42, doi:10.1016/j.conb.2017.05.015 (2017).

44  Reichert, C. F., Maire, M., Schmidt, C. and Cajochen, C. Sleep–wake regulation and its impact on working memory performance: the role of adenosine. *Biology (Basel)* **5**, doi:10.3390/biology5010011 (2016).

45  O'Callaghan, F., Muurlink, O. and Reid, N. Effects of caffeine on sleep quality and daytime functioning. *Risk Manag Healthc Policy* **11**, 263–71, doi:10.2147/RMHP.S156404 (2018).

46  Mets, M., Baas, D., van Boven, I., Olivier, B. and Verster, J. Effects of coffee on driving performance during prolonged simulated highway driving. *Psychopharmacology (Berl)* **222**, 337–42, doi:10.1007/s00213-012-2647-7 (2012).

47  Charron, G., Souloumiac, J., Fournier, M. C. and Canivenc, R. Pineal rhythm of N-acetyltransferase activity and melatonin in the male badger, Meles meles L, under natural daylight: relationship with the photoperiod. *J Pineal Res* **11**, 80–85, doi:10.1111/j.1600-079x.1991.tb00460.x (1991).

48  Verheggen, R. J. et al. Complete absence of evening melatonin increase in tetraplegics. *FASEB J* **26**, 3059–64, doi:10.1096/fj.12-205401 (2012).

49  Whelan, A., Halpine, M., Christie, S. D. and McVeigh, S. A. Systematic review of melatonin levels in individuals with complete cervical spinal cord injury. *J Spinal Cord Med*, 1–14, doi:10.1080/10790268.2018.1505312 (2018).

50  Spong, J., Kennedy, G. A., Brown, D. J., Armstrong, S. M. and Berlowitz, D. J. Melatonin supplementation in patients with complete tetraplegia and poor sleep. *Sleep Disord* **2013**, 128197, doi:10.1155/2013/128197 (2013).

51  Kostis, J. B. and Rosen, R. C. Central nervous system effects of beta-adrenergic-blocking drugs: the role of ancillary properties. *Circulation* **75**, 204–12, doi:10.1161/01.cir.75.1.204 (1987).

52  Scheer, F. A. et al. Repeated melatonin supplementation improves sleep in hypertensive patients treated with beta-blockers: a randomized controlled trial. *Sleep* **35**, 1395–1402, doi:10.5665/sleep.2122 (2012).

53  Ferracioli-Oda, E., Qawasmi, A. and Bloch, M. H. Meta-analysis: melatonin for the treatment of primary sleep disorders. *PLoS One* **8**, e63773, doi:10.1371/journal.pone.0063773 (2013).

54  Lockley, S. W. et al. Tasimelteon for non-24-hour sleep–wake disorder in totally blind people (SET and RESET): two multicentre, randomised, double-masked, placebo-controlled phase 3 trials. *Lancet* **386**, 1754–64, doi:10.1016/S0140–6736(15)60031–9 (2015).

55  Arendt, J. Melatonin in humans: it's about time. *J Neuroendocrinol* **17**, 537–8, doi:10.1111/j.1365–2826.2005.01333.x (2005).

56  Arendt, J. and Skene, D. J. Melatonin as a chronobiotic. *Sleep Med Rev* **9**, 25–39, doi:10.1016/j.smrv.2004.05.002 (2005).

57  Medeiros, S. L. S. et al. Cyclic alternation of quiet and active sleep states in the octopus. *iScience* **24**, 102223, doi:10.1016/j.isci.2021.102223 (2021).

58  Kanaya, H. J. et al. A sleep-like state in Hydra unravels conserved sleep mechanisms during the evolutionary development of the central nervous system. *Sci Adv* **6**, doi:10.1126/sciadv.abb9415 (2020).

59  Eelderink-Chen, Z. et al. A circadian clock in a nonphotosynthetic prokaryote. *Sci Adv* **7**, doi:10.1126/sciadv.abe2086 (2021).

60  Pittendrigh, C. S. Temporal organization: reflections of a Darwinian clock-watcher. *Annu Rev Physiol* **55**, 16–54, doi:10.1146/annurev.ph.55.030193.000313 (1993).

61  Laposky, A. D., Bass, J., Kohsaka, A. and Turek, F. W. Sleep and circadian rhythms: key components in the regulation of energy metabolism. *FEBS Lett* **582**, 142–51, doi:10.1016/j.febslet.2007.06.079 (2008).

62  Shokri-Kojori, E. et al. β-Amyloid accumulation in the human brain after one night of sleep deprivation. *Proc Natl Acad Sci USA* **115**, 4483–8, doi:10.1073/pnas.1721694115 (2018).

63  Walker, M. P. and Stickgold, R. Sleep, memory, and plasticity. *Annu Rev Psychol* **57**, 139–66, doi:10.1146/annurev.psych.56.091103.070307 (2006).

64  Foster, R. G. There is no mystery to sleep. *Psych J* **7**, 206–8, doi:10.1002/pchj.247 (2018).

65  Vyazovskiy, V. V. et al. Local sleep in awake rats. *Nature* **472**, 443–7, doi:10.1038/nature10009 (2011).

66  Shannon, S., Lewis, N., Lee, H. and Hughes, S. Cannabidiol in anxiety and sleep: a large case series. *Perm J* **23**, **18–041**, doi:10.7812/TPP/18–041 (2019).

67  Gray, S. L. et al. Cumulative use of strong anticholinergics and incident dementia: a prospective cohort study. *JAMA Intern Med* **175**, 401–7, doi:10.1001/jamainternmed.2014.7663 (2015).

68  Axelsson, J. et al. Beauty sleep: experimental study on the perceived health and attractiveness of sleep deprived people. *BMJ* **341**, c6614, doi:10.1136/bmj.c6614 (2010).

69  Mascetti, G. G. Unihemispheric sleep and asymmetrical sleep: behavioral, neurophysiological, and functional perspectives. *Nat Sci Sleep* **8**, 221–38, doi:10.2147/NSS.S71970 (2016).

70  Rattenborg, N. C. et al. Evidence that birds sleep in mid-flight. *Nat Commun* **7**, 12468, doi:10.1038/ncomms12468 (2016).

71  Winer, G. A., Cottrell, J. E., Gregg, V., Fournier, J. S. and Bica, L. A. Fundamentally misunderstanding visual perception. Adults' belief in visual emissions. *Am Psychol* **57**, 417–24, doi:10.1037//0003–066x.57.6–7.417 (2002).

72  Czeisler, C. A. et al. Stability, precision, and near-24-hour period of the human circadian pacemaker. *Science* **284**, 2177–81 (1999).

73  Campbell, S. S. and Murphy, P. J. Extraocular circadian phototransduction in humans. *Science* **279**, 396–9 (1998).

74  Foster, R. G. Shedding light on the biological clock. *Neuron* **20**, 829–32 (1998).

75  Lindblom, N. et al. Bright light exposure of a large skin area does not affect melatonin or bilirubin levels in humans. *Biol Psychiatry* **48**, 1098–1104 (2000).

76 Lindblom, N. et al. No evidence for extraocular light induced phase shifting of human melatonin, cortisol and thyrotropin rhythms. *Neuroreport* 11, 713–17 (2000).

77 Yamazaki, S., Goto, M. and Menaker, M. No evidence for extraocular photoreceptors in the circadian system of the Syrian hamster. *J Biol Rhythms* 14, 197–201, doi:10.1177/074873099129000605 (1999).

78 Wright, K. P., Jr and Czeisler, C. A. Absence of circadian phase resetting in response to bright light behind the knees. *Science* 297, 571, doi:10.1126/science.1071697 (2002).

79 Foster, R. G. et al. Circadian photoreception in the retinally degenerate mouse (rd/rd). *J Comp Physiol A* 169, 39–50 (1991).

80 Foster, R. G. et al. Photoreceptors regulating circadian behavior: a mouse model. *J Biol Rhythms* 8 **Suppl**, S17–23 (1993).

81 Freedman, M. S. et al. Regulation of mammalian circadian behavior by non-rod, non-cone, ocular photoreceptors. *Science* 284, 502–4 (1999).

82 Lucas, R. J., Freedman, M. S., Munoz, M., Garcia-Fernandez, J. M. and Foster, R. G. Regulation of the mammalian pineal by non-rod, non-cone, ocular photoreceptors. *Science* 284, 505–7 (1999).

83 Soni, B. G., Philp, A. R., Knox, B. E. and Foster, R. G. Novel retinal photoreceptors. *Nature* 394, 27–8, doi:10.1038/27794 (1998).

84 Berson, D. M., Dunn, F. A. and Takao, M. Phototransduction by retinal ganglion cells that set the circadian clock. *Science* 295, 1070–73, doi:10.1126/science.1067262 (2002).

85 Sekaran, S., Foster, R. G., Lucas, R. J. and Hankins, M. W. Calcium imaging reveals a network of intrinsically light-sensitive inner-retinal neurons. *Curr Biol* 13, 1290–98 (2003).

86 Lucas, R. J., Douglas, R. H. and Foster, R. G. Characterization of an ocular photopigment capable of driving pupillary constriction in mice. *Nat Neurosci* 4, 621–6, doi:10.1038/88443 (2001).

87 Hattar, S. et al. Melanopsin and rod-cone photoreceptive systems account for all major accessory visual functions in mice. *Nature* 424, 76–81, doi:10.1038/nature01761 (2003).

88 Provencio, I., Jiang, G., De Grip, W. J., Hayes, W. P. and Rollag, M. D. Melanopsin: an opsin in melanophores, brain, and eye. *Proc Natl Acad Sci USA* **95**, 340–45 (1998).

89 Foster, R. G., Hughes, S. and Peirson, S. N. Circadian photo-entrainment in mice and humans. *Biology (Basel)* **9**, doi:10.3390/biology9070180 (2020).

90 Honma, K., Honma, S. and Wada, T. Entrainment of human circadian rhythms by artificial bright light cycles. *Experientia* **43**, 572–4 (1987).

91 Randall, M. Labour in the agriculture industry, UK: February 2018. *Office for National Statistics, UK*, 1–11 (2018).

92 Porcheret, K. et al. Chronotype and environmental light exposure in a student population. *Chronobiol Int* **35**, 1365–74, doi:10.1080/07420528.2018.1482556 (2018).

93 Wright, K. P., Jr et al. Entrainment of the human circadian clock to the natural light-dark cycle. *Curr Biol* **23**, 1554–8, doi:10.1016/j.cub.2013.06.039 (2013).

94 Figueiro, M. G., Wood, B., Plitnick, B. and Rea, M. S. The impact of light from computer monitors on melatonin levels in college students. *Neuro Endocrinol Lett* **32**, 158–63 (2011).

95 Cajochen, C. et al. Evening exposure to a light-emitting diodes (LED)-backlit computer screen affects circadian physiology and cognitive performance. *J Appl Physiol (1985)* **110**, 1432–8, doi:10.1152/japplphysiol.00165.2011 (2011).

96 Chang, A. M., Aeschbach, D., Duffy, J. F. and Czeisler, C. A. Evening use of light-emitting eReaders negatively affects sleep, circadian timing, and next-morning alertness. *Proc Natl Acad Sci USA* **112**, 1232–7, doi:10.1073/pnas.1418490112 (2015).

97 Green, A., Cohen-Zion, M., Haim, A. and Dagan, Y. Evening light exposure to computer screens disrupts human sleep, biological rhythms, and attention abilities. *Chronobiol Int* **34**, 855–65, doi:10.1080/07420528.2017.1324878 (2017).

98 Kazemi, R., Alighanbari, N. and Zamanian, Z. The effects of screen light filtering software on cognitive performance and sleep among

night workers. *Health Promot Perspect* **9**, 233–40, doi:10.15171/hpp.2019.32 (2019).

99  Harbard, E., Allen, N. B., Trinder, J. and Bei, B. What's keeping teenagers up? Prebedtime behaviors and actigraphy-assessed sleep over school and vacation. *J Adolesc Health* **58**, 426–32, doi:10.1016/j.jadohealth.2015.12.011 (2016).

100  Zaidi, F. H. et al. Short-wavelength light sensitivity of circadian, pupillary, and visual awareness in humans lacking an outer retina. *Curr Biol* **17**, 2122–8, doi:10.1016/j.cub.2007.11.034 (2007).

101  Chellappa, S. L. et al. Non-visual effects of light on melatonin, alertness and cognitive performance: can blue-enriched light keep us alert? *PLoS One* **6**, e16429, doi:10.1371/journal.pone.0016429 (2011).

102  Mrosovsky, N. Masking: history, definitions, and measurement. *Chronobiol Int* **16**, 415–29, doi:10.3109/07420529908998717 (1999).

103  Hazelhoff, E. M., Dudink, J., Meijer, J. H. and Kervezee, L. Beginning to see the light: lessons learned from the development of the circadian system for optimizing light conditions in the neonatal intensive care unit. *Front Neurosci* **15**, 634034, doi:10.3389/fnins.2021.634034 (2021).

104  Kalafatakis, K., Russell, G. M. and Lightman, S. L. Mechanisms in endocrinology: does circadian and ultradian glucocorticoid exposure affect the brain? *Eur J Endocrinol* **180**, R73–R89, doi:10.1530/EJE-18-0853 (2019).

105  Andrews, R. C., Herlihy, O., Livingstone, D. E., Andrew, R. and Walker, B. R. Abnormal cortisol metabolism and tissue sensitivity to cortisol in patients with glucose intolerance. *J Clin Endocrinol Metab* **87**, 5587–93, doi:10.1210/jc.2002-020048 (2002).

106  Van der Valk, E. S., Savas, M. and van Rossum, E. F. C. Stress and obesity: are there more susceptible individuals? *Curr Obes Rep* **7**, 193–203, doi:10.1007/s13679-018-0306-y (2018).

107  Leal-Cerro, A., Soto, A., Martinez, M. A., Dieguez, C. and Casanueva, F. F. Influence of cortisol status on leptin secretion. *Pituitary* **4**, 111–16, doi:10.1023/a:1012903330944 (2001).

108  Spiegel, K., Leproult, R. and Van Cauter, E. Impact of sleep debt on metabolic and endocrine function. *Lancet* **354**, 1435–9, doi:10.1016/S0140-6736(99)01376-8 (1999).

109  Morey, J. N., Boggero, I. A., Scott, A. B. and Segerstrom, S. C. Current directions in stress and human immune function. *Curr Opin Psychol* **5**, 13–17, doi:10.1016/j.copsyc.2015.03.007 (2015).

110  Nojkov, B., Rubenstein, J. H., Chey, W. D. and Hoogerwerf, W. A. The impact of rotating shift work on the prevalence of irritable bowel syndrome in nurses. *Am J Gastroenterol* **105**, 842–7, doi:10.1038/ajg.2010.48 (2010).

111  Vyas, M. V. et al. Shift work and vascular events: systematic review and meta-analysis. *BMJ* **345**, e4800, doi:10.1136/bmj.e4800 (2012).

112  Ackermann, S., Hartmann, F., Papassotiropoulos, A., de Quervain, D. J. and Rasch, B. Associations between basal cortisol levels and memory retrieval in healthy young individuals. *J Cogn Neurosci* **25**, 1896–1907, doi:10.1162/jocn_a_00440 (2013).

113  Spira, A. P., Chen-Edinboro, L. P., Wu, M. N. and Yaffe, K. Impact of sleep on the risk of cognitive decline and dementia. *Curr Opin Psychiatry* **27**, 478–83, doi:10.1097/YCO.0000000000000106 (2014).

114  Ouanes, S. and Popp, J. High cortisol and the risk of dementia and Alzheimer's disease: a review of the literature. *Front Aging Neurosci* **11**, 43, doi:10.3389/fnagi.2019.00043 (2019).

115  Zankert, S., Bellingrath, S., Wust, S. and Kudielka, B. M. HPA axis responses to psychological challenge linking stress and disease: what do we know on sources of intra- and interindividual variability? *Psychoneuroendocrinology* **105**, 86–97, doi:10.1016/j.psyneuen.2018.10.027 (2019).

116  Lavretsky, H. and Newhouse, P. A. Stress, inflammation, and aging. *Am J Geriatr Psychiatry* **20**, 729–33, doi:10.1097/JGP.0b013e31826573cf (2012).

117  Costa, G. and Di Milia, L. Aging and shift work: a complex problem to face. *Chronobiol Int* **25**, 165–81, doi:10.1080/07420520802103410 (2008).

References

118 Dimitrov, S. et al. Cortisol and epinephrine control opposing circadian rhythms in T cell subsets. *Blood* **113**, 5134–43, doi:10.1182/blood-2008–11–190769 (2009).

119 Buckley, T. M. and Schatzberg, A. F. On the interactions of the hypothalamic-pituitary-adrenal (HPA) axis and sleep: normal HPA axis activity and circadian rhythm, exemplary sleep disorders. *J Clin Endocrinol Metab* **90**, 3106–14, doi:10.1210/jc.2004–1056 (2005).

120 Abell, J. G., Shipley, M. J., Ferrie, J. E., Kivimaki, M. and Kumari, M. Recurrent short sleep, chronic insomnia symptoms and salivary cortisol: a 10-year follow-up in the Whitehall II study. *Psychoneuroendocrinology* **68**, 91–9, doi:10.1016/j.psyneuen.2016.02.021 (2016).

121 Van Cauter, E. et al. Impact of sleep and sleep loss on neuroendocrine and metabolic function. *Horm Res* **67 Suppl** 1, 2–9, doi:10.1159/000097543 (2007).

122 Van Cauter, E., Spiegel, K., Tasali, E. and Leproult, R. Metabolic consequences of sleep and sleep loss. *Sleep Med* **9 Suppl** 1, S23–8, doi:10.1016/S1389–9457(08)70013–3 (2008).

123 Akerstedt, T. Psychosocial stress and impaired sleep. *Scand J Work Environ Health* **32**, 493–501 (2006).

124 Schwarz, J. et al. Does sleep deprivation increase the vulnerability to acute psychosocial stress in young and older adults? *Psychoneuroendocrinology* **96**, 155–65, doi:10.1016/j.psyneuen.2018.06.003 (2018).

125 Banks, S. and Dinges, D. F. Behavioral and physiological consequences of sleep restriction. *J Clin Sleep Med* **3**, 519–28 (2007).

126 Oginska, H. and Pokorski, J. Fatigue and mood correlates of sleep length in three age-social groups: school children, students, and employees. *Chronobiol Int* **23**, 1317–28, doi:10.1080/07420520601089349 (2006).

127 Scott, J. P., McNaughton, L. R. and Polman, R. C. Effects of sleep deprivation and exercise on cognitive, motor performance and mood. *Physiol Behav* **87**, 396–408, doi:10.1016/j.physbeh.2005.11.009 (2006).

128 Selvi, Y., Gulec, M., Agargun, M. Y. and Besiroglu, L. Mood changes after sleep deprivation in morningness–eveningness chronotypes

in healthy individuals. *J Sleep Res* **16**, 241–4, doi:10.1111/j.1365–2869.2007.00596.x (2007).

129 Dahl, R. E. and Lewin, D. S. Pathways to adolescent health: sleep regulation and behavior. *J Adolesc Health* **31**, 175–84 (2002).

130 Kelman, B. B. The sleep needs of adolescents. *J Sch Nurs* **15**, 14–19 (1999).

131 Muecke, S. Effects of rotating night shifts: literature review. *J Adv Nurs* **50**, 433–9, doi:10.1111/j.1365–2648.2005.03409.x (2005).

132 Acheson, A., Richards, J. B. and de Wit, H. Effects of sleep deprivation on impulsive behaviors in men and women. *Physiol Behav* **91**, 579–87, doi:10.1016/j.physbeh.2007.03.020 (2007).

133 McKenna, B. S., Dickinson, D. L., Orff, H. J. and Drummond, S. P. The effects of one night of sleep deprivation on known-risk and ambiguous-risk decisions. *J Sleep Res* **16**, 245–52, doi:10.1111/j.1365–2869.2007.00591.x (2007).

134 O'Brien, E. M. and Mindell, J. A. Sleep and risk-taking behavior in adolescents. *Behav Sleep Med* **3**, 113–33, doi:10.1207/s15402010bsm0303_1 (2005).

135 Venkatraman, V., Chuah, Y. M., Huettel, S. A. and Chee, M. W. Sleep deprivation elevates expectation of gains and attenuates response to losses following risky decisions. *Sleep* **30**, 603–9, doi:10.1093/sleep/30.5.603 (2007).

136 Baranski, J. V. and Pigeau, R. A. Self-monitoring cognitive performance during sleep deprivation: effects of modafinil, d-amphetamine and placebo. *J Sleep Res* **6**, 84–91 (1997).

137 Boivin, D. B., Tremblay, G. M. and James, F. O. Working on atypical schedules. *Sleep Med* **8**, 578–89, doi:10.1016/j.sleep.2007.03.015 (2007).

138 Killgore, W. D., Balkin, T. J. and Wesensten, N. J. Impaired decision making following 49 h of sleep deprivation. *J Sleep Res* **15**, 7–13, doi:10.1111/j.1365–2869.2006.00487.x (2006).

139 Roehrs, T. and Roth, T. Sleep, sleepiness, sleep disorders and alcohol use and abuse. *Sleep Med Rev* **5**, 287–97, doi:10.1053/smrv.2001.0162 (2001).

140  Roehrs, T. and Roth, T. Sleep, sleepiness, and alcohol use. *Alcohol Res Health* **25**, 101–9 (2001).

141  Mednick, S. C., Christakis, N. A. and Fowler, J. H. The spread of sleep loss influences drug use in adolescent social networks. *PLoS One* **5**, e9775, doi:10.1371/journal.pone.0009775 (2010).

142  Dinges, D. F. et al. Cumulative sleepiness, mood disturbance, and psychomotor vigilance performance decrements during a week of sleep restricted to 4–5 hours per night. *Sleep* **20**, 267–77 (1997).

143  Lamond, N. et al. The dynamics of neurobehavioural recovery following sleep loss. *J Sleep Res* **16**, 33–41, doi:10.1111/j.1365–2869.2007.00574.x (2007).

144  Pilcher, J. J. and Huffcutt, A. I. Effects of sleep deprivation on performance: a meta-analysis. *Sleep* **19**, 318–26, doi:10.1093/sleep/19.4.318 (1996).

145  Chee, M. W. and Chuah, L. Y. Functional neuroimaging insights into how sleep and sleep deprivation affect memory and cognition. *Curr Opin Neurol* **21**, 417–23, doi:10.1097/WCO.0b013e3283052cf7 (2008).

146  Dworak, M., Schierl, T., Bruns, T. and Struder, H. K. Impact of singular excessive computer game and television exposure on sleep patterns and memory performance of school-aged children. *Pediatrics* **120**, 978–85, doi:10.1542/peds.2007–0476 (2007).

147  Goder, R., Scharffetter, F., Aldenhoff, J. B. and Fritzer, G. Visual declarative memory is associated with non-rapid eye movement sleep and sleep cycles in patients with chronic non-restorative sleep. *Sleep Med* **8**, 503–8, doi:10.1016/j.sleep.2006.11.014 (2007).

148  Oken, B. S., Salinsky, M. C. and Elsas, S. M. Vigilance, alertness, or sustained attention: physiological basis and measurement. *Clin Neurophysiol* **117**, 1885–1901, doi:10.1016/j.clinph.2006.01.017 (2006).

149  Baranski, J. V. et al. Effects of sleep loss on team decision making: motivational loss or motivational gain? *Hum Factors* **49**, 646–60, doi:10.1518/001872007X215728 (2007).

150  Harrison, Y. and Horne, J. A. The impact of sleep deprivation on decision making: a review. *J Exp Psychol Appl* **6**, 236–49 (2000).

151 Killgore, W. D. et al. The effects of 53 hours of sleep deprivation on moral judgment. *Sleep* 30, 345–52, doi:10.1093/sleep/30.3.345 (2007).

152 Lucidi, F. et al. Sleep-related car crashes: risk perception and decision-making processes in young drivers. *Accid Anal Prev* 38, 302–9, doi:10.1016/j.aap.2005.09.013 (2006).

153 Horne, J. A. Sleep loss and 'divergent' thinking ability. *Sleep* 11, 528–36, doi:10.1093/sleep/11.6.528 (1988).

154 Jones, K. and Harrison, Y. Frontal lobe function, sleep loss and fragmented sleep. *Sleep Med Rev* 5, 463–75, doi:10.1053/smrv.2001.0203 (2001).

155 Killgore, W. D. et al. Sleep deprivation reduces perceived emotional intelligence and constructive thinking skills. *Sleep Med* 9, 517–26, doi:10.1016/j.sleep.2007.07.003 (2008).

156 Randazzo, A. C., Muehlbach, M. J., Schweitzer, P. K. and Walsh, J. K. Cognitive function following acute sleep restriction in children ages 10–14. *Sleep* 21, 861–8 (1998).

157 Kahol, K. et al. Effect of fatigue on psychomotor and cognitive skills. *Am J Surg* 195, 195–204, doi:10.1016/j.amjsurg.2007.10.004 (2008).

158 Tucker, A. M., Whitney, P., Belenky, G., Hinson, J. M. and Van Dongen, H. P. Effects of sleep deprivation on dissociated components of executive functioning. *Sleep* 33, 47–57, doi:10.1093/sleep/33.1.47 (2010).

159 Giesbrecht, T., Smeets, T., Leppink, J., Jelicic, M. and Merckelbach, H. Acute dissociation after 1 night of sleep loss. *J Abnorm Psychol* 116, 599–606, doi:10.1037/0021–843X.116.3.599 (2007).

160 Basner, M., Glatz, C., Griefahn, B., Penzel, T. and Samel, A. Aircraft noise: effects on macro- and microstructure of sleep. *Sleep Med* 9, 382–7, doi:10.1016/j.sleep.2007.07.002 (2008).

161 Philip, P. and Akerstedt, T. Transport and industrial safety, how are they affected by sleepiness and sleep restriction? *Sleep Med Rev* 10, 347–56, doi:10.1016/j.smrv.2006.04.002 (2006).

162 Pilcher, J. J., Lambert, B. J. and Huffcutt, A. I. Differential effects of permanent and rotating shifts on self-report sleep length: a meta-analytic review. *Sleep* 23, 155–63 (2000).

163  Scott, L. D. et al. The relationship between nurse work schedules, sleep duration, and drowsy driving. *Sleep* **30**, 1801–7, doi:10.1093/sleep/30.12.1801 (2007).

164  Meerlo, P., Sgoifo, A. and Suchecki, D. Restricted and disrupted sleep: effects on autonomic function, neuroendocrine stress systems and stress responsivity. *Sleep Med Rev* **12**, 197–210, doi:10.1016/j.smrv.2007.07.007 (2008).

165  Phan, T. X. and Malkani, R. G. Sleep and circadian rhythm disruption and stress intersect in Alzheimer's disease. *Neurobiol Stress* **10**, 100133, doi:10.1016/j.ynstr.2018.10.001 (2019).

166  Kundermann, B., Krieg, J. C., Schreiber, W. and Lautenbacher, S. The effect of sleep deprivation on pain. *Pain Res Manag* **9**, 25–32, doi:10.1155/2004/949187 (2004).

167  Landis, C. A., Savage, M. V., Lentz, M. J. and Brengelmann, G. L. Sleep deprivation alters body temperature dynamics to mild cooling and heating not sweating threshold in women. *Sleep* **21**, 101–8, doi:10.1093/sleep/21.1.101 (1998).

168  Roehrs, T., Hyde, M., Blaisdell, B., Greenwald, M. and Roth, T. Sleep loss and REM sleep loss are hyperalgesic. *Sleep* **29**, 145–51, doi:10.1093/sleep/29.2.145 (2006).

169  Irwin, M. Effects of sleep and sleep loss on immunity and cytokines. *Brain Behav Immun* **16**, 503–12 (2002).

170  Lorton, D. et al. Bidirectional communication between the brain and the immune system: implications for physiological sleep and disorders with disrupted sleep. *Neuroimmunomodulation* **13**, 357–74, doi:10.1159/000104864 (2006).

171  Davis, S. and Mirick, D. K. Circadian disruption, shift work and the risk of cancer: a summary of the evidence and studies in Seattle. *Cancer Causes Control* **17**, 539–45, doi:10.1007/s10552-005-9010-9 (2006).

172  Hansen, J. Risk of breast cancer after night- and shift work: current evidence and ongoing studies in Denmark. *Cancer Causes Control* **17**, 531–7, doi:10.1007/s10552-005-9006-5 (2006).

173  Kakizaki, M. et al. Sleep duration and the risk of breast cancer: the Ohsaki Cohort Study. *Br J Cancer* **99**, 1502–5, doi:10.1038/sj.bjc.6604684 (2008).

174  Gangwisch, J. E., Malaspina, D., Boden-Albala, B. and Heymsfield, S. B. Inadequate sleep as a risk factor for obesity: analyses of the NHANES I. *Sleep* **28**, 1289–96, doi:10.1093/sleep/28.10.1289 (2005).

175  Knutson, K. L., Spiegel, K., Penev, P. and Van Cauter, E. The metabolic consequences of sleep deprivation. *Sleep Med Rev* **11**, 163–78, doi:10.1016/j.smrv.2007.01.002 (2007).

176  Luyster, F. S. et al. Sleep: a health imperative. *Sleep* **35**, 727–34, doi:10.5665/sleep.1846 (2012).

177  Maemura, K., Takeda, N. and Nagai, R. Circadian rhythms in the CNS and peripheral clock disorders: role of the biological clock in cardiovascular diseases. *J Pharmacol Sci* **103**, 134–8 (2007).

178  Young, M. E. and Bray, M. S. Potential role for peripheral circadian clock dyssynchrony in the pathogenesis of cardiovascular dysfunction. *Sleep Med* **8**, 656–67, doi:10.1016/j.sleep.2006.12.010 (2007).

179  Johnson, E. O., Roth, T. and Breslau, N. The association of insomnia with anxiety disorders and depression: exploration of the direction of risk. *J Psychiatr Res* **40**, 700–708, doi:10.1016/j.jpsychires.2006.07.008 (2006).

180  Kahn-Greene, E. T., Killgore, D. B., Kamimori, G. H., Balkin, T. J. and Killgore, W. D. The effects of sleep deprivation on symptoms of psychopathology in healthy adults. *Sleep Med* **8**, 215–21, doi:10.1016/j.sleep.2006.08.007 (2007).

181  Riemann, D. and Voderholzer, U. Primary insomnia: a risk factor to develop depression? *J Affect Disord* **76**, 255–9 (2003).

182  Sharma, V. and Mazmanian, D. Sleep loss and postpartum psychosis. *Bipolar Disord* **5**, 98–105 (2003).

183  Carskadon, M. A. Sleep in adolescents: the perfect storm. *Pediatr Clin North Am* **58**, 637–47, doi:10.1016/j.pcl.2011.03.003 (2011).

184 Sleep Health Foundation. *Asleep on the job: costs of inadequate sleep in Australia*. August 2017. Report pdf available at: https://www.sleephealthfoundation.org.au/.

185 Hirotsu, C., Tufik, S. and Andersen, M. L. Interactions between sleep, stress, and metabolism: from physiological to pathological conditions. *Sleep Sci* **8**, 143–52, doi:10.1016/j.slsci.2015.09.002 (2015).

186 Ancoli-Israel, S., Ayalon, L. and Salzman, C. Sleep in the elderly: normal variations and common sleep disorders. *Harv Rev Psychiatry* **16**, 279–86, doi:10.1080/10673220802432210 (2008).

187 Dalziel, J. R. and Job, R. F. Motor vehicle accidents, fatigue and optimism bias in taxi drivers. *Accid Anal Prev* **29**, 489–94, doi:10.1016/s0001–4575(97)00028–6 (1997).

188 Folkard, S. Do permanent night workers show circadian adjustment? A review based on the endogenous melatonin rhythm. *Chronobiol Int* **25**, 215–24 (2008).

189 *The Lighting Handbook: Reference and Application (Illuminating Engineering Society of North America//Lighting Handbook)* 10th edn (Illuminating Engineering, 2019).

190 Czeisler, C. A. and Dijk, D. J. Use of bright light to treat maladaptation to night shift work and circadian rhythm sleep disorders. *J Sleep Res* **4**, 70–73 (1995).

191 Arendt, J. Shift work: coping with the biological clock. *Occup Med (Lond)* **60**, 10–20, doi:10.1093/occmed/kqp162 (2010).

192 Maidstone, R. et al. Shift work is associated with positive COVID-19 status in hospitalised patients. *Thorax* **76**, 601–6, doi:10.1136/thoraxjnl-2020–216651 (2021).

193 Hansen, J. Night shift work and risk of breast cancer. *Curr Environ Health Rep* **4**, 325–39, doi:10.1007/s40572–017–0155-y (2017).

194 Marquie, J. C., Tucker, P., Folkard, S., Gentil, C. and Ansiau, D. Chronic effects of shift work on cognition: findings from the VISAT longitudinal study. *Occup Environ Med* **72**, 258–64, doi:10.1136/oemed-2013–101993 (2015).

195 Wittmann, M., Dinich, J., Merrow, M. and Roenneberg, T. Social jetlag: misalignment of biological and social time. *Chronobiol Int* **23**, 497–509, doi:10.1080/07420520500545979 (2006).

196 Levandovski, R. et al. Depression scores associate with chronotype and social jetlag in a rural population. *Chronobiol Int* **28**, 771–8, doi: 10.3109/07420528.2011.602445 (2011).

197 Mitler, M. M. et al. Catastrophes, sleep, and public policy: consensus report. *Sleep* **11**, 100–109, doi:10.1093/sleep/11.1.100 (1988).

198 Cho, K. Chronic 'jet lag' produces temporal lobe atrophy and spatial cognitive deficits. *Nat Neurosci* **4**, 567–8, doi:10.1038/88384 (2001).

199 Cho, K., Ennaceur, A., Cole, J. C. and Suh, C. K. Chronic jet lag produces cognitive deficits. *J Neurosci* **20**, RC66 (2000).

200 Waterhouse, J. et al. Further assessments of the relationship between jet lag and some of its symptoms. *Chronobiol Int* **22**, 121–36, doi:10.1081/cbi-200036909 (2005).

201 Herxheimer, A. and Petrie, K. J. Melatonin for the prevention and treatment of jet lag. *Cochrane Database Syst Rev*, CD001520, doi:10.1002/14651858.CD001520 (2002).

202 Tortorolo, F., Farren, F. and Rada, G. Is melatonin useful for jet lag? *Medwave* **15 Suppl 3**, e6343, doi:10.5867/medwave.2015.6343 (2015).

203 Arendt, J. Does melatonin improve sleep? Efficacy of melatonin. *BMJ* **332**, 550, doi:10.1136/bmj.332.7540.550 (2006).

204 Wehrens, S. M. T. et al. Meal timing regulates the human circadian system. *Curr Biol* **27**, 1768–75 e1763, doi:10.1016/j.cub.2017.04.059 (2017).

205 Roenneberg, T., Kumar, C. J. and Merrow, M. The human circadian clock entrains to sun time. *Curr Biol* **17**, R44–5, doi:10.1016/j. cub.2006.12.011 (2007).

206 Roenneberg, T. et al. Why should we abolish daylight saving time? *J Biol Rhythms* **34**, 227–30, doi:10.1177/0748730419854197 (2019).

207 Hadlow, N. C., Brown, S., Wardrop, R. and Henley, D. The effects of season, daylight saving and time of sunrise on serum cortisol in a large population. *Chronobiol Int* **31**, 243–51, doi:10.3109/07420528.2 013.844162 (2014).

208 Harrison, Y. The impact of daylight saving time on sleep and related behaviours. *Sleep Med Rev* **17**, 285–92, doi:10.1016/j.smrv. 2012.10.001 (2013).

209 Zhang, H., Dahlen, T., Khan, A., Edgren, G. and Rzhetsky, A. Measurable health effects associated with the daylight saving time shift. *PLoS Comput Biol* **16**, e1007927, doi:10.1371/journal.pcbi.1007927 (2020).

210 Manfredini, R. et al. Daylight saving time and myocardial infarction: should we be worried? A review of the evidence. *Eur Rev Med Pharmacol Sci* **22**, 750–55, doi:10.26355/eurrev_201802_14306 (2018).

211 Sipilä, J. O., Ruuskanen, J. O., Rautava, P. and Kytö, V. Changes in ischemic stroke occurrence following daylight saving time transitions. *Sleep Med* **27–28**, 20–24, doi:10.1016/j.sleep.2016.10.009 (2016).

212 Barnes, C. M. and Wagner, D. T. Changing to daylight saving time cuts into sleep and increases workplace injuries. *J Appl Psychol* **94**, 1305–17, doi:10.1037/a0015320 (2009).

213 Fritz, J., VoPham, T., Wright, K. P., Jr and Vetter, C. A chronobiological evaluation of the acute effects of daylight saving time on traffic accident risk. *Curr Biol* **30**, 729–35 e722, doi:10.1016/j.cub.2019.12.045 (2020).

214 Todd, W. D. Potential pathways for circadian dysfunction and sundowning-related behavioral aggression in Alzheimer's disease and related dementias. *Front Neurosci* **14**, 910, doi:10.3389/fnins. 2020.00910 (2020).

215 Fabbian, F. et al. Chronotype, gender and general health. *Chronobiol Int* **33**, 863–82, doi:10.1080/07420528.2016.1176927 (2016).

216 Coppeta, L., Papa, F. and Magrini, A. Are shiftwork and indoor work related to D3 vitamin deficiency? A systematic review of current evidences. *J Environ Public Health* **2018**, 8468742, doi:10.1155/ 2018/8468742 (2018).

217 Perez-Lopez, F. R., Pilz, S. and Chedraui, P. Vitamin D supplementation during pregnancy: an overview. *Curr Opin Obstet Gynecol* **32**, 316–21, doi:10.1097/GCO.0000000000000641 (2020).

218 Friedman, M. Analysis, nutrition, and health benefits of tryptophan. *Int J Tryptophan Res* **11**, 1178646918802282, doi:10.1177/ 1178646918802282 (2018).

219  Casetta, G., Nolfo, A. P. and Palagi, E. Yawn contagion promotes motor synchrony in wild lions, *Panthera leo*. *Animal Behaviour* **174**, 149–59 (2021).

220  Giuntella, O. and Mazzonna, F. Sunset time and the economic effects of social jetlag: evidence from US time zone borders. *J Health Econ* **65**, 210–26, doi:10.1016/j.jhealeco.2019.03.007 (2019).

221  Dean, K. and Murray, R. M. Environmental risk factors for psychosis. *Dialogues Clin Neurosci* **7**, 69–80 (2005).

222  Gaine, M. E., Chatterjee, S. and Abel, T. Sleep deprivation and the epigenome. *Front Neural Circuits* **12**, 14, doi:10.3389/fncir.2018.00014 (2018).

223  Lindberg, E. et al. Sleep time and sleep-related symptoms across two generations – results of the community-based RHINE and RHINESSA studies. *Sleep Med* **69**, 8–13, doi:10.1016/j.sleep.2019.12.017 (2020).

224  Thorpy, M. J. Classification of sleep disorders. *Neurotherapeutics* **9**, 687–701, doi:10.1007/s13311–012–0145–6 (2012).

225  Greenberg, D. B. Clinical dimensions of fatigue. *Prim Care Companion J Clin Psychiatry* **4**, 90–93, doi:10.4088/pcc.v04n0301 (2002).

226  Marshall, M. The lasting misery of coronavirus long-haulers. *Nature* **585**, 339–41, doi:10.1038/d41586–020–02598–6 (2020).

227  Wehr, T. A. In short photoperiods, human sleep is biphasic. *J Sleep Res* **1**, 103–7 (1992).

228  Ekirch, A. R. Segmented sleep in pre-industrial societies. *Sleep* **39**, 715–16, doi:10.5665/sleep.5558 (2016).

229  Yetish, G. et al. Natural sleep and its seasonal variations in three pre-industrial societies. *Curr Biol* **25**, 2862–8, doi:10.1016/j.cub.2015.09.046 (2015).

230  Ekirch, A. R. *At Day's Close: A History of Nighttime* (W. W. Norton and Company, 2005).

231  Handley, S. *Sleep in Early Modern England* (Yale University Press, 2016).

232  Duncan, W. C., Barbato, G., Fagioli, I., Garcia-Borreguero, D. and Wehr, T. A. A biphasic daily pattern of slow wave activity during a

two-day 90-minute sleep–wake schedule. *Arch Ital Biol* **147**, 117–30 (2009).

233  Kleitman, N. Basic rest–activity cycle – 22 years later. *Sleep* **5**, 311–17, doi:10.1093/sleep/5.4.311 (1982).

234  Weaver, M. D. et al. Adverse impact of polyphasic sleep patterns in humans: Report of the National Sleep Foundation sleep timing and variability consensus panel. *Sleep Health*, doi:10.1016/j.sleh.2021. 02.009 (2021).

235  Shanware, N. P. et al. Casein kinase 1-dependent phosphorylation of familial advanced sleep phase syndrome-associated residues controls PERIOD 2 stability. *J Biol Chem* **286**, 12766–74, doi:10.1074/ jbc.M111.224014 (2011).

236  Toh, K. L. et al. An hPer2 phosphorylation site mutation in familial advanced sleep phase syndrome. *Science* **291**, 1040–43, doi:10.1126/ science.1057499 (2001).

237  Reid, K. J. et al. Familial advanced sleep phase syndrome. *Arch Neurol* **58**, 1089–94, doi:10.1001/archneur.58.7.1089 (2001).

238  Stepnowsky, C. J. and Ancoli-Israel, S. Sleep and its disorders in seniors. *Sleep Med Clin* **3**, 281–93, doi:10.1016/j.jsmc.2008.01.011 (2008).

239  Ancoli-Israel, S., Schnierow, B., Kelsoe, J. and Fink, R. A pedigree of one family with delayed sleep phase syndrome. *Chronobiol Int* **18**, 831–40, doi:10.1081/cbi-100107518 (2001).

240  Patke, A. et al. Mutation of the human circadian clock gene cry1 in familial delayed sleep phase disorder. *Cell* **169**, 203–15 e213, doi:10. 1016/j.cell.2017.03.027 (2017).

241  Crowley, S. J., Acebo, C. and Carskadon, M. A. Sleep, circadian rhythms, and delayed phase in adolescence. *Sleep Med* **8**, 602–12, doi:10.1016/j.sleep.2006.12.002 (2007).

242  Obeysekare, J. L. et al. Delayed sleep timing and circadian rhythms in pregnancy and transdiagnostic symptoms associated with postpartum depression. *Transl Psychiatry* **10**, 14, doi:10.1038/s41398-020-0683-3 (2020).

243  Turner, J. et al. A prospective study of delayed sleep phase syndrome in patients with severe resistant obsessive-compulsive disorder. *World Psychiatry* **6**, 108–11 (2007).

244  Esbensen, A. J. and Schwichtenberg, A. J. Sleep in neurodevelopmental disorders. *Int Rev Res Dev Disabil* **51**, 153–91, doi:10.1016/bs.irrdd.2016.07.005 (2016).

245  Andrews, C. D. et al. Sleep–wake disturbance related to ocular disease: a systematic review of phase-shifting pharmaceutical therapies. *Transl Vis Sci Technol* **8**, 49, doi:10.1167/tvst.8.3.49 (2019).

246  Wulff, K., Dijk, D. J., Middleton, B., Foster, R. G. and Joyce, E. M. Sleep and circadian rhythm disruption in schizophrenia. *Br J Psychiatry* **200**, 308–16, doi:10.1192/bjp.bp.111.096321 (2012).

247  Wulff, K., Gatti, S., Wettstein, J. G. and Foster, R. G. Sleep and circadian rhythm disruption in psychiatric and neurodegenerative disease. *Nat Rev Neurosci* **11**, 589–99, doi:10.1038/nrn2868 (2010).

248  Brennan, K. C. et al. Casein kinase Iẟ mutations in familial migraine and advanced sleep phase. *Sci Transl Med* **5**, 183ra156–11, doi:10.1126/scitranslmed.3005784 (2013).

249  Arendt, J. Melatonin: countering chaotic time cues. *Front Endocrinol (Lausanne)* **10**, 391, doi:10.3389/fendo.2019.00391 (2019).

250  Brown, M. A., Quan, S. F. and Eichling, P. S. Circadian rhythm sleep disorder, free-running type in a sighted male with severe depression, anxiety, and agoraphobia. *J Clin Sleep Med* **7**, 93–4 (2011).

251  Leng, Y., Musiek, E. S., Hu, K., Cappuccio, F. P. and Yaffe, K. Association between circadian rhythms and neurodegenerative diseases. *Lancet Neurol* **18**, 307–18, doi:10.1016/S1474-4422(18)30461-7 (2019).

252  American Academy of Sleep Medicine *International Classification of Sleep Disorders*, 3rd edn (2014).

253  Patel, D., Steinberg, J. and Patel, P. Insomnia in the elderly: a review. *J Clin Sleep Med* **14**, 1017–24, doi:10.5664/jcsm.7172 (2018).

254  Wennberg, A. M. V., Wu, M. N., Rosenberg, P. B. and Spira, A. P. Sleep disturbance, cognitive decline, and dementia: a review. *Semin Neurol* **37**, 395–406, doi:10.1055/s-0037–1604351 (2017).

255  Nutt, D., Wilson, S. and Paterson, L. Sleep disorders as core symptoms of depression. *Dialogues Clin Neurosci* **10**, 329–36 (2008).

256  Dauvilliers, Y. Insomnia in patients with neurodegenerative conditions. *Sleep Med* **8 Suppl 4**, S27–34, doi:10.1016/S1389–9457(08)70006–6 (2007).

257  Troxel, W. M. et al. Sleep symptoms predict the development of the metabolic syndrome. *Sleep* **33**, 1633–40, doi:10.1093/sleep/33.12.1633 (2010).

258  Kaneshwaran, K. et al. Sleep fragmentation, microglial aging, and cognitive impairment in adults with and without Alzheimer's dementia. *Sci Adv* **5**, eaax7331, doi:10.1126/sciadv.aax7331 (2019).

259  Abbott, S. M. and Videnovic, A. Chronic sleep disturbance and neural injury: links to neurodegenerative disease. *Nat Sci Sleep* **8**, 55–61, doi:10.2147/NSS.S78947 (2016).

260  Stamatakis, K. A. and Punjabi, N. M. Effects of sleep fragmentation on glucose metabolism in normal subjects. *Chest* **137**, 95–101, doi:10.1378/chest.09–0791 (2010).

261  Kim, A. M. et al. Tongue fat and its relationship to obstructive sleep apnea. *Sleep* **37**, 1639–48, doi:10.5665/sleep.4072 (2014).

262  Santos, M. and Hofmann, R. J. Ocular manifestations of obstructive sleep apnea. *J Clin Sleep Med* **13**, 1345–8, doi:10.5664/jcsm.6812 (2017).

263  Findley, L. J. and Suratt, P. M. Serious motor vehicle crashes: the cost of untreated sleep apnoea. *Thorax* **56**, 505, doi:10.1136/thorax.56.7.505 (2001).

264  Eckert, D. J. and Sweetman, A. Impaired central control of sleep depth propensity as a common mechanism for excessive overnight wake time: implications for sleep apnea, insomnia and beyond. *J Clin Sleep Med* **16**, 341–3, doi:10.5664/jcsm.8268 (2020).

265  Boing, S. and Randerath, W. J. Chronic hypoventilation syndromes and sleep-related hypoventilation. *J Thorac Dis* **7**, 1273–85, doi:10.3978/j.issn.2072–1439.2015.06.10 (2015).

266  Jen, R., Li, Y., Owens, R. L. and Malhotra, A. Sleep in chronic obstructive pulmonary disease: evidence gaps and challenges. *Can Respir J* **2016**, 7947198, doi:10.1155/2016/7947198 (2016).

267 Levy, P. et al. Intermittent hypoxia and sleep-disordered breathing: current concepts and perspectives. *Eur Respir J* **32**, 1082–95, doi:10.1183/09031936.00013308 (2008).

268 Mahoney, C. E., Cogswell, A., Koralnik, I. J. and Scammell, T. E. The neurobiological basis of narcolepsy. *Nat Rev Neurosci* **20**, 83–93, doi:10.1038/s41583–018–0097-x (2019).

269 Kaushik, M. K. et al. Continuous intrathecal orexin delivery inhibits cataplexy in a murine model of narcolepsy. *Proc Natl Acad Sci USA* **115**, 6046–51, doi:10.1073/pnas.1722686115 (2018).

270 Nellore, A. and Randall, T. D. Narcolepsy and influenza vaccination – the inappropriate awakening of immunity. *Ann Transl Med* **4**, S29, doi:10.21037/atm.2016.10.60 (2016).

271 Bonvalet, M., Ollila, H. M., Ambati, A. and Mignot, E. Autoimmunity in narcolepsy. *Curr Opin Pulm Med* **23**, 522–9, doi:10.1097/MCP.0000000000000426 (2017).

272 Luo, G. et al. Autoimmunity to hypocretin and molecular mimicry to flu in type 1 narcolepsy. *Proc Natl Acad Sci USA* **115**, E12323-E12332, doi:10.1073/pnas.1818150116 (2018).

273 Singh, S., Kaur, H., Singh, S. and Khawaja, I. Parasomnias: a comprehensive review. *Cureus* **10**, e3807, doi:10.7759/cureus.3807 (2018).

274 Tekriwal, A. et al. REM sleep behaviour disorder: prodromal and mechanistic insights for Parkinson's disease. *J Neurol Neurosurg Psychiatry* **88**, 445–51, doi:10.1136/jnnp-2016–314471 (2017).

275 Reddy, S. V., Kumar, M. P., Sravanthi, D., Mohsin, A. H. and Anuhya, V. Bruxism: a literature review. *J Int Oral Health* **6**, 105–9 (2014).

276 Walters, A. S. Clinical identification of the simple sleep-related movement disorders. *Chest* **131**, 1260–66, doi:10.1378/chest.06–1602 (2007).

277 Ferini-Strambi, L., Carli, G., Casoni, F. and Galbiati, A. Restless legs syndrome and Parkinson disease: a causal relationship between the two disorders? *Front Neurol* **9**, 551, doi:10.3389/fneur.2018.00551 (2018).

278 Patrick, L. R. Restless legs syndrome: pathophysiology and the role of iron and folate. *Altern Med Rev* **12**, 101–12 (2007).

279 Novak, M., Winkelman, J. W. and Unruh, M. Restless legs syndrome in patients with chronic kidney disease. *Semin Nephrol* **35**, 347–58, doi:10.1016/j.semnephrol.2015.06.006 (2015).

280 Sateia, M. J. International classification of sleep disorders – third edition: highlights and modifications. *Chest* **146**, 1387–94, doi:10.1378/chest.14–0970 (2014).

281 Pittler, M. H. and Ernst, E. Kava extract for treating anxiety. *Cochrane Database Syst Rev*, CD003383, doi:10.1002/14651858.CD003383 (2003).

282 Shinomiya, K. et al. Effects of kava-kava extract on the sleep–wake cycle in sleep-disturbed rats. *Psychopharmacology (Berl)* **180**, 564–9, doi:10.1007/s00213–005–2196–4 (2005).

283 Wick, J. Y. The history of benzodiazepines. *Consult Pharm* **28**, 538–48, doi:10.4140/TCP.n.2013.538 (2013).

284 Ferentinos, P. and Paparrigopoulos, T. Zopiclone and sleepwalking. *Int J Neuropsychopharmacol* **12**, 141–2, doi:10.1017/S1461145708009541 (2009).

285 Fernandez-Mendoza, J. et al. Sleep misperception and chronic insomnia in the general population: role of objective sleep duration and psychological profiles. *Psychosom Med* **73**, 88–97, doi:10.1097/PSY.0b013e3181fe365a (2011).

286 Van Maanen, A., Meijer, A. M., van der Heijden, K. B. and Oort, F. J. The effects of light therapy on sleep problems: A systematic review and meta-analysis. *Sleep Med Rev* **29**, 52–62, doi:10.1016/j.smrv.2015.08.009 (2016).

287 Milner, C. E. and Cote, K. A. Benefits of napping in healthy adults: impact of nap length, time of day, age, and experience with napping. *J Sleep Res* **18**, 272–81, doi:10.1111/j.1365–2869.2008.00718.x (2009).

288 Donskoy, I. and Loghmanee, D. Insomnia in adolescence. *Med Sci (Basel)* **6**, doi:10.3390/medsci6030072 (2018).

289 Fukuda, K. and Ishihara, K. Routine evening naps and night-time sleep patterns in junior high and high school students. *Psychiatry Clin Neurosci* **56**, 229–30, doi:10.1046/j.1440–1819.2002.00986.x (2002).

290 Dolezal, B. A., Neufeld, E. V., Boland, D. M., Martin, J. L. and Cooper, C. B. Interrelationship between sleep and exercise: a systematic review. *Adv Prev Med* **2017**, 1364387, doi:10.1155/2017/1364387 (2017).

291 Murray, K. et al. The relations between sleep, time of physical activity, and time outdoors among adult women. *PLoS One* **12**, e0182013, doi:10.1371/journal.pone.0182013 (2017).

292 Harding, E. C., Franks, N. P. and Wisden, W. The temperature dependence of sleep. *Front Neurosci* **13**, 336, doi:10.3389/fnins.2019.00336 (2019).

293 Stutz, J., Eiholzer, R. and Spengler, C. M. Effects of evening exercise on sleep in healthy participants: a systematic review and meta-analysis. *Sports Med* **49**, 269–87, doi:10.1007/s40279–018–1015–0 (2019).

294 Thomas, C., Jones, H., Whitworth-Turner, C. and Louis, J. High-intensity exercise in the evening does not disrupt sleep in endurance runners. *Eur J Appl Physiol* **120**, 359–68, doi:10.1007/s00421–019–04280-w (2020).

295 Dietrich, A. and McDaniel, W. F. Endocannabinoids and exercise. *Br J Sports Med* **38**, 536–41, doi:10.1136/bjsm.2004.011718 (2004).

296 McHill, A. W. et al. Later circadian timing of food intake is associated with increased body fat. *Am J Clin Nutr* **106**, 1213–19, doi:10.3945/ajcn.117.161588 (2017).

297 Beccuti, G. et al. Timing of food intake: Sounding the alarm about metabolic impairments? A systematic review. *Pharmacol Res* **125**, 132–41, doi:10.1016/j.phrs.2017.09.005 (2017).

298 Jehan, S. et al. Obesity, obstructive sleep apnea and type 2 diabetes mellitus: epidemiology and pathophysiologic insights. *Sleep Med Disord* **2**, 52–8 (2018).

299 Ruddick-Collins, L. C., Johnston, J. D., Morgan, P. J. and Johnstone, A. M. The big breakfast study: chrono-nutrition influence on energy expenditure and bodyweight. *Nutr Bull* **43**, 174–83, doi:10.1111/nbu.12323 (2018).

300 Fang, B., Liu, H., Yang, S., Xu, R. and Chen, G. Effect of subjective and objective sleep quality on subsequent peptic ulcer recurrence in older adults. *J Am Geriatr Soc* **67**, 1454–60, doi:10.1111/jgs.15871 (2019).

301  Verlander, L. A., Benedict, J. O. and Hanson, D. P. Stress and sleep patterns of college students. *Percept Mot Skills* **88**, 893–8, doi:10.2466/pms.1999.88.3.893 (1999).

302  Cajochen, C. et al. High sensitivity of human melatonin, alertness, thermoregulation, and heart rate to short wavelength light. *J Clin Endocrinol Metab* **90**, 1311–16, doi:10.1210/jc.2004–0957 (2005).

303  Mehta, R. and Zhu, R. J. Blue or red? Exploring the effect of color on cognitive task performances. *Science* **323**, 1226–9, doi:10.1126/science.1169144 (2009).

304  Lemmer, B. The sleep–wake cycle and sleeping pills. *Physiol Behav* **90**, 285–93, doi:10.1016/j.physbeh.2006.09.006 (2007).

305  He, Q., Chen, X., Wu, T., Li, L. and Fei, X. Risk of dementia in long-term benzodiazepine users: evidence from a meta-analysis of observational studies. *J Clin Neurol* **15**, 9–19, doi:10.3988/jcn.2019.15.1.9 (2019).

306  Osler, M. and Jorgensen, M. B. Associations of benzodiazepines, z-drugs, and other anxiolytics with subsequent dementia in patients with affective disorders: a nationwide cohort and nested case-control study. *Am J Psychiatry* **177**, 497–505, doi:10.1176/appi.ajp.2019.19030315 (2020).

307  Singleton, R. A., Jr and Wolfson, A. R. Alcohol consumption, sleep, and academic performance among college students. *J Stud Alcohol Drugs* **70**, 355–63, doi:10.15288/jsad.2009.70.355 (2009).

308  Raymann, R. J., Swaab, D. F. and Van Someren, E. J. Skin temperature and sleep-onset latency: changes with age and insomnia. *Physiol Behav* **90**, 257–66, doi:10.1016/j.physbeh.2006.09.008 (2007).

309  Krauchi, K., Cajochen, C., Werth, E. and Wirz-Justice, A. Functional link between distal vasodilation and sleep-onset latency? *Am J Physiol Regul Integr Comp Physiol* **278**, R741–8, doi:10.1152/ajpregu.2000.278.3.R741 (2000).

310  Fietze, I. et al. The effect of room acoustics on the sleep quality of healthy sleepers. *Noise Health* **18**, 240–46, doi:10.4103/1463–1741.192480 (2016).

311 Berk, M. Sleep and depression – theory and practice. *Aust Fam Physician* **38**, 302–4 (2009).

312 Cook, J. D., Eftekari, S. C., Dallmann, E., Sippy, M. and Plante, D. T. Ability of the Fitbit Alta HR to quantify and classify sleep in patients with suspected central disorders of hypersomnolence: a comparison against polysomnography. *J Sleep Res* **28**, e12789, doi:10.1111/jsr.12789 (2019).

313 Gavriloff, D. et al. Sham sleep feedback delivered via actigraphy biases daytime symptom reports in people with insomnia: Implications for insomnia disorder and wearable devices. *J Sleep Res* **27**, e12726, doi:10.1111/jsr.12726 (2018).

314 Fino, E. et al. (Not so) Smart sleep tracking through the phone: findings from a polysomnography study testing the reliability of four sleep applications. *J Sleep Res* **29**, e12935, doi:10.1111/jsr.12935 (2020).

315 Ko, P. R. et al. Consumer sleep technologies: a review of the landscape. *J Clin Sleep Med* **11**, 1455–61, doi:10.5664/jcsm.5288 (2015).

316 LeBourgeois, M. K., Giannotti, F., Cortesi, F., Wolfson, A. R. and Harsh, J. The relationship between reported sleep quality and sleep hygiene in Italian and American adolescents. *Pediatrics* **115**, 257–65, doi:10.1542/peds.2004–0815H (2005).

317 Kalmbach, D. A., Arnedt, J. T., Pillai, V. and Ciesla, J. A. The impact of sleep on female sexual response and behavior: a pilot study. *J Sex Med* **12**, 1221–32, doi:10.1111/jsm.12858 (2015).

318 Lastella, M., O'Mullan, C., Paterson, J. L. and Reynolds, A. C. Sex and sleep: perceptions of sex as a sleep promoting behavior in the general adult population. *Front Public Health* **7**, 33, doi:10.3389/fpubh.2019.00033 (2019).

319 Kroeger, M. Oxytocin: key hormone in sexual intercourse, parturition, and lactation. *Birth Gaz* **13**, 28–30 (1996).

320 Alley, J., Diamond, L. M., Lipschitz, D. L. and Grewen, K. Associations between oxytocin and cortisol reactivity and recovery in response to psychological stress and sexual arousal. *Psychoneuroendocrinology* **106**, 47–56, doi:10.1016/j.psyneuen.2019.03.031 (2019).

321 Kruger, T. H., Haake, P., Hartmann, U., Schedlowski, M. and Exton, M. S. Orgasm-induced prolactin secretion: feedback control of sexual drive? *Neurosci Biobehav Rev* **26**, 31–44, doi:10.1016/s0149-7634(01)00036-7 (2002).

322 Exton, M. S. et al. Coitus-induced orgasm stimulates prolactin secretion in healthy subjects. *Psychoneuroendocrinology* **26**, 287–94, doi:10.1016/s0306-4530(00)00053-6 (2001).

323 Bader, G. G. and Engdal, S. The influence of bed firmness on sleep quality. *Appl Ergon* **31**, 487–97 (2000).

324 Jacobson, B. H., Boolani, A. and Smith, D. B. Changes in back pain, sleep quality, and perceived stress after introduction of new bedding systems. *J Chiropr Med* **8**, 1–8, doi:10.1016/j.jcm.2008.09.002 (2009).

325 Krauchi, K. et al. Sleep on a high heat capacity mattress increases conductive body heat loss and slow wave sleep. *Physiol Behav* **185**, 23–30, doi:10.1016/j.physbeh.2017.12.014 (2018).

326 Chiba, S. et al. High rebound mattress toppers facilitate core body temperature drop and enhance deep sleep in the initial phase of nocturnal sleep. *PLoS One* **13**, e0197521, doi:10.1371/journal.pone.0197521 (2018).

327 Lytle, J., Mwatha, C. and Davis, K. K. Effect of lavender aromatherapy on vital signs and perceived quality of sleep in the intermediate care unit: a pilot study. *Am J Crit Care* **23**, 24–9, doi:10.4037/ajcc2014958 (2014).

328 Guadagna, S., Barattini, D. F., Rosu, S. and Ferini-Strambi, L. Plant extracts for sleep disturbances: a systematic review. *Evid Based Complement Alternat Med* **2020**, 3792390, doi:10.1155/2020/3792390 (2020).

329 Robertson, S., Loughran, S. and MacKenzie, K. Ear protection as a treatment for disruptive snoring: do ear plugs really work? *J Laryngol Otol* **120**, 381–4, doi:10.1017/S0022215106000363 (2006).

330 Blumen, M. et al. Effect of sleeping alone on sleep quality in female bed partners of snorers. *Eur Respir J* **34**, 1127–31, doi:10.1183/09031936.00012209 (2009).

331  Palagini, L. and Rosenlicht, N. Sleep, dreaming, and mental health: a review of historical and neurobiological perspectives. *Sleep Med Rev* **15**, 179–86, doi:10.1016/j.smrv.2010.07.003 (2011).

332  Paulson, S., Barrett, D., Bulkeley, K. and Naiman, R. Dreaming: a gateway to the unconscious? *Ann NY Acad Sci* **1406**, 28–45, doi:10.1111/nyas.13389 (2017).

333  Komasi, S., Soroush, A., Khazaie, H., Zakiei, A. and Saeidi, M. Dreams content and emotional load in cardiac rehabilitation patients and their relation to anxiety and depression. *Ann Card Anaesth* **21**, 388–92, doi:10.4103/aca.ACA_210_17 (2018).

334  Hartmann, E. and Brezler, T. A systematic change in dreams after 9/11/01. *Sleep* **31**, 213–18, doi:10.1093/sleep/31.2.213 (2008).

335  Revonsuo, A. The reinterpretation of dreams: an evolutionary hypothesis of the function of dreaming. *Behav Brain Sci* **23**, 877–901; discussion 904–1121, doi:10.1017/s0140525x00004015 (2000).

336  Rouder, J. N. and Morey, R. D. A Bayes factor meta-analysis of Bem's ESP claim. *Psychon Bull Rev* **18**, 682–9, doi:10.3758/s13423-011-0088-7 (2011).

337  Breslau, N. The epidemiology of trauma, PTSD, and other post-trauma disorders. *Trauma Violence Abuse* **10**, 198–210, doi:10.1177/1524838009334448 (2009).

338  Colvonen, P. J., Straus, L. D., Acheson, D. and Gehrman, P. A Review of the relationship between emotional learning and memory, sleep, and PTSD. *Curr Psychiatry Rep* **21**, 2, doi:10.1007/s11920-019-0987-2 (2019).

339  Porcheret, K., Holmes, E. A., Goodwin, G. M., Foster, R. G. and Wulff, K. Psychological effect of an analogue traumatic event reduced by sleep deprivation. *Sleep* **38**, 1017–25, doi:10.5665/sleep.4802 (2015).

340  Porcheret, K. et al. Investigation of the impact of total sleep deprivation at home on the number of intrusive memories to an analogue trauma. *Transl Psychiatry* **9**, 104, doi:10.1038/s41398-019-0403-z (2019).

341  Lal, S. K. and Craig, A. A critical review of the psychophysiology of driver fatigue. *Biol Psychol* **55**, 173–94 (2001).

342  Cai, Q., Gao, Z. K., Yang, Y. X., Dang, W. D. and Grebogi, C. Multiplex limited penetrable horizontal visibility graph from EEG signals for driver fatigue detection. *Int J Neural Syst* **29**, 1850057, doi:10.1142/S0129065718500570 (2019).

343  Lok, R., Smolders, K., Beersma, D. G. M. and de Kort, Y. A. W. Light, alertness, and alerting effects of white light: a literature overview. *J Biol Rhythms* **33**, 589–601, doi:10.1177/0748730418796443 (2018).

344  Perry-Jenkins, M., Goldberg, A. E., Pierce, C. P. and Sayer, A. G. Shift work, role overload, and the transition to parenthood. *J Marriage Fam* **69**, 123–38, doi: 10.1111/j.1741–3737.2006.00349.x (2007).

345  Roenneberg, T., Allebrandt, K. V., Merrow, M. and Vetter, C. Social jetlag and obesity. *Curr Biol* **22**, 939–43, doi:10.1016/j.cub.2012.03.038 (2012).

346  Baird, B., Castelnovo, A., Gosseries, O. and Tononi, G. Frequent lucid dreaming associated with increased functional connectivity between frontopolar cortex and temporoparietal association areas. *Sci Rep* **8**, 17798, doi:10.1038/s41598–018–36190-w (2018).

347  Lawson, C. C. et al. Rotating shift work and menstrual cycle characteristics. *Epidemiology* **22**, 305–12, doi:10.1097/EDE.0b013e3182130016 (2011).

348  Kang, W., Jang, K. H., Lim, H. M., Ahn, J. S. and Park, W. J. The menstrual cycle associated with insomnia in newly employed nurses performing shift work: a 12-month follow-up study. *Int Arch Occup Environ Health* **92**, 227–35, doi:10.1007/s00420–018–1371-y (2019).

349  Garcia, J. E., Jones, G. S. and Wright, G. L., Jr. Prediction of the time of ovulation. *Fertil Steril* **36**, 308–15 (1981).

350  Wilcox, A. J., Weinberg, C. R. and Baird, D. D. Timing of sexual intercourse in relation to ovulation. Effects on the probability of conception, survival of the pregnancy, and sex of the baby. *N Engl J Med* **333**, 1517–21, doi:10.1056/NEJM199512073332301 (1995).

351  Kerdelhue, B. et al. Timing of initiation of the preovulatory luteinizing hormone surge and its relationship with the circadian cortisol rhythm in the human. *Neuroendocrinology* **75**, 158–63, doi:10.1159/000048233 (2002).

352  Sellix, M. T. and Menaker, M. Circadian clocks in the ovary. *Trends Endocrinol Metab* **21**, 628–36, doi:10.1016/j.tem.2010.06.002 (2010).

353  Miller, B. H. and Takahashi, J. S. Central circadian control of female reproductive function. *Front Endocrinol (Lausanne)* **4**, 195, doi:10.3389/fendo.2013.00195 (2013).

354  Baker, F. C. and Driver, H. S. Circadian rhythms, sleep, and the menstrual cycle. *Sleep Med* **8**, 613–22, doi:10.1016/j.sleep.2006.09.011 (2007).

355  Nurminen, T. Shift work and reproductive health. *Scand J Work Environ Health* **24 Suppl 3**, 28–34 (1998).

356  Lemmer, B. No correlation between lunar and menstrual cycle – an early report by the French physician J. A. Murat in 1806. *Chronobiol Int* **36**, 587–90, doi:10.1080/07420528.2019.1583669 (2019).

357  Ilias, I., Spanoudi, F., Koukkou, E., Adamopoulos, D. A. and Nikopoulou, S. C. Do lunar phases influence menstruation? A year-long retrospective study. *Endocr Regul* **47**, 121–2, doi:10.4149/endo_2013_03_121 (2013).

358  Helfrich-Forster, C. et al. Women temporarily synchronize their menstrual cycles with the luminance and gravimetric cycles of the Moon. *Sci Adv* **7**, doi:10.1126/sciadv.abe1358 (2021).

359  Vyazovskiy, V. V. and Foster, R. G. Sleep: a biological stimulus from our nearest celestial neighbor? *Curr Biol* **24**, R557–60, doi:10.1016/j.cub.2014.05.027 (2014).

360  Staboulidou, I., Soergel, P., Vaske, B. and Hillemanns, P. The influence of lunar cycle on frequency of birth, birth complications, neonatal outcome and the gender: a retrospective analysis. *Acta Obstet Gynecol Scand* **87**, 875–9, doi:10.1080/00016340802233090 (2008).

361  Naylor, E. Tidally rhythmic behaviour of marine animals. *Symp Soc Exp Biol* **39**, 63–93 (1985).

362  Bulla, M., Oudman, T., Bijleveld, A. I., Piersma, T. and Kyriacou, C. P. Marine biorhythms: bridging chronobiology and ecology. *Philos Trans R Soc Lond B Biol Sci* **372**, doi:10.1098/rstb.2016.0253 (2017).

363 Palmer, J. D., Udry, J. R. and Morris, N. M. Diurnal and weekly, but no lunar rhythms in human copulation. *Hum Biol* **54**, 111–21 (1982).

364 Refinetti, R. Time for sex: nycthemeral distribution of human sexual behavior. *J Circadian Rhythms* **3**, 4, doi:10.1186/1740-3391-3-4 (2005).

365 Junger, J. et al. Do women's preferences for masculine voices shift across the ovulatory cycle? *Horm Behav* **106**, 122–34, doi:10.1016/j.yhbeh.2018.10.008 (2018).

366 Little, A. C., Jones, B. C. and Burriss, R. P. Preferences for masculinity in male bodies change across the menstrual cycle. *Horm Behav* **51**, 633–9, doi:10.1016/j.yhbeh.2007.03.006 (2007).

367 DeBruine, L. et al. Evidence for menstrual cycle shifts in women's preferences for masculinity: a response to Harris (in press), 'Menstrual cycle and facial preferences reconsidered'. *Evol Psychol* **8**, 768–75 (2010).

368 Gildersleeve, K., Haselton, M. G. and Fales, M. R. Do women's mate preferences change across the ovulatory cycle? A meta-analytic review. *Psychol Bull* **140**, 1205–59, doi:10.1037/a0035438 (2014).

369 Williams, M. N. and Jacobson, A. Effect of copulins on rating of female attractiveness, mate-guarding, and self-perceived sexual desirability. *Evolutionary Psychology* **14** (2016).

370 Kuukasjärvi, S. et al. Attractiveness of women's body odors over the menstrual cycle: the role of oral contraceptives and receiver sex. *Behavioral Ecology* **15**, 579–84 (2004).

371 Doty, R. L., Ford, M., Preti, G. and Huggins, G. R. Changes in the intensity and pleasantness of human vaginal odors during the menstrual cycle. *Science* **190**, 1316–18, doi:10.1126/science.1239080 (1975).

372 Su, H. W., Yi, Y. C., Wei, T. Y., Chang, T. C. and Cheng, C. M. Detection of ovulation, a review of currently available methods. *Bioeng Transl Med* **2**, 238–46, doi:10.1002/btm2.10058 (2017).

373 Winters, S. J. Diurnal rhythm of testosterone and luteinizing hormone in hypogonadal men. *J Androl* **12**, 185–90 (1991).

374 Xie, M., Utzinger, K. S., Blickenstorfer, K. and Leeners, B. Diurnal and seasonal changes in semen quality of men in subfertile partnerships. *Chronobiol Int* **35**, 1375–84, doi:10.1080/07420528.2018.14839 42 (2018).

375 Kaiser, I. H. and Halberg, F. Circadian periodic aspects of birth. *Ann NY Acad Sci* **98**, 1056–68, doi:10.1111/j.1749–6632.1962.tb30618.x (1962).

376 Chaney, C., Goetz, T. G. and Valeggia, C. A time to be born: variation in the hour of birth in a rural population of Northern Argentina. *Am J Phys Anthropol* **166**, 975–8, doi:10.1002/ajpa.23483 (2018).

377 Sharkey, J. T., Cable, C. and Olcese, J. Melatonin sensitizes human myometrial cells to oxytocin in a protein kinase C alpha/extracellular-signal regulated kinase-dependent manner. *J Clin Endocrinol Metab* **95**, 2902–8, doi:10.1210/jc.2009–2137 (2010).

378 Millar, L. J., Shi, L., Hoerder-Suabedissen, A. and Molnar, Z. Neonatal hypoxia ischaemia: mechanisms, models, and therapeutic challenges. *Front Cell Neurosci* **11**, 78, doi:10.3389/fncel.2017.00078 (2017).

379 Anderson, S. T. and FitzGerald, G. A. Sexual dimorphism in body clocks. *Science* **369**, 1164–5, doi:10.1126/science.abd4964 (2020).

380 Boivin, D. B., Shechter, A., Boudreau, P., Begum, E. A. and Ng Ying-Kin, N. M. Diurnal and circadian variation of sleep and alertness in men vs. naturally cycling women. *Proc Natl Acad Sci USA* **113**, 10980–85, doi:10.1073/pnas.1524484113 (2016).

381 Roenneberg, T. et al. A marker for the end of adolescence. *Curr Biol* **14**, R1038–9, doi:10.1016/j.cub.2004.11.039 (2004).

382 Fischer, D., Lombardi, D. A., Marucci-Wellman, H. and Roenneberg, T. Chronotypes in the US – influence of age and sex. *PLoS One* **12**, e0178782, doi:10.1371/journal.pone.0178782 (2017).

383 Feillet, C. et al. Sexual dimorphism in circadian physiology is altered in LXRalpha deficient mice. *PLoS One* **11**, e0150665, doi:10.1371/journal.pone.0150665 (2016).

384 Meers, J. M. and Nowakowski, S. Sleep, premenstrual mood disorder, and women's health. *Curr Opin Psychol* **34**, 43–9, doi:10.1016/j.copsyc.2019.09.003 (2020).

385 Yonkers, K. A., O'Brien, P. M. and Eriksson, E. Premenstrual syndrome. *Lancet* **371**, 1200–1210, doi:10.1016/S0140–6736(08)60527–9 (2008).

386 Kruijver, F. P. and Swaab, D. F. Sex hormone receptors are present in the human suprachiasmatic nucleus. *Neuroendocrinology* **75**, 296–305, doi:10.1159/000057339 (2002).

387 Wollnik, F. and Turek, F. W. Estrous correlated modulations of circadian and ultradian wheel-running activity rhythms in LEW/Ztm rats. *Physiol Behav* **43**, 389–96, doi:10.1016/0031–9384(88)90204–1 (1988).

388 Parry, B. L. et al. Reduced phase-advance of plasma melatonin after bright morning light in the luteal, but not follicular, menstrual cycle phase in premenstrual dysphoric disorder: an extended study. *Chronobiol Int* **28**, 415–24, doi:10.3109/07420528.2011.567365 (2011).

389 Van Reen, E. and Kiesner, J. Individual differences in self-reported difficulty sleeping across the menstrual cycle. *Arch Womens Ment Health* **19**, 599–608, doi:10.1007/s00737–016–0621–9 (2016).

390 Tempesta, D. et al. Lack of sleep affects the evaluation of emotional stimuli. *Brain Res Bull* **82**, 104–8, doi:10.1016/j.brainresbull.2010.01.014 (2010).

391 Schwarz, J. F. et al. Shortened night sleep impairs facial responsiveness to emotional stimuli. *Biol Psychol* **93**, 41–4, doi:10.1016/j.biopsycho.2013.01.008 (2013).

392 Meers, J. M., Bower, J. L. and Alfano, C. A. Poor sleep and emotion dysregulation mediate the association between depressive and premenstrual symptoms in young adult women. *Arch Womens Ment Health* **23**, 351–9, doi:10.1007/s00737–019–00984–2 (2020).

393 Hollander, L. E. et al. Sleep quality, estradiol levels, and behavioral factors in late reproductive age women. *Obstet Gynecol* **98**, 391–7, doi:10.1016/s0029–7844(01)01485–5 (2001).

394 Manber, R. and Armitage, R. Sex, steroids, and sleep: a review. *Sleep* **22**, 540–55 (1999).

395 Proserpio, P. et al. Insomnia and menopause: a narrative review on mechanisms and treatments. *Climacteric* **23**, 539–49, doi:10.1080/13697137.2020.1799973 (2020).

396 Shaver, J. L. and Woods, N. F. Sleep and menopause: a narrative review. *Menopause* **22**, 899–915, doi:10.1097/GME.0000000000000499 (2015).

397 Kravitz, H. M. et al. Sleep difficulty in women at midlife: a community survey of sleep and the menopausal transition. *Menopause* **10**, 19–28, doi:10.1097/00042192–200310010–00005 (2003).

398 Walters, J. F., Hampton, S. M., Ferns, G. A. and Skene, D. J. Effect of menopause on melatonin and alertness rhythms investigated in constant routine conditions. *Chronobiol Int* **22**, 859–72, doi:10.1080/07420520500263193 (2005).

399 Freedman, R. R. Hot flashes: behavioral treatments, mechanisms, and relation to sleep. *Am J Med* **118 Suppl 12B**, 124–30, doi:10.1016/j.amjmed.2005.09.046 (2005).

400 Baker, F. C., de Zambotti, M., Colrain, I. M. and Bei, B. Sleep problems during the menopausal transition: prevalence, impact, and management challenges. *Nat Sci Sleep* **10**, 73–95, doi:10.2147/NSS.S125807 (2018).

401 Kravitz, H. M. and Joffe, H. Sleep during the perimenopause: a SWAN story. *Obstet Gynecol Clin North Am* **38**, 567–86, doi:10.1016/j.ogc.2011.06.002 (2011).

402 Moe, K. E. Hot flashes and sleep in women. *Sleep Med Rev* **8**, 487–97, doi:10.1016/j.smrv.2004.07.005 (2004).

403 Franklin, K. A., Sahlin, C., Stenlund, H. and Lindberg, E. Sleep apnoea is a common occurrence in females. *Eur Respir J* **41**, 610–15, doi:10.1183/09031936.00212711 (2013).

404 Kapsimalis, F. and Kryger, M. H. Gender and obstructive sleep apnea syndrome, part 2: mechanisms. *Sleep* **25**, 499–506 (2002).

405 Cintron, D. et al. Efficacy of menopausal hormone therapy on sleep quality: systematic review and meta-analysis. *Endocrine* **55**, 702–11, doi:10.1007/s12020–016–1072–9 (2017).

406 McCurry, S. M. et al. Telephone-based cognitive behavioral therapy for insomnia in perimenopausal and postmenopausal women with vasomotor symptoms: a MsFLASH randomized clinical trial. *JAMA Intern Med* **176**, 913–20, doi:10.1001/jamainternmed.2016.1795 (2016).

407 Salk, R. H., Hyde, J. S. and Abramson, L. Y. Gender differences in depression in representative national samples: meta-analyses of diagnoses and symptoms. *Psychol Bull* **143**, 783–822, doi:10.1037/bul0000102 (2017).

408 Barrett-Connor, E. et al. The association of testosterone levels with overall sleep quality, sleep architecture, and sleep-disordered breathing. *J Clin Endocrinol Metab* **93**, 2602–9, doi:10.1210/jc.2007–2622 (2008).

409 Swaab, D. F., Gooren, L. J. and Hofman, M. A. Brain research, gender and sexual orientation. *J Homosex* **28**, 283–301, doi:10.1300/J082v28n03_07 (1995).

410 Swaab, D. F. and Hofman, M. A. An enlarged suprachiasmatic nucleus in homosexual men. *Brain Res* **537**, 141–8, doi:10.1016/0006–8993(90)90350-k (1990).

411 Román-Gálvez, R. M. et al. Factors associated with insomnia in pregnancy: a prospective Cohort Study. *Eur J Obstet Gynecol Reprod Biol* **221**, 70–75, doi:10.1016/j.ejogrb.2017.12.007 (2018).

412 Kivela, L., Papadopoulos, M. R. and Antypa, N. Chronotype and psychiatric disorders. *Curr Sleep Med Rep* **4**, 94–103, doi:10.1007/s40675–018–0113–8 (2018).

413 Goyal, D., Gay, C. L. and Lee, K. A. Patterns of sleep disruption and depressive symptoms in new mothers. *J Perinat Neonatal Nurs* **21**, 123–9, doi:10.1097/01.JPN.0000270629.58746.96 (2007).

414 Doan, T., Gay, C. L., Kennedy, H. P., Newman, J. and Lee, K. A. Nighttime breastfeeding behavior is associated with more nocturnal sleep among first-time mothers at one month postpartum. *J Clin Sleep Med* **10**, 313–19, doi:10.5664/jcsm.3538 (2014).

415 Wulff, K. and Siegmund, R. Emergence of circadian rhythms in infants before and after birth: evidence for variations by parental influence. *Z Geburtshilfe Neonatol* **206**, 166–71, doi:10.1055/s-2002–34963 (2002).

416 Gay, C. L., Lee, K. A. and Lee, S. Y. Sleep patterns and fatigue in new mothers and fathers. *Biol Res Nurs* **5**, 311–18, doi:10.1177/1099800403262142 (2004).

417 Stremler, R. et al. A behavioral-educational intervention to promote maternal and infant sleep: a pilot randomized, controlled
trial. *Sleep* **29**, 1609–15, doi:10.1093/sleep/29.12.1609 (2006).

418 Hunter, L. P., Rychnovsky, J. D. and Yount, S. M. A selective review
of maternal sleep characteristics in the postpartum period. *J Obstet
Gynecol Neonatal Nurs* **38**, 60–68, doi:10.1111/j.1552–6909.2008.00309.x
(2009).

419 Kennedy, H. P., Gardiner, A., Gay, C. and Lee, K. A. Negotiating
sleep: a qualitative study of new mothers. *J Perinat Neonatal Nurs*
**21**, 114–22, doi:10.1097/01.JPN.0000270628.51122.1d (2007).

420 Cronin, R. S. et al. An individual participant data meta-analysis of
maternal going-to-sleep position, interactions with fetal vulnerability, and the risk of late stillbirth. *EClinicalMedicine* **10**, 49–57,
doi:10.1016/j.eclinm.2019.03.014 (2019).

421 Condon, R. G. and Scaglion, R. The ecology of human birth seasonality. *Hum Ecol* **10**, 495–511, doi:10.1007/BF01531169 (1982).

422 Lundin, C. et al. Combined oral contraceptive use is associated
with both improvement and worsening of mood in the different
phases of the treatment cycle – a double-blind, placebo-controlled
randomized trial. *Psychoneuroendocrinology* **76**, 135–43, doi:10.1016/j.
psyneuen.2016.11.033 (2017).

423 Yonkers, K. A., Cameron, B., Gueorguieva, R., Altemus, M. and
Kornstein, S. G. The influence of cyclic hormonal contraception
on expression of premenstrual syndrome. *J Womens Health
(Larchmt)* **26**, 321–8, doi:10.1089/jwh.2016.5941 (2017).

424 Simmons, R. G. et al. Predictors of contraceptive switching and
discontinuation within the first 6 months of use among Highly
Effective Reversible Contraceptive Initiative Salt Lake study participants. *Am J Obstet Gynecol* **220**, 376 e371–6 e312, doi:10.1016/j.
ajog.2018.12.022 (2019).

425 Smith, K. et al. Do progestin-only contraceptives contribute to the
risk of developing depression as implied by Beta-Arrestin 1 levels in
leukocytes? A pilot study. *Int J Environ Res Public Health* **15**,
doi:10.3390/ijerph15091966 (2018).

426 Lewis, C. A. et al. Effects of hormonal contraceptives on mood: a focus on emotion recognition and reactivity, reward processing, and stress response. *Curr Psychiatry Rep* **21**, 115, doi:10.1007/s11920–019–1095-z (2019).

427 Jocz, P., Stolarski, M. and Jankowski, K. S. Similarity in chronotype and preferred time for sex and its role in relationship quality and sexual satisfaction. *Front Psychol* **9**, 443, doi:10.3389/fpsyg.2018.00443 (2018).

428 Richter, K., Adam, S., Geiss, L., Peter, L. and Niklewski, G. Two in a bed: the influence of couple sleeping and chronotypes on relationship and sleep. An overview. *Chronobiol Int* **33**, 1464–72, doi:10.10 80/07420528.2016.1220388 (2016).

429 Cooke, P. S., Nanjappa, M. K., Ko, C., Prins, G. S. and Hess, R. A. Estrogens in Male Physiology. *Physiol Rev* **97**, 995–1043, doi:10.1152/physrev.00018.2016 (2017).

430 Fillinger, L., Janussen, D., Lundalv, T. and Richter, C. Rapid glass sponge expansion after climate-induced Antarctic ice shelf collapse. *Curr Biol* **23**, 1330–34, doi:10.1016/j.cub.2013.05.051 (2013).

431 Poblano, A., Haro, R. and Arteaga, C. Neurophysiologic measurement of continuity in the sleep of fetuses during the last week of pregnancy and in newborns. *Int J Biol Sci* **4**, 23–8, doi:10.7150/ijbs.4.23 (2007).

432 Lancel, M., Faulhaber, J., Holsboer, F. and Rupprecht, R. Progesterone induces changes in sleep comparable to those of agonistic GABAA receptor modulators. *Am J Physiol* **271**, E763–72, doi:10.1152/ajpendo.1996.271.4.E763 (1996).

433 Silvestri, R. and Arico, I. Sleep disorders in pregnancy. *Sleep Sci* **12**, 232–9, doi:10.5935/1984–0063.20190098 (2019).

434 Bell, A. V., Hinde, K. and Newson, L. Who was helping? The scope for female cooperative breeding in early Homo. *PLoS One* **8**, e83667, doi:10.1371/journal.pone.0083667 (2013).

435 Bruni, O. et al. Longitudinal study of sleep behavior in normal infants during the first year of life. *J Clin Sleep Med* **10**, 1119–27, doi:10.5664/jcsm.4114 (2014).

436 Tham, E. K., Schneider, N. and Broekman, B. F. Infant sleep and its relation with cognition and growth: a narrative review. *Nat Sci Sleep* **9**, 135–49, doi:10.2147/NSS.S125992 (2017).

437 Mindell, J. A. et al. Behavioral treatment of bedtime problems and night wakings in infants and young children. *Sleep* **29**, 1263–76 (2006).

438 Rivkees, S. A. Developing circadian rhythmicity in infants. *Pediatrics* **112**, 373–81, doi:10.1542/peds.112.2.373 (2003).

439 Bateson, P. et al. Developmental plasticity and human health. *Nature* **430**, 419–21, doi:10.1038/nature02725 (2004).

440 Burnham, M. M., Goodlin-Jones, B. L., Gaylor, E. E. and Anders, T. F. Nighttime sleep–wake patterns and self-soothing from birth to one year of age: a longitudinal intervention study. *J Child Psychol Psychiatry* **43**, 713–25, doi:10.1111/1469–7610.00076 (2002).

441 Gaylor, E. E., Burnham, M. M., Goodlin-Jones, B. L. and Anders, T. F. A longitudinal follow-up study of young children's sleep patterns using a developmental classification system. *Behav Sleep Med* **3**, 44–61, doi:10.1207/s15402010bsm0301_6 (2005).

442 Matricciani, L., Paquet, C., Galland, B., Short, M. and Olds, T. Children's sleep and health: A meta-review. *Sleep Med Rev* **46**, 136–50, doi:10.1016/j.smrv.2019.04.011 (2019).

443 Stormark, K. M., Fosse, H. E., Pallesen, S. and Hysing, M. The association between sleep problems and academic performance in primary school-aged children: findings from a Norwegian longitudinal population-based study. *PLoS One* **14**, e0224139, doi:10.1371/journal.pone.0224139 (2019).

444 Sluggett, L., Wagner, S. L. and Harris, R. L. Sleep duration and obesity in children and adolescents. *Can J Diabetes* **43**, 146–52, doi:10.1016/j.jcjd.2018.06.006 (2019).

445 Meltzer, L. J. and Montgomery-Downs, H. E. Sleep in the family. *Pediatr Clin North Am* **58**, 765–74, doi:10.1016/j.pcl.2011.03.010 (2011).

446 Mindell, J. A. and Williamson, A. A. Benefits of a bedtime routine in young children: sleep, development, and beyond. *Sleep Med Rev* **40**, 93–108, doi:10.1016/j.smrv.2017.10.007 (2018).

447 Moturi, S. and Avis, K. Assessment and treatment of common pediatric sleep disorders. *Psychiatry (Edgmont)* 7, 24–37 (2010).

448 Akacem, L. D., Wright, K. P., Jr and LeBourgeois, M. K. Bedtime and evening light exposure influence circadian timing in preschool-age children: a field study. *Neurobiol Sleep Circadian Rhythms* 1, 27–31, doi:10.1016/j.nbscr.2016.11.002 (2016).

449 Patton, G. C. et al. Our future: a *Lancet* commission on adolescent health and wellbeing. *Lancet* 387, 2423–78, doi:10.1016/S0140-6736(16)00579–1 (2016).

450 Crowley, S. J., Wolfson, A. R., Tarokh, L. and Carskadon, M. A. An update on adolescent sleep: new evidence informing the perfect storm model. *J Adolesc* 67, 55–65, doi:10.1016/j.adolescence.2018.06.001 (2018).

451 Keyes, K. M., Maslowsky, J., Hamilton, A. and Schulenberg, J. The great sleep recession: changes in sleep duration among US adolescents, 1991–2012. *Pediatrics* 135, 460–68, doi:10.1542/peds.2014-2707 (2015).

452 Matricciani, L., Olds, T. and Petkov, J. In search of lost sleep: secular trends in the sleep time of school-aged children and adolescents. *Sleep Med Rev* 16, 203–11, doi:10.1016/j.smrv.2011.03.005 (2012).

453 Hirshkowitz, M. et al. National Sleep Foundation's sleep time duration recommendations: methodology and results summary. *Sleep Health* 1, 40–43, doi:10.1016/j.sleh.2014.12.010 (2015).

454 Paruthi, S. et al. Recommended amount of sleep for pediatric populations: a consensus statement of the American Academy of Sleep Medicine. *J Clin Sleep Med* 12, 785–6, doi:10.5664/jcsm.5866 (2016).

455 Gradisar, M., Gardner, G. and Dohnt, H. Recent worldwide sleep patterns and problems during adolescence: a review and meta-analysis of age, region, and sleep. *Sleep Med* 12, 110–18, doi:10.1016/j.sleep.2010.11.008 (2011).

456 Basch, C. E., Basch, C. H., Ruggles, K. V. and Rajan, S. Prevalence of sleep duration on an average school night among 4 nationally representative successive samples of American high school

students, 2007–2013. *Prev Chronic Dis* **11**, E216, doi:10.5888/pcd11. 140383 (2014).

457 Owens, J., Adolescent Sleep Working Group and Committee on Adolescence. Insufficient sleep in adolescents and young adults: an update on causes and consequences. *Pediatrics* **134**, e921–32, doi:10.1542/peds.2014–1696 (2014).

458 Chaput, J. P. et al. Systematic review of the relationships between sleep duration and health indicators in school-aged children and youth. *Appl Physiol Nutr Metab* **41**, S266–82, doi:10.1139/apnm-2015–0627 (2016).

459 McKnight-Eily, L. R. et al. Relationships between hours of sleep and health-risk behaviors in US adolescent students. *Prev Med* **53**, 271–3, doi:10.1016/j.ypmed.2011.06.020 (2011).

460 Shochat, T., Cohen-Zion, M. and Tzischinsky, O. Functional consequences of inadequate sleep in adolescents: a systematic review. *Sleep Med Rev* **18**, 75–87, doi:10.1016/j.smrv.2013.03.005 (2014).

461 Hysing, M., Harvey, A. G., Linton, S. J., Askeland, K. G. and Sivertsen, B. Sleep and academic performance in later adolescence: results from a large population-based study. *J Sleep Res* **25**, 318–24, doi:10.1111/jsr.12373 (2016).

462 Beebe, D. W., Field, J., Miller, M. M., Miller, L. E. and LeBlond, E. Impact of multi-night experimentally induced short sleep on adolescent performance in a simulated classroom. *Sleep* **40**, doi:10.1093/sleep/zsw035 (2017).

463 Godsell, S. and White, J. Adolescent perceptions of sleep and influences on sleep behaviour: a qualitative study. *J Adolesc* **73**, 18–25, doi:10.1016/j.adolescence.2019.03.010 (2019).

464 Van Dyk, T. R., Becker, S. P. and Byars, K. C. Rates of mental health symptoms and associations with self-reported sleep quality and sleep hygiene in adolescents presenting for insomnia treatment. *J Clin Sleep Med* **15**, 1433–42, doi:10.5664/jcsm.7970 (2019).

465 Jankowski, K. S., Fajkowska, M., Domaradzka, E. and Wytykowska, A. Chronotype, social jetlag and sleep loss in relation to sex

steroids. *Psychoneuroendocrinology* **108**, 87–93, doi:10.1016/j.psyn-euen.2019.05.027 (2019).

466 Jenni, O. G., Achermann, P. and Carskadon, M. A. Homeostatic sleep regulation in adolescents. *Sleep* **28**, 1446–54, doi:10.1093/sleep/28.11.1446 (2005).

467 Taylor, D. J., Jenni, O. G., Acebo, C. and Carskadon, M. A. Sleep tendency during extended wakefulness: insights into adolescent sleep regulation and behavior. *J Sleep Res* **14**, 239–44, doi:10.1111/j.1365-2869.2005.00467.x (2005).

468 Basheer, R., Strecker, R. E., Thakkar, M. M. and McCarley, R. W. Adenosine and sleep–wake regulation. *Prog Neurobiol* **73**, 379–96, doi:10.1016/j.pneurobio.2004.06.004 (2004).

469 Illingworth, G. The challenges of adolescent sleep. *Interface Focus* **10**, 20190080, doi:10.1098/rsfs.2019.0080 (2020).

470 Cain, N. and Gradisar, M. Electronic media use and sleep in school-aged children and adolescents: a review. *Sleep Med* **11**, 735–42, doi:10.1016/j.sleep.2010.02.006 (2010).

471 Twenge, J. M., Krizan, Z. and Hisler, G. Decreases in self-reported sleep duration among U.S. adolescents 2009–2015 and association with new media screen time. *Sleep Med* **39**, 47–53, doi:10.1016/j.sleep.2017.08.013 (2017).

472 Bartel, K. A., Gradisar, M. and Williamson, P. Protective and risk factors for adolescent sleep: a meta-analytic review. *Sleep Med Rev* **21**, 72–85, doi:10.1016/j.smrv.2014.08.002 (2015).

473 Vernon, L., Modecki, K. L. and Barber, B. L. Mobile phones in the bedroom: trajectories of sleep habits and subsequent adolescent psychosocial development. *Child Dev* **89**, 66–77, doi:10.1111/cdev.12836 (2018).

474 Orzech, K. M., Grandner, M. A., Roane, B. M. and Carskadon, M. A. Digital media use in the 2 h before bedtime is associated with sleep variables in university students. *Comput Human Behav* **55**, 43–50, doi:10.1016/j.chb.2015.08.049 (2016).

475 Perrault, A. A. et al. Reducing the use of screen electronic devices in the evening is associated with improved sleep and

daytime vigilance in adolescents. *Sleep* **42**, doi:10.1093/sleep/zsz125 (2019).

476 Crowley, S. J. et al. A longitudinal assessment of sleep timing, circadian phase, and phase angle of entrainment across human adolescence. *PLoS One* **9**, e112199, doi:10.1371/journal.pone.0112199 (2014).

477 Troxel, W. M. and Wolfson, A. R. The intersection between sleep science and policy: introduction to the special issue on school start times. *Sleep Health* **3**, 419–22, doi:10.1016/j.sleh.2017.10.001 (2017).

478 Minges, K. E. and Redeker, N. S. Delayed school start times and adolescent sleep: a systematic review of the experimental evidence. *Sleep Med Rev* **28**, 86–95, doi:10.1016/j.smrv.2015.06.002 (2016).

479 Bowers, J. M. and Moyer, A. Effects of school start time on students' sleep duration, daytime sleepiness, and attendance: a meta-analysis. *Sleep Health* **3**, 423–31, doi:10.1016/j.sleh.2017.08.004 (2017).

480 Wheaton, A. G., Chapman, D. P. and Croft, J. B. School start times, sleep, behavioral, health, and academic outcomes: a review of the literature. *J Sch Health* **86**, 363–81, doi:10.1111/josh.12388 (2016).

481 Foster, R. G. Sleep, circadian rhythms and health. *Interface Focus* **10**, 20190098, doi:10.1098/rsfs.2019.0098 (2020).

482 Kobak, R., Abbott, C., Zisk, A. and Bounoua, N. Adapting to the changing needs of adolescents: parenting practices and challenges to sensitive attunement. *Curr Opin Psychol* **15**, 137–42, doi:10.1016/j.copsyc.2017.02.018 (2017).

483 Blunden, S. L., Chapman, J. and Rigney, G. A. Are sleep education programs successful? The case for improved and consistent research efforts. *Sleep Med Rev* **16**, 355–70, doi:10.1016/j.smrv.2011.08.002 (2012).

484 Blunden. S. and Rigney, G. Lessons learned from sleep education in schools: a review of do's and don'ts. *J Clin Sleep Med* **11**, 671–80, doi:10.5664/jcsm.4782 (2015).

485 Facer-Childs, E. R., Middleton, B., Skene, D. J. and Bagshaw, A. P. Resetting the late timing of 'night owls' has a positive impact on mental health and performance. *Sleep Med* **60**, 236–47, doi:10.1016/j.sleep.2019.05.001 (2019).

486  Van Dyk, T. R. et al. Feasibility and emotional impact of experimentally extending sleep in short-sleeping adolescents. *Sleep* **40**, doi:10.1093/sleep/zsx123 (2017).

487  Livingston, G. et al. Dementia prevention, intervention, and care: 2020 report of the *Lancet* Commission. *Lancet* **396**, 413–46, doi:10.1016/S0140–6736(20)30367–6 (2020).

488  Ohayon, M. M., Carskadon, M. A., Guilleminault, C. and Vitiello, M. V. Meta-analysis of quantitative sleep parameters from childhood to old age in healthy individuals: developing normative sleep values across the human lifespan. *Sleep* **27**, 1255–73, doi:10.1093/sleep/27.7.1255 (2004).

489  Dijk, D. J., Duffy, J. F. and Czeisler, C. A. Age-related increase in awakenings: impaired consolidation of nonREM sleep at all circadian phases. *Sleep* **24**, 565–77, doi:10.1093/sleep/24.5.565 (2001).

490  Bliwise, D. L. Sleep in normal aging and dementia. *Sleep* **16**, 40–81, doi:10.1093/sleep/16.1.40 (1993).

491  Czeisler, C. A. et al. Association of sleep–wake habits in older people with changes in output of circadian pacemaker. *Lancet* **340**, 933–96, doi:10.1016/0140–6736(92)92817-y (1992).

492  Duffy, J. F. et al. Peak of circadian melatonin rhythm occurs later within the sleep of older subjects. *Am J Physiol Endocrinol Metab* **282**, E297–303, doi:10.1152/ajpendo.00268.2001 (2002).

493  Sherman, B., Wysham, C. and Pfohl, B. Age-related changes in the circadian rhythm of plasma cortisol in man. *J Clin Endocrinol Metab* **61**, 439–43, doi:10.1210/jcem-61-3-439 (1985).

494  Van Someren, E. J. Circadian and sleep disturbances in the elderly. *Exp Gerontol* **35**, 1229–37, doi:10.1016/s0531–5565(00)00191–1 (2000).

495  Munch, M. et al. Age-related attenuation of the evening circadian arousal signal in humans. *Neurobiol Aging* **26**, 1307–19, doi:10.1016/j.neurobiolaging.2005.03.004 (2005).

496  Zeitzer, J. M. et al. Do plasma melatonin concentrations decline with age? *Am J Med* **107**, 432–6, doi:10.1016/s0002–9343(99)00266–1 (1999).

497  Farajnia, S. et al. Evidence for neuronal desynchrony in the aged suprachiasmatic nucleus clock. *J Neurosci* **32**, 5891–9, doi:10.1523/JNEUROSCI.0469-12.2012 (2012).

498  Zhou, J. N., Hofman, M. A. and Swaab, D. F. VIP neurons in the human SCN in relation to sex, age, and Alzheimer's disease. *Neurobiol Aging* **16**, 571–6, doi:10.1016/0197-4580(95)00043-e (1995).

499  Pagani, L. et al. Serum factors in older individuals change cellular clock properties. *Proc Natl Acad Sci USA* **108**, 7218–23, doi:10.1073/pnas.1008882108 (2011).

500  Crowley, S. J., Cain, S. W., Burns, A. C., Acebo, C. and Carskadon, M. A. Increased sensitivity of the circadian system to light in early/mid-puberty. *J Clin Endocrinol Metab* **100**, 4067–73, doi:10.1210/jc.2015-2775 (2015).

501  Duffy, J. F., Zeitzer, J. M. and Czeisler, C. A. Decreased sensitivity to phase-delaying effects of moderate intensity light in older subjects. *Neurobiol Aging* **28**, 799–807, doi:10.1016/j.neurobiolaging.2006.03.005 (2007).

502  Cuthbertson, F. M., Peirson, S. N., Wulff, K., Foster, R. G. and Downes, S. M. Blue light-filtering intraocular lenses: review of potential benefits and side effects. *J Cataract Refract Surg* **35**, 1281–97, doi:10.1016/j.jcrs.2009.04.017 (2009).

503  Alexander, I. et al. Impact of cataract surgery on sleep in patients receiving either ultraviolet-blocking or blue-filtering intraocular lens implants. *Invest Ophthalmol Vis Sci* **55**, 4999–5004, doi:10.1167/iovs.14-14054 (2014).

504  Dijk, D. J. and Czeisler, C. A. Contribution of the circadian pacemaker and the sleep homeostat to sleep propensity, sleep structure, electroencephalographic slow waves, and sleep spindle activity in humans. *J Neurosci* **15**, 3526–38 (1995).

505  Gadie, A., Shafto, M., Leng, Y., Kievit, R. A. and Cam, C. A. N. How are age-related differences in sleep quality associated with health outcomes? An epidemiological investigation in a UK cohort of 2406 adults. *BMJ Open* **7**, e014920, doi:10.1136/bmjopen-2016-014920 (2017).

506  Schmidt, C., Peigneux, P. and Cajochen, C. Age-related changes in sleep and circadian rhythms: impact on cognitive performance and underlying neuroanatomical networks. *Front Neurol* **3**, 118, doi: 10.3389/fneur.2012.00118 (2012).

507  Pengo, M. F., Won, C. H. and Bourjeily, G. Sleep in women across the life span. *Chest* **154**, 196–206, doi:10.1016/j.chest.2018.04.005 (2018).

508  Cheung, S. S. Responses of the hands and feet to cold exposure. *Temperature (Austin)* **2**, 105–20, doi:10.1080/23328940.2015.1008890 (2015).

509  Oshima-Saeki, C., Taniho, Y., Arita, H. and Fujimoto, E. Lower-limb warming improves sleep quality in elderly people living in nursing homes. *Sleep Sci* **10**, 87–91, doi:10.5935/1984–0063.20170016 (2017).

510  Middelkoop, H. A., Smilde-van den Doel, D. A., Neven, A. K., Kamphuisen, H. A. and Springer, C. P. Subjective sleep characteristics of 1,485 males and females aged 50–93: effects of sex and age, and factors related to self-evaluated quality of sleep. *J Gerontol A Biol Sci Med Sci* **51**, M108–15, doi:10.1093/gerona/51a.3.m108 (1996).

511  Fonda, D. Nocturia: a disease or normal ageing? *BJU Int* **84 Suppl** 1, 13–15, doi:10.1046/j.1464–410X.1999.00055.x (1999).

512  Van Dijk, L., Kooij, D. G. and Schellevis, F. G. Nocturia in the Dutch adult population. *BJU Int* **90**, 644–8, doi:10.1046/j.1464–410X.2002.03011.x (2002).

513  Duffy, J. F., Scheuermaier, K. and Loughlin, K. R. Age-related sleep disruption and reduction in the circadian rhythm of urine output: contribution to nocturia? *Curr Aging Sci* **9**, 34–43, doi:10.2174/187460 9809666151130220343 (2016).

514  Sugaya, K., Nishijima, S., Miyazato, M., Kadekawa, K. and Ogawa, Y. Effects of melatonin and rilmazafone on nocturia in the elderly. *J Int Med Res* **35**, 685–91, doi:10.1177/147323000703500513 (2007).

515  Homma, Y. et al. Nocturia in the adult: classification on the basis of largest voided volume and nocturnal urine production. *J Urol* **163**, 777–81, doi:10.1016/s0022–5347(05)67802–0 (2000).

516  Jin, M. H. and Moon, G. du. Practical management of nocturia in urology. *Indian J Urol* **24**, 289–94, doi:10.4103/0970–1591.42607 (2008).

517 Moon, D. G. et al. Antidiuretic hormone in elderly male patients with severe nocturia: a circadian study. *BJU Int* **94**, 571–5, doi:10.1111/j.1464–410X.2004.05003.x (2004).

518 Asplund, R., Sundberg, B. and Bengtsson, P. Oral desmopressin for nocturnal polyuria in elderly subjects: a double-blind, placebo-controlled randomized exploratory study. *BJU Int* **83**, 591–5, doi:10.1046/j.1464–410x.1999.00012.x (1999).

519 Oelke, M., Fangmeyer, B., Zinke, J. and Witt, J. H. Nocturia in men with benign prostatic hyperplasia. *Aktuelle Urol* **49**, 319–27, doi:10.1055/a-0650–3700 (2018).

520 Umlauf, M. G. et al. Obstructive sleep apnea, nocturia and polyuria in older adults. *Sleep* **27**, 139–44, doi:10.1093/sleep/27.1.139 (2004).

521 Margel, D., Shochat, T., Getzler, O., Livne, P. M. and Pillar, G. Continuous positive airway pressure reduces nocturia in patients with obstructive sleep apnea. *Urology* **67**, 974–7, doi:10.1016/j.urology.2005.11.054 (2006).

522 Charloux, A., Gronfier, C., Lonsdorfer-Wolf, E., Piquard, F. and Brandenberger, G. Aldosterone release during the sleep–wake cycle in humans. *Am J Physiol* **276**, E43–9, doi:10.1152/ajpendo.1999.276.1.E43 (1999).

523 Stewart, R. B., Moore, M. T., May, F. E., Marks, R. G. and Hale, W. E. Nocturia: a risk factor for falls in the elderly. *J Am Geriatr Soc* **40**, 1217–20, doi:10.1111/j.1532–5415.1992.tb03645.x (1992).

524 Asplund, R., Johansson, S., Henriksson, S. and Isacsson, G. Nocturia, depression and antidepressant medication. *BJU Int* **95**, 820–23, doi:10.1111/j.1464–410X.2005.05408.x (2005).

525 Asplund, R. Nocturia in relation to sleep, health, and medical treatment in the elderly. *BJU Int* **96 Suppl** 1, 15–21, doi:10.1111/j.1464–410X.2005.05653.x (2005).

526 Hall, S. A. et al. Commonly used antihypertensives and lower urinary tract symptoms: results from the Boston Area Community Health (BACH) Survey. *BJU Int* **109**, 1676–84, doi:10.1111/j.1464–410X.2011.10593.x (2012).

527 Washino, S., Ugata, Y., Saito, K. and Miyagawa, T. Calcium channel blockers are associated with nocturia in men aged 40 years or older. *J Clin Med* **10**, doi:10.3390/jcm10081603 (2021).

528 Salman, M. et al. Effect of calcium channel blockers on lower urinary tract symptoms: a systematic review. *Biomed Res Int* **2017**, 4269875, doi:10.1155/2017/4269875 (2017).

529 Rongve, A., Boeve, B. F. and Aarsland, D. Frequency and correlates of caregiver-reported sleep disturbances in a sample of persons with early dementia. *J Am Geriatr Soc* **58**, 480–86, doi:10.1111/j.1532–5415.2010.02733.x (2010).

530 Naismith, S. L. et al. Sleep disturbance relates to neuropsychological functioning in late-life depression. *J Affect Disord* **132**, 139–45, doi:10.1016/j.jad.2011.02.027 (2011).

531 Ancoli-Israel, S., Klauber, M. R., Butters, N., Parker, L. and Kripke, D. F. Dementia in institutionalized elderly: relation to sleep apnea. *J Am Geriatr Soc* **39**, 258–63, doi:10.1111/j.1532–5415.1991.tb01647.x (1991).

532 Jaussent, I. et al. Excessive sleepiness is predictive of cognitive decline in the elderly. *Sleep* **35**, 1201–7, doi:10.5665/sleep.2070 (2012).

533 Ayalon, L. et al. Adherence to continuous positive airway pressure treatment in patients with Alzheimer's disease and obstructive sleep apnea. *Am J Geriatr Psychiatry* **14**, 176–80, doi:10.1097/01. JGP.0000192484.12684.cd (2006).

534 Alzheimer's Association. 2016 Alzheimer's disease facts and figures. *Alzheimers Dement* **12**, 459–509, doi:10.1016/j.jalz.2016.03.001 (2016).

535 Jack, C. R., Jr et al. Tracking pathophysiological processes in Alzheimer's disease: an updated hypothetical model of dynamic biomarkers. *Lancet Neurol* **12**, 207–16, doi:10.1016/S1474–4422(12) 70291–0 (2013).

536 Kar, S. and Quirion, R. Amyloid beta peptides and central cholinergic neurons: functional interrelationship and relevance to Alzheimer's disease pathology. *Prog Brain Res* **145**, 261–74, doi:10.1016/S0079–6123(03)45018–8 (2004).

537 Kar, S., Slowikowski, S. P., Westaway, D. and Mount, H. T. Interactions between beta-amyloid and central cholinergic neurons:

implications for Alzheimer's disease. *J Psychiatry Neurosci* **29**, 427–41 (2004).

538 Song, H. R., Woo, Y. S., Wang, H. R., Jun, T. Y. and Bahk, W. M. Effect of the timing of acetylcholinesterase inhibitor ingestion on sleep. *Int Clin Psychopharmacol* **28**, 346–8, doi:10.1097/YIC.0b013e328364f58d (2013).

539 Hatfield, C. F., Herbert, J., van Someren, E. J., Hodges, J. R. and Hastings, M. H. Disrupted daily activity/rest cycles in relation to daily cortisol rhythms of home-dwelling patients with early Alzheimer's dementia. *Brain* **127**, 1061–74, doi:10.1093/brain/awh129 (2004).

540 Hahn, E. A., Wang, H. X., Andel, R. and Fratiglioni, L. A change in sleep pattern may predict Alzheimer disease. *Am J Geriatr Psychiatry* **22**, 1262–71, doi:10.1016/j.jagp.2013.04.015 (2014).

541 Benito-Leon, J., Bermejo-Pareja, F., Vega, S. and Louis, E. D. Total daily sleep duration and the risk of dementia: a prospective population-based study. *Eur J Neurol* **16**, 990–97, doi:10.1111/j.1468-1331.2009.02618.x (2009).

542 Lim, A. S., Kowgier, M., Yu, L., Buchman, A. S. and Bennett, D. A. Sleep fragmentation and the risk of incident Alzheimer's disease and cognitive decline in older persons. *Sleep* **36**, 1027–32, doi:10.5665/sleep.2802 (2013).

543 Kang, J. E. et al. Amyloid-beta dynamics are regulated by orexin and the sleep–wake cycle. *Science* **326**, 1005–7, doi:10.1126/science.1180962 (2009).

544 Xie, L. et al. Sleep drives metabolite clearance from the adult brain. *Science* **342**, 373–7, doi:10.1126/science.1241224 (2013).

545 Reeves, B. C. et al. Glymphatic system impairment in Alzheimer's disease and idiopathic normal pressure hydrocephalus. *Trends Mol Med* **26**, 285–95, doi:10.1016/j.molmed.2019.11.008 (2020).

546 Cordone, S., Annarumma, L., Rossini, P. M. and De Gennaro, L. Sleep and beta-amyloid deposition in Alzheimer disease: insights on mechanisms and possible innovative treatments. *Front Pharmacol* **10**, 695, doi:10.3389/fphar.2019.00695 (2019).

547 Sundaram, S. et al. Inhibition of casein kinase 1delta/epsilon improves cognitive-affective behavior and reduces amyloid load in the APP-PS1 mouse model of Alzheimer's disease. *Sci Rep* **9**, 13743, doi:10.1038/s41598-019-50197-x (2019).

548 Tandberg, E., Larsen, J. P. and Karlsen, K. A community-based study of sleep disorders in patients with Parkinson's disease. *Mov Disord* **13**, 895–9, doi:10.1002/mds.870130606 (1998).

549 Nussbaum, R. L. and Ellis, C. E. Alzheimer's disease and Parkinson's disease. *N Engl J Med* **348**, 1356–64, doi:10.1056/NEJM2003ra020003 (2003).

550 Kudo, T., Loh, D. H., Truong, D., Wu, Y. and Colwell, C. S. Circadian dysfunction in a mouse model of Parkinson's disease. *Exp Neurol* **232**, 66–75, doi:10.1016/j.expneurol.2011.08.003 (2011).

551 Willison, L. D., Kudo, T., Loh, D. H., Kuljis, D. and Colwell, C. S. Circadian dysfunction may be a key component of the non-motor symptoms of Parkinson's disease: insights from a transgenic mouse model. *Exp Neurol* **243**, 57–66, doi:10.1016/j.expneurol.2013.01.014 (2013).

552 McCurry, S. M. et al. Increasing walking and bright light exposure to improve sleep in community-dwelling persons with Alzheimer's disease: results of a randomized, controlled trial. *J Am Geriatr Soc* **59**, 1393–1402, doi:10.1111/j.1532-5415.2011.03519.x (2011).

553 Ettcheto, M. et al. Benzodiazepines and related drugs as a risk factor in Alzheimer's disease dementia. *Front Aging Neurosci* **11**, 344, doi:10.3389/fnagi.2019.00344 (2019).

554 Mendelson, W. B. A review of the evidence for the efficacy and safety of trazodone in insomnia. *J Clin Psychiatry* **66**, 469–76, doi:10.4088/jcp.v66n0409 (2005).

555 Molano, J. and Vaughn, B. V. Approach to insomnia in patients with dementia. *Neurol Clin Pract* **4**, 7–15, doi:10.1212/CPJ.0b013e3182a78edf (2014).

556 Shochat, T., Martin, J., Marler, M. and Ancoli-Israel, S. Illumination levels in nursing home patients: effects on sleep and activity rhythms. *J Sleep Res* **9**, 373–9, doi:10.1046/j.1365-2869.2000.00221.x (2000).

557 Martins da Silva, R., Afonso, P., Fonseca, M. and Teodoro, T. Comparing sleep quality in institutionalized and non-institutionalized elderly individuals. *Aging Ment Health* **24**, 1452–8, doi:10.1080/13607 863.2019.1619168 (2020).

558 Riemersma-van der Lek, R. F. et al. Effect of bright light and melatonin on cognitive and noncognitive function in elderly residents of group care facilities: a randomized controlled trial. *JAMA* **299**, 2642–55, doi:10.1001/jama.299.22.2642 (2008).

559 Figueiro, M. G. Light, sleep and circadian rhythms in older adults with Alzheimer's disease and related dementias. *Neurodegener Dis Manag* **7**, 119–45, doi:10.2217/nmt-2016–0060 (2017).

560 Chapell, M. et al. Myopia and night-time lighting during sleep in children and adults. *Percept Mot Skills* **92**, 640–42, doi:10.2466/ pms.2001.92.3.640 (2001).

561 Guggenheim, J. A., Hill, C. and Yam, T. F. Myopia, genetics, and ambient lighting at night in a UK sample. *Br J Ophthalmol* **87**, 580– 82, doi:10.1136/bjo.87.5.580 (2003).

562 Gee, B. M., Lloyd, K., Sutton, J. and McOmber, T. Weighted blankets and sleep quality in children with autism spectrum disorders: a single-subject design. *Children (Basel)* **8**, doi:10.3390/children8010010 (2020).

563 Dawson, D. and Reid, K. Fatigue, alcohol and performance impairment. *Nature* **388**, 235, doi:10.1038/40775 (1997).

564 Goldstein, D., Hahn, C. S., Hasher, L., Wiprzycka, U. J. and Zelazo, P. D. Time of day, intellectual performance, and behavioral problems in morning versus evening type adolescents: is there a synchrony effect? *Pers Individ Dif* **42**, 431–40, doi:10.1016/j.paid.2006.07.008 (2007).

565 Zerbini, G. and Merrow, M. Time to learn: how chronotype impacts education. *Psych J* **6**, 263–76, doi:10.1002/pchj.178 (2017).

566 Van der Vinne, V. et al. Timing of examinations affects school performance differently in early and late chronotypes. *J Biol Rhythms* **30**, 53–60, doi:10.1177/0748730414564786 (2015).

567 Van Dongen, H. P., Maislin, G., Mullington, J. M. and Dinges, D. F. The cumulative cost of additional wakefulness: dose-response

effects on neurobehavioral functions and sleep physiology from chronic sleep restriction and total sleep deprivation. *Sleep* **26**, 117–26, doi:10.1093/sleep/26.2.117 (2003).

568 Belenky, G. et al. Patterns of performance degradation and restoration during sleep restriction and subsequent recovery: a sleep dose-response study. *J Sleep Res* **12**, 1–12, doi:10.1046/j.1365-2869.2003.00337.x (2003).

569 Lim, J. and Dinges, D. F. A meta-analysis of the impact of short-term sleep deprivation on cognitive variables. *Psychol Bull* **136**, 375–89, doi:10.1037/a0018883 (2010).

570 Durmer, J. S. and Dinges, D. F. Neurocognitive consequences of sleep deprivation. *Semin Neurol* **25**, 117–29, doi:10.1055/s-2005-867080 (2005).

571 Bioulac, S. et al. Risk of motor vehicle accidents related to sleepiness at the wheel: a systematic review and meta-analysis. *Sleep* **41**, doi:10.1093/sleep/zsy075 (2018).

572 Saper, C. B., Fuller, P. M., Pedersen, N. P., Lu, J. and Scammell, T. E. Sleep state switching. *Neuron* **68**, 1023–42, doi:10.1016/j.neuron.2010.11.032 (2010).

573 Bendor, D. and Wilson, M. A. Biasing the content of hippocampal replay during sleep. *Nat Neurosci* **15**, 1439–44, doi:10.1038/nn.3203 (2012).

574 Yoo, S. S., Hu, P. T., Gujar, N., Jolesz, F. A. and Walker, M. P. A deficit in the ability to form new human memories without sleep. *Nat Neurosci* **10**, 385–92, doi:10.1038/nn1851 (2007).

575 Ong, J. L. et al. Auditory stimulation of sleep slow oscillations modulates subsequent memory encoding through altered hippocampal function. *Sleep* **41**, doi:10.1093/sleep/zsy031 (2018).

576 Marshall, L. and Born, J. The contribution of sleep to hippocampus-dependent memory consolidation. *Trends Cogn Sci* **11**, 442–50, doi:10.1016/j.tics.2007.09.001 (2007).

577 Schmid, D., Erlacher, D., Klostermann, A., Kredel, R. and Hossner, E. J. Sleep-dependent motor memory consolidation in healthy adults: a meta-analysis. *Neurosci Biobehav Rev* **118**, 270–81, doi:10.1016/j.neubiorev.2020.07.028 (2020).

578  Schonauer, M., Geisler, T. and Gais, S. Strengthening procedural memories by reactivation in sleep. *J Cogn Neurosci* **26**, 143–53, doi:10.1162/jocn_a_00471 (2014).

579  Kurniawan, I. T., Cousins, J. N., Chong, P. L. and Chee, M. W. Procedural performance following sleep deprivation remains impaired despite extended practice and an afternoon nap. *Sci Rep* **6**, 36001, doi:10.1038/srep36001 (2016).

580  Wagner, U., Gais, S., Haider, H., Verleger, R. and Born, J. Sleep inspires insight. *Nature* **427**, 352–5, doi:10.1038/nature02223 (2004).

581  Murray, G. Diurnal mood variation in depression: a signal of disturbed circadian function? *J Affect Disord* **102**, 47–53, doi:10.1016/j.jad.2006.12.001 (2007).

582  Wirz-Justice, A. Diurnal variation of depressive symptoms. *Dialogues Clin Neurosci* **10**, 337–43 (2008).

583  Roiser, J. P., Howes, O. D., Chaddock, C. A., Joyce, E. M. and McGuire, P. Neural and behavioral correlates of aberrant salience in individuals at risk for psychosis. *Schizophr Bull* **39**, 1328–36, doi:10.1093/schbul/sbs147 (2013).

584  Benca, R. M., Obermeyer, W. H., Thisted, R. A. and Gillin, J. C. Sleep and psychiatric disorders. A meta-analysis. *Arch Gen Psychiatry* **49**, 651–68; discussion 669–70, doi:10.1001/archpsyc.1992.01820080059010 (1992).

585  Kessler, R. C. et al. Lifetime prevalence and age-of-onset distributions of mental disorders in the World Health Organization's World Mental Health Survey Initiative. *World Psychiatry* **6**, 168–76 (2007).

586  Manoach, D. S. and Stickgold, R. Does abnormal sleep impair memory consolidation in schizophrenia? *Front Hum Neurosci* **3**, 21, doi:10.3389/neuro.09.021.2009 (2009).

587  Cohrs, S. Sleep disturbances in patients with schizophrenia: impact and effect of antipsychotics. *CNS Drugs* **22**, 939–62, doi:10.2165/00023210-200822110-00004 (2008).

588  Martin, J. et al. Actigraphic estimates of circadian rhythms and sleep/wake in older schizophrenia patients. *Schizophr Res* **47**, 77–86 (2001).

589 Martin, J. L., Jeste, D. V. and Ancoli-Israel, S. Older schizophrenia patients have more disrupted sleep and circadian rhythms than age-matched comparison subjects. *J Psychiatr Res* **39**, 251–9, doi:10.1016/j.jpsychires.2004.08.011 (2005).

590 Wulff, K., Joyce, E., Middleton, B., Dijk, D. J. and Foster, R. G. The suitability of actigraphy, diary data, and urinary melatonin profiles for quantitative assessment of sleep disturbances in schizophrenia: a case report. *Chronobiol Int* **23**, 485–95, doi:10.1080/07420520500545987 (2006).

591 Wulff, K., Porcheret, K., Cussans, E. and Foster, R. G. Sleep and circadian rhythm disturbances: multiple genes and multiple phenotypes. *Curr Opin Genet Dev* **19**, 237–46, doi:10.1016/j.gde.2009.03.007 (2009).

592 Goldman, M. et al. Biological predictors of 1-year outcome in schizophrenia in males and females. *Schizophr Res* **21**, 65–73 (1996).

593 Hofstetter, J. R., Lysaker, P. H. and Mayeda, A. R. Quality of sleep in patients with schizophrenia is associated with quality of life and coping. *BMC Psychiatry* **5**, 13, doi:10.1186/1471-244X-5-13 (2005).

594 Auslander, L. A. and Jeste, D. V. Perceptions of problems and needs for service among middle-aged and elderly outpatients with schizophrenia and related psychotic disorders. *Community Ment Health J* **38**, 391–402 (2002).

595 Ehlers, C. L., Frank, E. and Kupfer, D. J. Social zeitgebers and biological rhythms. A unified approach to understanding the etiology of depression. *Arch Gen Psychiatry* **45**, 948–52, doi:10.1001/archpsyc.1988.01800340076012 (1988).

596 Chemerinski, E. et al. Insomnia as a predictor for symptom worsening following antipsychotic withdrawal in schizophrenia. *Compr Psychiatry* **43**, 393–6 (2002).

597 Pritchett, D. et al. Evaluating the links between schizophrenia and sleep and circadian rhythm disruption. *J Neural Transm (Vienna)* **119**, 1061–75, doi:10.1007/s00702-012-0817-8 (2012).

598 Oliver, P. L. et al. Disrupted circadian rhythms in a mouse model of schizophrenia. *Curr Biol* **22**, 314–19, doi:10.1016/j.cub.2011.12.051 (2012).

599 Pritchett, D. et al. Deletion of metabotropic glutamate receptors 2 and 3 (mGlu2 and mGlu3) in mice disrupts sleep and wheel-running activity, and increases the sensitivity of the circadian system to light. *PLoS One* **10**, e0125523, doi:10.1371/journal.pone.0125523 (2015).

600 Uhlhaas, P. J. and Singer, W. Neural synchrony in brain disorders: relevance for cognitive dysfunctions and pathophysiology. *Neuron* **52**, 155–68, doi:10.1016/j.neuron.2006.09.020 (2006).

601 Richardson, G. and Wang-Weigand, S. Effects of long-term exposure to ramelteon, a melatonin receptor agonist, on endocrine function in adults with chronic insomnia. *Hum Psychopharmacol* **24**, 103–11, doi:10.1002/hup.993 (2009).

602 Jagannath, A., Peirson, S. N. and Foster, R. G. Sleep and circadian rhythm disruption in neuropsychiatric illness. *Curr Opin Neurobiol* **23**, 888–94, doi:10.1016/j.conb.2013.03.008 (2013).

603 Freeman, D. et al. The effects of improving sleep on mental health (OASIS): a randomised controlled trial with mediation analysis. *Lancet Psychiatry* **4**, 749–58, doi:10.1016/S2215–0366(17)30328–0 (2017).

604 Alvaro, P. K., Roberts, R. M. and Harris, J. K. A systematic review assessing bidirectionality between sleep disturbances, anxiety, and depression. *Sleep* **36**, 1059–68, doi:10.5665/sleep.2810 (2013).

605 Goldstein, T. R., Bridge, J. A. and Brent, D. A. Sleep disturbance preceding completed suicide in adolescents. *J Consult Clin Psychol* **76**, 84–91, doi:10.1037/0022–006X.76.1.84 (2008).

606 Rumble, M. E. et al. The relationship of person-specific eveningness chronotype, greater seasonality, and less rhythmicity to suicidal behavior: a literature review. *J Affect Disord* **227**, 721–30, doi:10.1016/j.jad.2017.11.078 (2018).

607 Gold, A. K. and Sylvia, L. G. The role of sleep in bipolar disorder. *Nat Sci Sleep* **8**, 207–14, doi:10.2147/NSS.S85754 (2016).

608 Monk, T. H., Germain, A. and Reynolds, C. F. Sleep disturbance in bereavement. *Psychiatr Ann* **38**, 671–5, doi:10.3928/00485713–20081001–06 (2008).

609 Noguchi, T., Lo, K., Diemer, T. and Welsh, D. K. Lithium effects on circadian rhythms in fibroblasts and suprachiasmatic nucleus slices

from Cry knockout mice. *Neurosci Lett* **619**, 49–53, doi:10.1016/j. neulet.2016.02.030 (2016).

610  Sanghani, H. R. et al. Patient fibroblast circadian rhythms predict lithium sensitivity in bipolar disorder. *Mol Psychiatry*, doi:10.1038/ s41380-020-0769-6 (2020).

611  Desborough, M. J. R. and Keeling, D. M. The aspirin story – from willow to wonder drug. *Br J Haematol* **177**, 674–83, doi:10.1111/ bjh.14520 (2017).

612  Levi, F., Le Louarn, C. and Reinberg, A. Timing optimizes sustained-release indomethacin treatment of osteoarthritis. *Clin Pharmacol Ther* **37**, 77–84, doi:10.1038/clpt.1985.15 (1985).

613  Maurer, M., Ortonne, J. P. and Zuberbier, T. Chronic urticaria: an internet survey of health behaviours, symptom patterns and treatment needs in European adult patients. *Br J Dermatol* **160**, 633–41, doi:10.1111/j.1365-2133.2008.08920.x (2009).

614  Labrecque, G. and Vanier, M. C. Biological rhythms in pain and in the effects of opioid analgesics. *Pharmacol Ther* **68**, 129–47, doi:10.1016/ 0163-7258(95)02003-9 (1995).

615  Rund, S. S., O'Donnell, A. J., Gentile, J. E. and Reece, S. E. Daily rhythms in mosquitoes and their consequences for malaria transmission. *Insects* **7**, doi:10.3390/insects7020014 (2016).

616  Smolensky, M. H. et al. Diurnal and twenty-four hour patterning of human diseases: acute and chronic common and uncommon medical conditions. *Sleep Med Rev* **21**, 12–22, doi:10.1016/j.smrv.2014. 06.005 (2015).

617  Jamieson, R. A. Acute perforated peptic ulcer; frequency and incidence in the West of Scotland. *Br Med J* **2**, 222–7, doi:10.1136/ bmj.2.4933.222 (1955).

618  Kujubu, D. A. and Aboseif, S. R. An overview of nocturia and the syndrome of nocturnal polyuria in the elderly. *Nat Clin Pract Nephrol* **4**, 426–35, doi:10.1038/ncpneph0856 (2008).

619  Barloese, M. C., Jennum, P. J., Lund, N. T. and Jensen, R. H. Sleep in cluster headache – beyond a temporal rapid eye movement relationship? *Eur J Neurol* **22**, 656–64, doi:10.1111/ene.12623 (2015).

620  Durrington, H. J., Farrow, S. N., Loudon, A. S. and Ray, D. W. The circadian clock and asthma. *Thorax* **69**, 90–92, doi:10.1136/thoraxjnl-2013-203482 (2014).

621  Nihei, T. et al. Circadian variation of Rho-kinase activity in circulating leukocytes of patients with vasospastic angina. *Circ J* **78**, 1183–90, doi:10.1253/circj.cj-13-1458 (2014).

622  Truong, K. K., Lam, M. T., Grandner, M. A., Sassoon, C. S. and Malhotra, A. Timing matters: circadian rhythm in sepsis, obstructive lung disease, obstructive sleep apnea, and cancer. *Ann Am Thorac Soc* **13**, 1144–54, doi:10.1513/AnnalsATS.201602-125FR (2016).

623  Scott, J. T. Morning stiffness in rheumatoid arthritis. *Ann Rheum Dis* **19**, 361–8, doi:10.1136/ard.19.4.361 (1960).

624  Cutolo, M. Chronobiology and the treatment of rheumatoid arthritis. *Curr Opin Rheumatol* **24**, 312–18, doi:10.1097/BOR.0b013e3283521c78 (2012).

625  Smolensky, M. H., Reinberg, A. and Labrecque, G. Twenty-four hour pattern in symptom intensity of viral and allergic rhinitis: treatment implications. *J Allergy Clin Immunol* **95**, 1084–96, doi:10.1016/s0091-6749(95)70212-1 (1995).

626  Van Oosterhout, W. et al. Chronotypes and circadian timing in migraine. *Cephalalgia* **38**, 617–25, doi:10.1177/0333102417698953 (2018).

627  Elliott, W. J. Circadian variation in the timing of stroke onset: a meta-analysis. *Stroke* **29**, 992–6, doi:10.1161/01.str.29.5.992 (1998).

628  Suarez-Barrientos, A. et al. Circadian variations of infarct size in acute myocardial infarction. *Heart* **97**, 970–76, doi:10.1136/hrt.2010.212621 (2011).

629  Muller, J. E. et al. Circadian variation in the frequency of sudden cardiac death. *Circulation* **75**, 131–8, doi:10.1161/01.cir.75.1.131 (1987).

630  Khachiyants, N., Trinkle, D., Son, S. J. and Kim, K. Y. Sundown syndrome in persons with dementia: an update. *Psychiatry Investig* **8**, 275–87, doi:10.4306/pi.2011.8.4.275 (2011).

631  Gallerani, M. et al. The time for suicide. *Psychol Med* **26**, 867–70, doi:10.1017/s0033291700037909 (1996).

632  Allada, R. and Bass, J. Circadian mechanisms in medicine. *N Engl J Med* **384**, 550–61, doi:10.1056/NEJMra1802337 (2021).

633  Kaur, G., Phillips, C. L., Wong, K., McLachlan, A. J. and Saini, B. Timing of administration: for commonly-prescribed medicines in Australia. *Pharmaceutics* **8**, doi:10.3390/pharmaceutics8020013 (2016).

634  Mangoni, A. A. and Jackson, S. H. Age-related changes in pharmacokinetics and pharmacodynamics: basic principles and practical applications. *Br J Clin Pharmacol* **57**, 6–14, doi:10.1046/j.1365-2125.2003.02007.x (2004).

635  Coleman, J. J. and Pontefract, S. K. Adverse drug reactions. *Clin Med (Lond)* **16**, 481–5, doi:10.7861/clinmedicine.16-5-481 (2016).

636  Asher, G. N., Corbett, A. H. and Hawke, R. L. Common herbal dietary supplement-drug interactions. *Am Fam Physician* **96**, 101–7 (2017).

637  Baraldo, M. The influence of circadian rhythms on the kinetics of drugs in humans. *Expert Opin Drug Metab Toxicol* **4**, 175–92, doi:10.1517/17425255.4.2.175 (2008).

638  Mehta, S. R. et al. The circadian pattern of ischaemic heart disease events in Indian population. *J Assoc Physicians India* **46**, 767–71 (1998).

639  Stubblefield, J. J. and Lechleiter, J. D. Time to target stroke: examining the circadian system in stroke. *Yale J Biol Med* **92**, 349–57 (2019).

640  Scheer, F. A. et al. The human endogenous circadian system causes greatest platelet activation during the biological morning independent of behaviors. *PLoS One* **6**, e24549, doi:10.1371/journal.pone.0024549 (2011).

641  McLoughlin, S. C., Haines, P. and FitzGerald, G. A. Clocks and cardiovascular function. *Methods Enzymol* **552**, 211–28, doi:10.1016/bs.mie.2014.11.029 (2015).

642  Wong, P. M., Hasler, B. P., Kamarck, T. W., Muldoon, M. F. and Manuck, S. B. Social jetlag, chronotype, and cardiometabolic risk. *J Clin Endocrinol Metab* **100**, 4612–20, doi:10.1210/jc.2015-2923 (2015).

643  Morris, C. J., Purvis, T. E., Hu, K. and Scheer, F. A. Circadian misalignment increases cardiovascular disease risk factors in humans. *Proc Natl Acad Sci USA* **113**, E1402–11, doi:10.1073/pnas.1516953113 (2016).

644 Thosar, S. S., Butler, M. P. and Shea, S. A. Role of the circadian system in cardiovascular disease. *J Clin Invest* **128**, 2157–67, doi:10.1172/JCI80590 (2018).

645 Manfredini, R. et al. Circadian variation in stroke onset: identical temporal pattern in ischemic and hemorrhagic events. *Chronobiol Int* **22**, 417–53, doi:10.1081/CBI-200062927 (2005).

646 Butt, M. U., Zakaria, M. and Hussain, H. M. Circadian pattern of onset of ischaemic and haemorrhagic strokes, and their relation to sleep/wake cycle. *J Pak Med Assoc* **59**, 129–32 (2009).

647 Duss, S. B. et al. The role of sleep in recovery following ischemic stroke: a review of human and animal data. *Neurobiol Sleep Circadian Rhythms* **2**, 94–105, doi:10.1016/j.nbscr.2016.11.003 (2017).

648 Hodor, A., Palchykova, S., Baracchi, F., Noain, D. and Bassetti, C. L. Baclofen facilitates sleep, neuroplasticity, and recovery after stroke in rats. *Ann Clin Transl Neurol* **1**, 765–77, doi:10.1002/acn3.115 (2014).

649 Parra, O. et al. Early treatment of obstructive apnoea and stroke outcome: a randomised controlled trial. *Eur Respir J* **37**, 1128–36, doi:10.1183/09031936.00034410 (2011).

650 Zunzunegui, C., Gao, B., Cam, E., Hodor, A. and Bassetti, C. L. Sleep disturbance impairs stroke recovery in the rat. *Sleep* **34**, 1261–9, doi:10.5665/SLEEP.1252 (2011).

651 Fleming, M. K. et al. Sleep disruption after brain injury is associated with worse motor outcomes and slower functional recovery. *Neurorehabil Neural Repair* **34**, 661–71, doi:10.1177/1545968320929669 (2020).

652 Bowles, N. P., Thosar, S. S., Herzig, M. X. and Shea, S. A. Chronotherapy for hypertension. *Curr Hypertens Rep* **20**, 97, doi:10.1007/s11906-018-0897-4 (2018).

653 Hackam, D. G. and Spence, J. D. Antiplatelet therapy in ischemic stroke and transient ischemic attack. *Stroke* **50**, 773–8, doi:10.1161/STROKEAHA.118.023954 (2019).

654 Bonten, T. N. et al. Time-dependent effects of aspirin on blood pressure and morning platelet reactivity: a randomized cross-over

trial. *Hypertension* **65**, 743–50, doi:10.1161/HYPERTENSIONAHA. 114.04980 (2015).

655 Buurma, M., van Diemen, J. J. K., Thijs, A., Numans, M. E. and Bonten, T. N. Circadian rhythm of cardiovascular disease: the potential of chronotherapy with aspirin. *Front Cardiovasc Med* **6**, 84, doi:10.3389/fcvm.2019.00084 (2019).

656 Hermida, R. C. et al. Bedtime hypertension treatment improves cardiovascular risk reduction: the Hygia Chronotherapy Trial. *Eur Heart J*, doi:10.1093/eurheartj/ehz754 (2019).

657 Mayor, S. Taking antihypertensives at bedtime nearly halves cardiovascular deaths when compared with morning dosing, study finds. *BMJ* **367**, l6173, doi:10.1136/bmj.l6173 (2019).

658 Sanders, G. D. et al. in *Angiotensin-Converting Enzyme Inhibitors (ACEIs), Angiotensin II Receptor Antagonists (ARBs), and Direct Renin Inhibitors for Treating Essential Hypertension: An Update; AHRQ Comparative Effectiveness Reviews* (2011).

659 Altman, R., Luciardi, H. L., Muntaner, J. and Herrera, R. N. The antithrombotic profile of aspirin. Aspirin resistance, or simply failure? *Thromb J* **2**, 1, doi:10.1186/1477-9560-2-1 (2004).

660 Zhu, L. L., Xu, L. C., Chen, Y., Zhou, Q. and Zeng, S. Poor awareness of preventing aspirin-induced gastrointestinal injury with combined protective medications. *World J Gastroenterol* **18**, 3167–72, doi:10.3748/wjg.v18.i24.3167 (2012).

661 Plakogiannis, R. and Cohen, H. Optimal low-density lipoprotein cholesterol lowering – morning versus evening statin administration. *Ann Pharmacother* **41**, 106–10, doi:10.1345/aph.1G659 (2007).

662 Peirson, S. N. and Foster, R. G. Bad light stops play. *EMBO Rep* **12**, 380, doi:10.1038/embor.2011.70 (2011).

663 Ede, M. C. Circadian rhythms of drug effectiveness and toxicity. *Clin Pharmacol Ther* **14**, 925–35, doi:10.1002/cpt1973146925 (1973).

664 Esposito, E. et al. Potential circadian effects on translational failure for neuroprotection. *Nature* **582**, 395–8, doi:10.1038/s41586-020-2348-z (2020).

665  Shankar, A. and Williams, C. T. The darkness and the light: diurnal rodent models for seasonal affective disorder. *Dis Model Mech* **14**, doi:10.1242/dmm.047217 (2021).

666  Segal, J. P., Tresidder, K. A., Bhatt, C., Gilron, I. and Ghasemlou, N. Circadian control of pain and neuroinflammation. *J Neurosci Res* **96**, 1002–20, doi:10.1002/jnr.24150 (2018).

667  Buttgereit, F., Smolen, J. S., Coogan, A. N. and Cajochen, C. Clocking in: chronobiology in rheumatoid arthritis. *Nat Rev Rheumatol* **11**, 349–56, doi:10.1038/nrrheum.2015.31 (2015).

668  Broner, S. W. and Cohen, J. M. Epidemiology of cluster headache. *Curr Pain Headache Rep* **13**, 141–6, doi:10.1007/s11916–009–0024-y (2009).

669  Burish, M. J., Chen, Z. and Yoo, S. H. Emerging relevance of circadian rhythms in headaches and neuropathic pain. *Acta Physiol (Oxf)* **225**, e13161, doi:10.1111/apha.13161 (2019).

670  Noseda, R. and Burstein, R. Migraine pathophysiology: anatomy of the trigeminovascular pathway and associated neurological symptoms, CSD, sensitization and modulation of pain. *Pain* **154** Suppl 1, doi:10.1016/j.pain.2013.07.021 (2013).

671  Rozen, T. D. and Fishman, R. S. Cluster headache in the United States of America: demographics, clinical characteristics, triggers, suicidality, and personal burden. *Headache* **52**, 99–113, doi:10.1111/j.1526–4610.2011.02028.x (2012).

672  Headache Classification Committee of the International Headache Society (IHS). The International Classification of Headache Disorders, 3rd edn. *Cephalalgia* **38**, 1–211, doi:10.1177/0333102417738202 (2018).

673  Stewart, W. F., Lipton, R. B., Celentano, D. D. and Reed, M. L. Prevalence of migraine headache in the United States. Relation to age, income, race, and other sociodemographic factors. *JAMA* **267**, 64–9 (1992).

674  Hemelsoet, D., Hemelsoet, K. and Devreese, D. The neurological illness of Friedrich Nietzsche. *Acta Neurol Belg* **108**, 9–16 (2008).

675  Borsook, D. et al. Sex and the migraine brain. *Neurobiol Dis* **68**, 200–214, doi:10.1016/j.nbd.2014.03.008 (2014).

676  Ong, J. C. et al. Can circadian dysregulation exacerbate migraines? *Headache* **58**, 1040–51, doi:10.1111/head.13310 (2018).

677  Leso, V. et al. Shift work and migraine: a systematic review. *J Occup Health* **62**, e12116, doi:10.1002/1348-9585.12116 (2020).

678  Chen, Z. What's next for chronobiology and drug discovery. *Expert Opin Drug Discov* **12**, 1181–5, doi:10.1080/17460441.2017.1378179 (2017).

679  Johansson, A. S., Brask, J., Owe-Larsson, B., Hetta, J. and Lundkvist, G. B. Valproic acid phase shifts the rhythmic expression of Period2::Luciferase. *J Biol Rhythms* **26**, 541–51, doi:10.1177/0748730411419775 (2011).

680  Biggs, K. R. and Prosser, R. A. GABAB receptor stimulation phase-shifts the mammalian circadian clock in vitro. *Brain Res* **807**, 250–54, doi:10.1016/s0006-8993(98)00820-8 (1998).

681  Glasser, S. P. Circadian variations and chronotherapeutic implications for cardiovascular management: a focus on COER verapamil. *Heart Dis* **1**, 226–32 (1999).

682  Mwamburi, M., Liebler, E. J. and Tenaglia, A. T. Review of non-invasive vagus nerve stimulation (gammaCore): efficacy, safety, potential impact on comorbidities, and economic burden for episodic and chronic cluster headache. *Am J Manag Care* **23**, S317–25 (2017).

683  Gilron, I., Bailey, J. M. and Vandenkerkhof, E. G. Chronobiological characteristics of neuropathic pain: clinical predictors of diurnal pain rhythmicity. *Clin J Pain* **29**, 755–9, doi:10.1097/AJP.0b013e318275f287 (2013).

684  Zhang, J. et al. Regulation of peripheral clock to oscillation of substance P contributes to circadian inflammatory pain. *Anesthesiology* **117**, 149–60, doi:10.1097/ALN.0b013e31825b4fc1 (2012).

685  Gong, J., Chehrazi-Raffle, A., Reddi, S. and Salgia, R. Development of PD-1 and PD-L1 inhibitors as a form of cancer immunotherapy: a comprehensive review of registration trials and future considerations. *J Immunother Cancer* **6**, 8, doi:10.1186/s40425-018-0316-z (2018).

686 Filipski, E. et al. Disruption of circadian coordination accelerates malignant growth in mice. *Pathol Biol (Paris)* **51**, 216–19, doi:10.1016/s0369-8114(03)00034-8 (2003).

687 Fu, L., Pelicano, H., Liu, J., Huang, P. and Lee, C. The circadian gene Period2 plays an important role in tumor suppression and DNA damage response in vivo. *Cell* **111**, 41–50, doi:10.1016/s0092-8674(02)00961-3 (2002).

688 Mteyrek, A., Filipski, E., Guettier, C., Okyar, A. and Lévi, F. Clock gene Per2 as a controller of liver carcinogenesis. *Oncotarget* **7**, 85832–47, doi:10.18632/oncotarget.11037 (2016).

689 Altman, B. J. et al. MYC disrupts the circadian clock and metabolism in cancer cells. *Cell Metab* **22**, 1009–19, doi:10.1016/j.cmet.2015.09.003 (2015).

690 Dakup, P. P. et al. The circadian clock protects against ionizing radiation-induced cardiotoxicity. *FASEB J* **34**, 3347–58, doi:10.1096/fj.201901850RR (2020).

691 Schernhammer, E. S. et al. Rotating night shifts and risk of breast cancer in women participating in the nurses' health study. *J Natl Cancer Inst* **93**, 1563–8, doi:10.1093/jnci/93.20.1563 (2001).

692 Lozano-Lorca, M. et al. Night shift work, chronotype, sleep duration, and prostate cancer risk: CAPLIFE study. *Int J Environ Res Public Health* **17**, doi:10.3390/ijerph17176300 (2020).

693 Erren, T. C., Morfeld, P. and Gross, V. J. Night shift work, chronotype, and prostate cancer risk: incentives for additional analyses and prevention. *Int J Cancer* **137**, 1784–5, doi:10.1002/ijc.29524 (2015).

694 Papantoniou, K. et al. Night shift work, chronotype and prostate cancer risk in the MCC-Spain case-control study. *Int J Cancer* **137**, 1147–57, doi:10.1002/ijc.29400 (2015).

695 Viswanathan, A. N., Hankinson, S. E. and Schernhammer, E. S. Night shift work and the risk of endometrial cancer. *Cancer Res* **67**, 10618–22, doi:10.1158/0008-5472.CAN-07-2485 (2007).

696 Schernhammer, E. S. et al. Night-shift work and risk of colorectal cancer in the nurses' health study. *J Natl Cancer Inst* **95**, 825–8, doi:10.1093/jnci/95.11.825 (2003).

697 Papantoniou, K. et al. Rotating night shift work and colorectal cancer risk in the nurses' health studies. *Int J Cancer* **143**, 2709–17, doi:10.1002/ijc.31655 (2018).

698 Wegrzyn, L. R. et al. Rotating night-shift work and the risk of breast cancer in the nurses' health studies. *Am J Epidemiol* **186**, 532–40, doi:10.1093/aje/kwx140 (2017).

699 Papantoniou, K. et al. Breast cancer risk and night shift work in a case-control study in a Spanish population. *Eur J Epidemiol* **31**, 867–78, doi:10.1007/s10654-015-0073-y (2016).

700 Hansen, J. Light at night, shiftwork, and breast cancer risk. *J Natl Cancer Inst* **93**, 1513–15, doi:10.1093/jnci/93.20.1513 (2001).

701 Cordina-Duverger, E. et al. Night shift work and breast cancer: a pooled analysis of population-based case-control studies with complete work history. *Eur J Epidemiol* **33**, 369–79, doi:10.1007/s10654-018-0368-x (2018).

702 Straif, K. et al. Carcinogenicity of shift-work, painting, and fire-fighting. *Lancet Oncol* **8**, 1065–6, doi:10.1016/S1470-2045(07)70373-X (2007).

703 Tokumaru, O. et al. Incidence of cancer among female flight attendants: a meta-analysis. *J Travel Med* **13**, 127–32, doi:10.1111/j.1708-8305.2006.00029.x (2006).

704 Pukkala, E. et al. Cancer incidence among 10,211 airline pilots: a Nordic study. *Aviat Space Environ Med* **74**, 699–706 (2003).

705 Band, P. R. et al. Cohort study of Air Canada pilots: mortality, cancer incidence, and leukemia risk. *Am J Epidemiol* **143**, 137–43, doi:10.1093/oxfordjournals.aje.a008722 (1996).

706 Kettner, N. M. et al. Circadian homeostasis of liver metabolism suppresses hepatocarcinogenesis. *Cancer Cell* **30**, 909–24, doi:10.1016/j.ccell.2016.10.007 (2016).

707 Koritala, B. S. C. et al. Night shift schedule causes circadian dysregulation of DNA repair genes and elevated DNA damage in humans. *J Pineal Res* **70**, e12726, doi:10.1111/jpi.12726 (2021).

708 Fu, L. and Kettner, N. M. The circadian clock in cancer development and therapy. *Prog Mol Biol Transl Sci* **119**, 221–82, doi:10.1016/B978-0-12-396971-2.00009-9 (2013).

709 Yang, M. Y. et al. Downregulation of circadian clock genes in chronic myeloid leukemia: alternative methylation pattern of hPER3. *Cancer Sci* **97**, 1298–1307, doi:10.1111/j.1349–7006.2006.00331.x (2006).

710 Samulin Erdem, J. et al. Mechanisms of breast cancer in shift workers: DNA methylation in five core circadian genes in nurses working night shifts. *J Cancer* **8**, 2876–84, doi:10.7150/jca.21064 (2017).

711 Sulli, G. et al. Pharmacological activation of REV-ERBs is lethal in cancer and oncogene-induced senescence. *Nature* **553**, 351–5, doi:10.1038/nature25170 (2018).

712 Oshima, T. et al. Cell-based screen identifies a new potent and highly selective CK2 inhibitor for modulation of circadian rhythms and cancer cell growth. *Sci Adv* **5**, eaau9060, doi:10.1126/sciadv.aau9060 (2019).

713 Bu, Y. et al. A PERK-miR-211 axis suppresses circadian regulators and protein synthesis to promote cancer cell survival. *Nat Cell Biol* **20**, 104–15, doi:10.1038/s41556–017–0006-y (2018).

714 Grutsch, J. F. et al. Validation of actigraphy to assess circadian organization and sleep quality in patients with advanced lung cancer. *J Circadian Rhythms* **9**, 4, doi:10.1186/1740–3391–9–4 (2011).

715 Steur, L. M. H. et al. Sleep–wake rhythm disruption is associated with cancer-related fatigue in pediatric acute lymphoblastic leukemia. *Sleep* **43**, doi:10.1093/sleep/zsz320 (2020).

716 Palesh, O. et al. Relationship between subjective and actigraphy-measured sleep in 237 patients with metastatic colorectal cancer. *Qual Life Res* **26**, 2783–91, doi:10.1007/s11136–017–1617–2 (2017).

717 Innominato, P. F. et al. Circadian rhythm in rest and activity: a biological correlate of quality of life and a predictor of survival in patients with metastatic colorectal cancer. *Cancer Res* **69**, 4700–4707, doi:10.1158/0008–5472.CAN-08–4747 (2009).

718 Lévi, F. et al. Wrist actimetry circadian rhythm as a robust predictor of colorectal cancer patients survival. *Chronobiol Int* **31**, 891–900, doi:10.3109/07420528.2014.924523 (2014).

719  Nurse, P. A journey in science: cell-cycle control. *Mol Med* **22**, 112–19, doi:10.2119/molmed.2016.00189 (2017).

720  Li, S., Balmain, A. and Counter, C. M. A model for RAS mutation patterns in cancers: finding the sweet spot. *Nat Rev Cancer* **18**, 767–77, doi:10.1038/s41568-018-0076-6 (2018).

721  Tsuchiya, Y., Minami, I., Kadotani, H., Todo, T. and Nishida, E. Circadian clock-controlled diurnal oscillation of Ras/ERK signaling in mouse liver. *Proc Jpn Acad Ser B Phys Biol Sci* **89**, 59–65, doi:10.2183/pjab.89.59 (2013).

722  Relogio, A. et al. Ras-mediated deregulation of the circadian clock in cancer. *PLoS Genet* **10**, e1004338, doi:10.1371/journal.pgen.1004338 (2014).

723  Jacob, L., Freyn, M., Kalder, M., Dinas, K. and Kostev, K. Impact of tobacco smoking on the risk of developing 25 different cancers in the UK: a retrospective study of 422,010 patients followed for up to 30 years. *Oncotarget* **9**, 17420–29, doi:10.18632/oncotarget.24724 (2018).

724  Zienolddiny, S. et al. Analysis of polymorphisms in the circadian-related genes and breast cancer risk in Norwegian nurses working night shifts. *Breast Cancer Res* **15**, R53, doi:10.1186/bcr3445 (2013).

725  Levy-Lahad, E. and Friedman, E. Cancer risks among BRCA1 and BRCA2 mutation carriers. *Br J Cancer* **96**, 11–15, doi:10.1038/sj.bjc.6603535 (2007).

726  Buchi, K. N., Moore, J. G., Hrushesky, W. J., Sothern, R. B. and Rubin, N. H. Circadian rhythm of cellular proliferation in the human rectal mucosa. *Gastroenterology* **101**, 410–15, doi:10.1016/0016-5085(91)90019-h (1991).

727  Frentz, G., Moller, U., Holmich, P. and Christensen, I. J. On circadian rhythms in human epidermal cell proliferation. *Acta Derm Venereol* **71**, 85–7 (1991).

728  Hrushesky, W. J. Circadian timing of cancer chemotherapy. *Science* **228**, 73–5, doi:10.1126/science.3883493 (1985).

729  Rivard, G. E., Infante-Rivard, C., Hoyoux, C. and Champagne, J. Maintenance chemotherapy for childhood acute lymphoblastic

leukaemia: better in the evening. *Lancet* **2**, 1264–6, doi:10.1016/s0140–6736(85)91551-x (1985).

730 Lévi, F. et al. Chronotherapy of colorectal cancer metastases. *Hepatogastroenterology* **48**, 320–22 (2001).

731 Lévi, F., Okyar, A., Dulong, S., Innominato, P. F. and Clairambault, J. Circadian timing in cancer treatments. *Annu Rev Pharmacol Toxicol* **50**, 377–421, doi:10.1146/annurev.pharmtox.48.113006.094626 (2010).

732 Hill, R. J. W., Innominato, P. F., Lévi, F. and Ballesta, A. Optimizing circadian drug infusion schedules towards personalized cancer chronotherapy. *PLoS Comput Biol* **16**, e1007218, doi:10.1371/journal.pcbi.1007218 (2020).

733 Chan, S. et al. Could time of whole brain radiotherapy delivery impact overall survival in patients with multiple brain metastases? *Ann Palliat Med* **5**, 267–79, doi:10.21037/apm.2016.09.05 (2016).

734 Lim, G. B. Surgery: circadian rhythms influence surgical outcomes. *Nat Rev Cardiol* **15**, 5, doi:10.1038/nrcardio.2017.186 (2018).

735 Czeisler, C. A., Pellegrini, C. A. and Sade, R. M. Should sleep-deprived surgeons be prohibited from operating without patients' consent? *Ann Thorac Surg* **95**, 757–66, doi:10.1016/j.athoracsur.2012.11.052 (2013).

736 Lévi, F. and Okyar, A. Circadian clocks and drug delivery systems: impact and opportunities in chronotherapeutics. *Expert Opin Drug Deliv* **8**, 1535–41, doi:10.1517/17425247.2011.618184 (2011).

737 Man, K., Loudon, A. and Chawla, A. Immunity around the clock. *Science* **354**, 999–1003, doi:10.1126/science.aah4966 (2016).

738 Scheiermann, C., Kunisaki, Y. and Frenette, P. S. Circadian control of the immune system. *Nat Rev Immunol* **13**, 190–98, doi:10.1038/nri3386 (2013).

739 Lyons, A. B., Moy, L., Moy, R. and Tung, R. Circadian rhythm and the skin: a review of the literature. *J Clin Aesthet Dermatol* **12**, 42–5 (2019).

740 Chen, S., Fuller, K. K., Dunlap, J. C. and Loros, J. J. A pro- and anti-inflammatory axis modulates the macrophage circadian clock. *Front Immunol* **11**, 867, doi:10.3389/fimmu.2020.00867 (2020).

741  Edgar, R. S. et al. Cell autonomous regulation of herpes and influenza virus infection by the circadian clock. *Proc Natl Acad Sci USA* **113**, 10085–90, doi:10.1073/pnas.1601895113 (2016).

742  Sengupta, S. et al. Circadian control of lung inflammation in influenza infection. *Nat Commun* **10**, 4107, doi:10.1038/s41467-019-11400-9 (2019).

743  Long, J. E. et al. Morning vaccination enhances antibody response over afternoon vaccination: a cluster-randomised trial. *Vaccine* **34**, 2679–85, doi:10.1016/j.vaccine.2016.04.032 (2016).

744  Vinciguerra, M. et al. Exploitation of host clock gene machinery by hepatitis viruses B and C. *World J Gastroenterol* **19**, 8902–9, doi:10.3748/wjg.v19.i47.8902 (2013).

745  Benegiamo, G. et al. Mutual antagonism between circadian protein period 2 and hepatitis C virus replication in hepatocytes. *PLoS One* **8**, e60527, doi:10.1371/journal.pone.0060527 (2013).

746  Zhuang, X., Rambhatla, S. B., Lai, A. G. and McKeating, J. A. Interplay between circadian clock and viral infection. *J Mol Med (Berl)* **95**, 1283–9, doi:10.1007/s00109-017-1592-7 (2017).

747  Spiegel, K., Sheridan, J. F. and Van Cauter, E. Effect of sleep deprivation on response to immunization. *JAMA* **288**, 1471–2, doi:10.1001/jama.288.12.1471-a (2002).

748  Taylor, D. J., Kelly, K., Kohut, M. L. and Song, K. S. Is insomnia a risk factor for decreased influenza vaccine response? *Behav Sleep Med* **15**, 270–87, doi:10.1080/15402002.2015.1126596 (2017).

749  Prather, A. A. et al. Sleep and antibody response to hepatitis B vaccination. *Sleep* **35**, 1063–9, doi:10.5665/sleep.1990 (2012).

750  Lange, T., Perras, B., Fehm, H. L. and Born, J. Sleep enhances the human antibody response to hepatitis A vaccination. *Psychosom Med* **65**, 831–5, doi:10.1097/01.psy.0000091382.61178.fi (2003).

751  Glaser, R. and Kiecolt-Glaser, J. K. Stress-induced immune dysfunction: implications for health. *Nat Rev Immunol* **5**, 243–51, doi:10.1038/nri1571 (2005).

752  Segerstrom, S. C. and Miller, G. E. Psychological stress and the human immune system: a meta-analytic study of 30 years of

inquiry. *Psychol Bull* **130**, 601–30, doi:10.1037/0033–2909.130.4.601 (2004).

753 Irwin, M. et al. Partial sleep deprivation reduces natural killer cell activity in humans. *Psychosom Med* **56**, 493–8, doi:10.1097/00006842–199411000–00004 (1994).

754 Straub, R. H. and Cutolo, M. Involvement of the hypothalamic–pituitary–adrenal/gonadal axis and the peripheral nervous system in rheumatoid arthritis: viewpoint based on a systemic pathogenetic role. *Arthritis Rheum* **44**, 493–507, doi:10.1002/1529–0131 (200103)44:3<493::AID-ANR95>3.0.CO;2-U (2001).

755 Hotez, P. J. and Herricks, J. R. Impact of the neglected tropical diseases on human development in the organisation of islamic cooperation nations. *PLoS Negl Trop Dis* **9**, e0003782, doi:10.1371/journal.pntd.0003782 (2015).

756 Waite, J. L., Suh, E., Lynch, P. A. and Thomas, M. B. Exploring the lower thermal limits for development of the human malaria parasite, *Plasmodium falciparum*. *Biol Lett* **15**, 20190275, doi:10.1098/rsbl.2019.0275 (2019).

757 Dobson, M. J. Malaria in England: a geographical and historical perspective. *Parassitologia* **36**, 35–60 (1994).

758 Nixon, C. P. *Plasmodium falciparum* gametocyte transit through the cutaneous microvasculature: a new target for malaria transmission blocking vaccines? *Hum Vaccin Immunother* **12**, 3189–95, doi:10.1080/21645515.2016.1183076 (2016).

759 Meibalan, E. and Marti, M. Biology of malaria transmission. *Cold Spring Harb Perspect Med* **7**, doi:10.1101/cshperspect.a025452 (2017).

760 Long, C. A. and Zavala, F. Immune responses in malaria. *Cold Spring Harb Perspect Med* **7**, doi:10.1101/cshperspect.a025577 (2017).

761 Lell, B., Brandts, C. H., Graninger, W. and Kremsner, P. G. The circadian rhythm of body temperature is preserved during malarial fever. *Wien Klin Wochenschr* **112**, 1014–15 (2000).

762 Reece, S. E., Prior, K. F. and Mideo, N. The life and times of parasites: rhythms in strategies for within-host survival and between-host

transmission. *J Biol Rhythms* **32**, 516–33, doi:10.1177/0748730417718904 (2017).

763  Rijo-Ferreira, F. et al. The malaria parasite has an intrinsic clock. *Science* **368**, 746–53, doi:10.1126/science.aba2658 (2020).

764  Carvalho Cabral, P., Olivier, M. and Cermakian, N. The complex interplay of parasites, their hosts, and circadian clocks. *Front Cell Infect Microbiol* **9**, 425, doi:10.3389/fcimb.2019.00425 (2019).

765  O'Donnell, A. J., Schneider, P., McWatters, H. G. and Reece, S. E. Fitness costs of disrupting circadian rhythms in malaria parasites. *Proc Biol Sci* **278**, 2429–36, doi:10.1098/rspb.2010.2457 (2011).

766  Prior, K. F. et al. Timing of host feeding drives rhythms in parasite replication. *PLoS Pathog* **14**, e1006900, doi:10.1371/journal.ppat.1006900 (2018).

767  Hirako, I. C. et al. Daily rhythms of TNFalpha expression and food intake regulate synchrony of plasmodium stages with the host circadian cycle. *Cell Host Microbe* **23**, 796–808 e796, doi:10.1016/j.chom.2018.04.016 (2018).

768  Descamps, S. Breeding synchrony and predator specialization: a test of the predator swamping hypothesis in seabirds. *Ecol Evol* **9**, 1431–6, doi:10.1002/ece3.4863 (2019).

769  Lavtar, P. et al. Association of circadian rhythm genes ARNTL/ BMAL1 and CLOCK with multiple sclerosis. *PLoS One* **13**, e0190601, doi:10.1371/journal.pone.0190601 (2018).

770  Gustavsen, S. et al. Shift work at young age is associated with increased risk of multiple sclerosis in a Danish population. *Mult Scler Relat Disord* **9**, 104–9, doi:10.1016/j.msard.2016.06.010 (2016).

771  Dowell, S. F. and Ho, M. S. Seasonality of infectious diseases and severe acute respiratory syndrome – what we don't know can hurt us. *Lancet Infect Dis* **4**, 704–8, doi:10.1016/S1473-3099(04)01177-6 (2004).

772  Babcock, J. and Krouse, H. J. Evaluating the sleep/wake cycle in persons with asthma: three case scenarios. *J Am Acad Nurse Pract* **22**, 270–77, doi:10.1111/j.1745-7599.2010.00505.x (2010).

773 Ray, S. and Reddy, A. B. COVID-19 management in light of the circadian clock. *Nat Rev Mol Cell Biol* **21**, 494–5, doi:10.1038/s41580-020-0275-3 (2020).

774 Collaborators, G. B. D. O. et al. Health effects of overweight and obesity in 195 countries over 25 years. *N Engl J Med* **377**, 13–27, doi:10.1056/NEJMoa1614362 (2017).

775 Eknoyan, G. A history of obesity, or how what was good became ugly and then bad. *Adv Chronic Kidney Dis* **13**, 421–7, doi:10.1053/j.ackd.2006.07.002 (2006).

776 Yaffe, K. et al. Cardiovascular risk factors across the life course and cognitive decline: a pooled cohort study. *Neurology*, doi:10.1212/WNL.0000000000011747 (2021).

777 Kwok, S. et al. Obesity: a critical risk factor in the COVID-19 pandemic. *Clin Obes* **10**, e12403, doi:10.1111/cob.12403 (2020).

778 Kalsbeek, A., la Fleur, S. and Fliers, E. Circadian control of glucose metabolism. *Mol Metab* **3**, 372–83, doi:10.1016/j.molmet.2014.03.002 (2014).

779 Adeva-Andany, M. M., Funcasta-Calderon, R., Fernandez-Fernandez, C., Castro-Quintela, E. and Carneiro-Freire, N. Metabolic effects of glucagon in humans. *J Clin Transl Endocrinol* **15**, 45–53, doi:10.1016/j.jcte.2018.12.005 (2019).

780 Weeke, J. and Gundersen, H. J. Circadian and 30 minutes variations in serum TSH and thyroid hormones in normal subjects. *Acta Endocrinol (Copenh)* **89**, 659–72, doi:10.1530/acta.0.0890659 (1978).

781 Lieb, K., Reincke, M., Riemann, D. and Voderholzer, U. Sleep deprivation and growth-hormone secretion. *Lancet* **356**, 2096–7, doi:10.1016/S0140-6736(05)74304-X (2000).

782 Tsai, M., Asakawa, A., Amitani, H. and Inui, A. Stimulation of leptin secretion by insulin. *Indian J Endocrinol Metab* **16**, S543–8, doi:10.4103/2230-8210.105570 (2012).

783 Thie, N. M., Kato, T., Bader, G., Montplaisir, J. Y. and Lavigne, G. J. The significance of saliva during sleep and the relevance of oromotor movements. *Sleep Med Rev* **6**, 213–27, doi:10.1053/smrv.2001.0183 (2002).

784 Duboc, H., Coffin, B. and Siproudhis, L. Disruption of circadian rhythms and gut motility: an overview of underlying mechanisms and associated pathologies. *J Clin Gastroenterol* **54**, 405–14, doi:10.1097/MCG.0000000000001333 (2020).

785 Vaughn, B., Rotolo, S. and Roth, H. Circadian rhythm and sleep influences on digestive physiology and disorders. *ChronoPhysiology and Therapy* **4**, 67–77, doi:https://doi.org/10.2147/CPT.S44806 (2014).

786 Yamamoto, H., Nagai, K. and Nakagawa, H. Role of SCN in daily rhythms of plasma glucose, FFA, insulin and glucagon. *Chronobiol Int* **4**, 483–91, doi:10.3109/07420528709078539 (1987).

787 Van den Pol, A. N. and Powley, T. A fine-grained anatomical analysis of the role of the rat suprachiasmatic nucleus in circadian rhythms of feeding and drinking. *Brain Res* **160**, 307–26, doi:10.1016/0006–8993(79)90427-x (1979).

788 Turek, F. W. et al. Obesity and metabolic syndrome in circadian Clock mutant mice. *Science* **308**, 1043–5, doi:10.1126/science.1108750 (2005).

789 Stokkan, K. A., Yamazaki, S., Tei, H., Sakaki, Y. and Menaker, M. Entrainment of the circadian clock in the liver by feeding. *Science* **291**, 490–93, doi:10.1126/science.291.5503.490 (2001).

790 Chaix, A., Lin, T., Le, H. D., Chang, M. W. and Panda, S. Time-restricted feeding prevents obesity and metabolic syndrome in mice lacking a circadian clock. *Cell Metab* **29**, 303–19 e304, doi:10.1016/j.cmet.2018.08.004 (2019).

791 Kroenke, C. H. et al. Work characteristics and incidence of type 2 diabetes in women. *Am J Epidemiol* **165**, 175–83, doi:10.1093/aje/kwj355 (2007).

792 Suwazono, Y. et al. Shiftwork and impaired glucose metabolism: a 14-year cohort study on 7104 male workers. *Chronobiol Int* **26**, 926–41, doi:10.1080/07420520903044422 (2009).

793 Pan, A., Schernhammer, E. S., Sun, Q. and Hu, F. B. Rotating night shift work and risk of type 2 diabetes: two prospective cohort studies in women. *PLoS Med* **8**, e1001141, doi:10.1371/journal.pmed.1001141 (2011).

794  Shan, Z. et al. Rotating night shift work and adherence to unhealthy lifestyle in predicting risk of type 2 diabetes: results from two large US cohorts of female nurses. *BMJ* **363**, k4641, doi:10.1136/bmj.k4641 (2018).

795  Meisinger, C., Heier, M., Loewel, H. and Study, M. K. A. C. Sleep disturbance as a predictor of type 2 diabetes mellitus in men and women from the general population. *Diabetologia* **48**, 235–41, doi:10.1007/s00125-004-1634-x (2005).

796  Gangwisch, J. E. et al. Sleep duration as a risk factor for diabetes incidence in a large U.S. sample. *Sleep* **30**, 1667–73, doi:10.1093/sleep/30.12.1667 (2007).

797  Beihl, D. A., Liese, A. D. and Haffner, S. M. Sleep duration as a risk factor for incident type 2 diabetes in a multiethnic cohort. *Ann Epidemiol* **19**, 351–7, doi:10.1016/j.annepidem.2008.12.001 (2009).

798  Cummings, D. E. et al. A preprandial rise in plasma ghrelin levels suggests a role in meal initiation in humans. *Diabetes* **50**, 1714–19, doi:10.2337/diabetes.50.8.1714 (2001).

799  Schmid, S. M., Hallschmid, M., Jauch-Chara, K., Born, J. and Schultes, B. A single night of sleep deprivation increases ghrelin levels and feelings of hunger in normal-weight healthy men. *J Sleep Res* **17**, 331–4, doi:10.1111/j.1365-2869.2008.00662.x (2008).

800  Froy, O. Metabolism and circadian rhythms – implications for obesity. *Endocr Rev* **31**, 1–24, doi:10.1210/er.2009-0014 (2010).

801  Spiegel, K., Tasali, E., Penev, P. and Van Cauter, E. Brief communication: sleep curtailment in healthy young men is associated with decreased leptin levels, elevated ghrelin levels, and increased hunger and appetite. *Ann Intern Med* **141**, 846–50, doi:10.7326/0003-4819-141-11-200412070-00008 (2004).

802  Van Drongelen, A., Boot, C. R., Merkus, S. L., Smid, T. and van der Beek, A. J. The effects of shift work on body weight change – a systematic review of longitudinal studies. *Scand J Work Environ Health* **37**, 263–75, doi:10.5271/sjweh.3143 (2011).

803  Licinio, J. Longitudinally sampled human plasma leptin and cortisol concentrations are inversely correlated. *J Clin Endocrinol Metab* **83**, 1042, doi:10.1210/jcem.83.3.4668-3 (1998).

804 Heptulla, R. et al. Temporal patterns of circulating leptin levels in lean and obese adolescents: relationships to insulin, growth hormone, and free fatty acids rhythmicity. *J Clin Endocrinol Metab* **86**, 90–96, doi:10.1210/jcem.86.1.7136 (2001).

805 Izquierdo, A. G., Crujeiras, A. B., Casanueva, F. F. and Carreira, M. C. Leptin, obesity, and leptin resistance: where are we 25 years later? *Nutrients* **11**, doi:10.3390/nu11112704 (2019).

806 Cohen, P. and Spiegelman, B. M. Cell biology of fat storage. *Mol Biol Cell* **27**, 2523–7, doi:10.1091/mbc.E15–10–0749 (2016).

807 Maury, E., Hong, H. K. and Bass, J. Circadian disruption in the pathogenesis of metabolic syndrome. *Diabetes Metab* **40**, 338–46, doi:10.1016/j.diabet.2013.12.005 (2014).

808 Gnocchi, D., Pedrelli, M., Hurt-Camejo, E. and Parini, P. Lipids around the clock: focus on circadian rhythms and lipid metabolism. *Biology (Basel)* **4**, 104–32, doi:10.3390/biology4010104 (2015).

809 Gottlieb, D. J. et al. Association of sleep time with diabetes mellitus and impaired glucose tolerance. *Arch Intern Med* **165**, 863–7, doi:10.1001/archinte.165.8.863 (2005).

810 Cappuccio, F. P. et al. Meta-analysis of short sleep duration and obesity in children and adults. *Sleep* **31**, 619–26, doi:10.1093/sleep/31.5.619 (2008).

811 Belyavskiy, E., Pieske-Kraigher, E. and Tadic, M. Obstructive sleep apnea, hypertension, and obesity: a dangerous triad. *J Clin Hypertens (Greenwich)* **21**, 1591–3, doi:10.1111/jch.13688 (2019).

812 Driver, H. S., Shulman, I., Baker, F. C. and Buffenstein, R. Energy content of the evening meal alters nocturnal body temperature but not sleep. *Physiol Behav* **68**, 17–23, doi:10.1016/s0031–9384(99)00145–6 (1999).

813 Strand, D. S., Kim, D. and Peura, D. A. 25 years of proton pump inhibitors: a comprehensive review. *Gut Liver* **11**, 27–37, doi:10.5009/gnl15502 (2017).

814 Hatlebakk, J. G., Katz, P. O., Camacho-Lobato, L. and Castell, D. O. Proton pump inhibitors: better acid suppression when taken before a meal than without a meal. *Aliment Pharmacol Ther* **14**, 1267–72, doi:10.1046/j.1365–2036.2000.00829.x (2000).

815  Jung, H. K., Choung, R. S. and Talley, N. J. Gastroesophageal reflux disease and sleep disorders: evidence for a causal link and therapeutic implications. *J Neurogastroenterol Motil* **16**, 22–9, doi:10.5056/jnm.2010.16.1.22 (2010).

816  Hatlebakk, J. G., Katz, P. O., Kuo, B. and Castell, D. O. Nocturnal gastric acidity and acid breakthrough on different regimens of omeprazole 40 mg daily. *Aliment Pharmacol Ther* **12**, 1235–40, doi:10.1046/j.1365-2036.1998.00426.x (1998).

817  Syed, A. U. et al. Adenylyl cyclase 5-generated cAMP controls cerebral vascular reactivity during diabetic hyperglycemia. *J Clin Invest* **129**, 3140–52, doi:10.1172/JCI124705 (2019).

818  Sanchez, A. et al. Role of sugars in human neutrophilic phagocytosis. *Am J Clin Nutr* **26**, 1180–84, doi:10.1093/ajcn/26.11.1180 (1973).

819  Rotimi, C. N., Tekola-Ayele, F., Baker, J. L. and Shriner, D. The African diaspora: history, adaptation and health. *Curr Opin Genet Dev* **41**, 77–84, doi:10.1016/j.gde.2016.08.005 (2016).

820  Yang, Q. et al. Added sugar intake and cardiovascular diseases mortality among US adults. *JAMA Intern Med* **174**, 516–24, doi:10.1001/jamainternmed.2013.13563 (2014).

821  Stanhope, K. L. Sugar consumption, metabolic disease and obesity: the state of the controversy. *Crit Rev Clin Lab Sci* **53**, 52–67, doi:10.3109/10408363.2015.1084990 (2016).

822  Sherwani, S. I., Khan, H. A., Ekhzaimy, A., Masood, A. and Sakharkar, M. K. Significance of HbA1c test in diagnosis and prognosis of diabetic patients. *Biomark Insights* **11**, 95–104, doi:10.4137/BMI.S38440 (2016).

823  Fildes, A. et al. Probability of an obese person attaining normal body weight: cohort study using electronic health records. *Am J Public Health* **105**, e54–9, doi:10.2105/AJPH.2015.302773 (2015).

824  Shea, S. A., Hilton, M. F., Hu, K. and Scheer, F. A. Existence of an endogenous circadian blood pressure rhythm in humans that peaks in the evening. *Circ Res* **108**, 980–84, doi:10.1161/CIRCRESAHA.110.233668 (2011).

825  Selfridge, J. M., Moyer, K., Capelluto, D. G. and Finkielstein, C. V. Opening the debate: how to fulfill the need for physicians' training

in circadian-related topics in a full medical school curriculum. *J Circadian Rhythms* 13, 7, doi:10.5334/jcr.ah (2015).

826 Atkinson, G. and Reilly, T. Circadian variation in sports performance. *Sports Med* 21, 292–312, doi:10.2165/00007256–199621040–00005 (1996).

827 De Goede, P., Wefers, J., Brombacher, E. C., Schrauwen, P. and Kalsbeek, A. Circadian rhythms in mitochondrial respiration. *J Mol Endocrinol* 60, R115–R130, doi:10.1530/JME-17–0196 (2018).

828 Van Moorsel, D. et al. Demonstration of a day–night rhythm in human skeletal muscle oxidative capacity. *Mol Metab* 5, 635–45, doi:10.1016/j.molmet.2016.06.012 (2016).

829 Kline, C. E. et al. Circadian variation in swim performance. *J Appl Physiol (1985)* 102, 641–9, doi:10.1152/japplphysiol.00910.2006 (2007).

830 Zitting, K. M. et al. Human resting energy expenditure varies with circadian phase. *Curr Biol* 28, 3685–90 e3683, doi:10.1016/j.cub.2018.10.005 (2018).

831 Facer-Childs, E. and Brandstaetter, R. The impact of circadian phenotype and time since awakening on diurnal performance in athletes. *Curr Biol* 25, 518–22, doi:10.1016/j.cub.2014.12.036 (2015).

832 Vieira, A. F., Costa, R. R., Macedo, R. C., Coconcelli, L. and Kruel, L. F. Effects of aerobic exercise performed in fasted v. fed state on fat and carbohydrate metabolism in adults: a systematic review and meta-analysis. *Br J Nutr* 116, 1153–64, doi:10.1017/S0007114516003160 (2016).

833 Iwayama, K. et al. Exercise increases 24–h fat oxidation only when it is performed before breakfast. *EBioMedicine* 2, 2003–9, doi:10.1016/j.ebiom.2015.10.029 (2015).

834 Colberg, S. R., Grieco, C. R. and Somma, C. T. Exercise effects on postprandial glycemia, mood, and sympathovagal balance in type 2 diabetes. *J Am Med Dir Assoc* 15, 261–6, doi:10.1016/j.jamda.2013.11.026 (2014).

835 Borror, A., Zieff, G., Battaglini, C. and Stoner, L. The effects of postprandial exercise on glucose control in individuals with type 2 diabetes: a systematic review. *Sports Med* 48, 1479–91, doi:10.1007/s40279–018–0864-x (2018).

836  Reebs, S. G. and Mrosovsky, N. Effects of induced wheel running on the circadian activity rhythms of Syrian hamsters: entrainment and phase response curve. *J Biol Rhythms* 4, 39–48, doi:10.1177/074873048900400103 (1989).

837  Youngstedt, S. D., Elliott, J. A. and Kripke, D. F. Human circadian phase-response curves for exercise. *J Physiol* 597, 2253–68, doi:10.1113/JP276943 (2019).

838  Lewis, P., Korf, H. W., Kuffer, L., Gross, J. V. and Erren, T. C. Exercise time cues (zeitgebers) for human circadian systems can foster health and improve performance: a systematic review. *BMJ Open Sport Exerc Med* 4, e000443, doi:10.1136/bmjsem-2018–000443 (2018).

839  Nedeltcheva, A. V. and Scheer, F. A. Metabolic effects of sleep disruption, links to obesity and diabetes. *Curr Opin Endocrinol Diabetes Obes* 21, 293–8, doi:10.1097/MED.0000000000000082 (2014).

840  Zimberg, I. Z. et al. Short sleep duration and obesity: mechanisms and future perspectives. *Cell Biochem Funct* 30, 524–9, doi:10.1002/cbf.2832 (2012).

841  Depner, C. M., Stothard, E. R. and Wright, K. P., Jr. Metabolic consequences of sleep and circadian disorders. *Curr Diab Rep* 14, 507, doi:10.1007/s11892–014–0507-z (2014).

842  Shi, S. Q., Ansari, T. S., McGuinness, O. P., Wasserman, D. H. and Johnson, C. H. Circadian disruption leads to insulin resistance and obesity. *Curr Biol* 23, 372–81, doi:10.1016/j.cub.2013.01.048 (2013).

843  Stenvers, D. J., Scheer, F., Schrauwen, P., la Fleur, S. E. and Kalsbeek, A. Circadian clocks and insulin resistance. *Nat Rev Endocrinol* 15, 75–89, doi:10.1038/s41574–018–0122–1 (2019).

844  Virtanen, M. et al. Long working hours and alcohol use: systematic review and meta-analysis of published studies and unpublished individual participant data. *BMJ* 350, g7772, doi:10.1136/bmj.g7772 (2015).

845  Summa, K. C. et al. Disruption of the circadian clock in mice increases intestinal permeability and promotes alcohol-induced hepatic pathology and inflammation. *PLoS One* 8, e67102, doi:10.1371/journal.pone.0067102 (2013).

846 Bailey, S. M. Emerging role of circadian clock disruption in alcohol-induced liver disease. *Am J Physiol Gastrointest Liver Physiol* **315**, G364–G373, doi:10.1152/ajpgi.00010.2018 (2018).

847 Eastman, C. I., Stewart, K. T. and Weed, M. R. Evening alcohol consumption alters the circadian rhythm of body temperature. *Chronobiol Int* **11**, 141–2, doi:10.3109/07420529409055901 (1994).

848 Danel, T., Libersa, C. and Touitou, Y. The effect of alcohol consumption on the circadian control of human core body temperature is time dependent. *Am J Physiol Regul Integr Comp Physiol* **281**, R52–5, doi:10.1152/ajpregu.2001.281.1.R52 (2001).

849 Daimon, K., Yamada, N., Tsujimoto, T. and Takahashi, S. Circadian rhythm abnormalities of deep body temperature in depressive disorders. *J Affect Disord* **26**, 191–8, doi:10.1016/0165-0327(92)90015-x (1992).

850 Lack, L. C. and Lushington, K. The rhythms of human sleep propensity and core body temperature. *J Sleep Res* **5**, 1–11, doi:10.1046/j.1365-2869.1996.00005.x (1996).

851 Miyata, S. et al. REM sleep is impaired by a small amount of alcohol in young women sensitive to alcohol. *Intern Med* **43**, 679–84, doi:10.2169/internalmedicine.43.679 (2004).

852 Simou, E., Britton, J. and Leonardi-Bee, J. Alcohol and the risk of sleep apnoea: a systematic review and meta-analysis. *Sleep Med* **42**, 38–46, doi:10.1016/j.sleep.2017.12.005 (2018).

853 Stenvers, D. J., Jonkers, C. F., Fliers, E., Bisschop, P. and Kalsbeek, A. Nutrition and the circadian timing system. *Prog Brain Res* **199**, 359–76, doi:10.1016/B978-0-444-59427-3.00020-4 (2012).

854 Kara, Y., Tuzun, S., Oner, C. and Simsek, E. E. Night eating syndrome according to obesity groups and the related factors. *J Coll Physicians Surg Pak* **30**, 833–8, doi:10.29271/jcpsp.2020.08.833 (2020).

855 Gallant, A. R., Lundgren, J. and Drapeau, V. The night-eating syndrome and obesity. *Obes Rev* **13**, 528–36, doi:10.1111/j.1467-789X.2011.00975.x (2012).

856 Bo, S. et al. Consuming more of daily caloric intake at dinner predisposes to obesity. A 6-year population-based prospective cohort study. *PLoS One* **9**, e108467, doi:10.1371/journal.pone.0108467 (2014).

857  Garaulet, M. et al. Timing of food intake predicts weight loss effectiveness. *Int J Obes (Lond)* **37**, 604–11, doi:10.1038/ijo.2012.229 (2013).

858  Jakubowicz, D., Barnea, M., Wainstein, J. and Froy, O. High caloric intake at breakfast vs. dinner differentially influences weight loss of overweight and obese women. *Obesity (Silver Spring)* **21**, 2504–12, doi:10.1002/oby.20460 (2013).

859  Jakubowicz, D. et al. High-energy breakfast with low-energy dinner decreases overall daily hyperglycaemia in type 2 diabetic patients: a randomised clinical trial. *Diabetologia* **58**, 912–19, doi:10.1007/s00125–015–3524–9 (2015).

860  Ekmekcioglu, C. and Touitou, Y. Chronobiological aspects of food intake and metabolism and their relevance on energy balance and weight regulation. *Obes Rev* **12**, 14–25, doi:10.1111/j.1467–789X. 2010.00716.x (2011).

861  Hutchison, A. T., Wittert, G. A. and Heilbronn, L. K. Matching meals to body clocks – impact on weight and glucose metabolism. *Nutrients* **9**, doi:10.3390/nu9030222 (2017).

862  Morris, C. J. et al. Endogenous circadian system and circadian misalignment impact glucose tolerance via separate mechanisms in humans. *Proc Natl Acad Sci USA* **112**, E2225–34, doi:10.1073/pnas. 1418955112 (2015).

863  Sender, R., Fuchs, S. and Milo, R. Revised estimates for the number of human and bacteria cells in the body. *PLoS Biol* **14**, e1002533, doi:10.1371/journal.pbio.1002533 (2016).

864  Saklayen, M. G. The global epidemic of the metabolic syndrome. *Curr Hypertens Rep* **20**, 12, doi:10.1007/s11906–018–0812-z (2018).

865  Paulose, J. K., Wright, J. M., Patel, A. G. and Cassone, V. M. Human gut bacteria are sensitive to melatonin and express endogenous circadian rhythmicity. *PLoS One* **11**, e0146643, doi:10.1371/journal. pone.0146643 (2016).

866  Thaiss, C. A. et al. Microbiota diurnal rhythmicity programs host transcriptome oscillations. *Cell* **167**, 1495–1510 e1412, doi:10.1016/j. cell.2016.11.003 (2016).

867  Liang, X., Bushman, F. D. and FitzGerald, G. A. Rhythmicity of the intestinal microbiota is regulated by gender and the host circadian clock. *Proc Natl Acad Sci USA* **112**, 10479–84, doi:10.1073/pnas.1501305112 (2015).

868  Leone, V. et al. Effects of diurnal variation of gut microbes and high-fat feeding on host circadian clock function and metabolism. *Cell Host Microbe* **17**, 681–9, doi:10.1016/j.chom.2015.03.006 (2015).

869  Parkar, S. G., Kalsbeek, A. and Cheeseman, J. F. Potential role for the gut microbiota in modulating host circadian rhythms and metabolic health. *Microorganisms* **7**, doi:10.3390/microorganisms7020041 (2019).

870  Kuang, Z. et al. The intestinal microbiota programs diurnal rhythms in host metabolism through histone deacetylase 3. *Science* **365**, 1428–34, doi:10.1126/science.aaw3134 (2019).

871  Rinninella, E. et al. What is the healthy gut microbiota composition? A changing ecosystem across age, environment, diet, and diseases. *Microorganisms* **7**, doi:10.3390/microorganisms7010014 (2019).

872  Depommier, C. et al. Supplementation with Akkermansia muciniphila in overweight and obese human volunteers: a proof-of-concept exploratory study. *Nat Med* **25**, 1096–1103, doi:10.1038/s41591–019–0495–2 (2019).

873  Janssen, A. W. and Kersten, S. The role of the gut microbiota in metabolic health. *FASEB J* **29**, 3111–23, doi:10.1096/fj.14–269514 (2015).

874  Rosselot, A. E., Hong, C. I. and Moore, S. R. Rhythm and bugs: circadian clocks, gut microbiota, and enteric infections. *Curr Opin Gastroenterol* **32**, 7–11, doi:10.1097/MOG.0000000000000227 (2016).

875  Voigt, R. M. et al. The circadian clock mutation promotes intestinal dysbiosis. *Alcohol Clin Exp Res* **40**, 335–47, doi:10.1111/acer.12943 (2016).

876  Mukherji, A., Kobiita, A., Ye, T. and Chambon, P. Homeostasis in intestinal epithelium is orchestrated by the circadian clock and microbiota cues transduced by TLRs. *Cell* **153**, 812–27, doi:10.1016/j.cell.2013.04.020 (2013).

## References

877 Butler, T. D. and Gibbs, J. E. Circadian host-microbiome interactions in immunity. *Front Immunol* **11**, 1783, doi:10.3389/fimmu.2020.01783 (2020).

878 Murakami, M. et al. Gut microbiota directs PPARgamma-driven reprogramming of the liver circadian clock by nutritional challenge. *EMBO Rep* **17**, 1292–1303, doi:10.15252/embr.201642463 (2016).

879 Round, J. L. and Mazmanian, S. K. The gut microbiota shapes intestinal immune responses during health and disease. *Nat Rev Immunol* **9**, 313–23, doi:10.1038/nri2515 (2009).

880 Zheng, D., Ratiner, K. and Elinav, E. Circadian influences of diet on the microbiome and immunity. *Trends Immunol* **41**, 512–30, doi:10.1016/j.it.2020.04.005 (2020).

881 Li, Y., Hao, Y., Fan, F. and Zhang, B. The role of microbiome in insomnia, circadian disturbance and depression. *Front Psychiatry* **9**, 669, doi:10.3389/fpsyt.2018.00669 (2018).

882 Hill, L. V. and Embil, J. A. Vaginitis: current microbiologic and clinical concepts. *CMAJ* **134**, 321–31 (1986).

883 Bahijri, S. et al. Relative metabolic stability, but disrupted circadian cortisol secretion during the fasting month of Ramadan. *PLoS One* **8**, e60917, doi:10.1371/journal.pone.0060917 (2013).

884 BaHammam, A. S. and Almeneessier, A. S. Recent evidence on the impact of Ramadan diurnal intermittent fasting, mealtime, and circadian rhythm on cardiometabolic risk: a review. *Front Nutr* **7**, 28, doi:10.3389/fnut.2020.00028 (2020).

885 Institute of Medicine (IOM). *To Err is Human: Building a Safer Health System*. National Academy Press, Washington, DC (2000).

886 Brensilver, J. M., Smith, L. and Lyttle, C. S. Impact of the Libby Zion case on graduate medical education in internal medicine. *Mt Sinai J Med* **65**, 296–300 (1998).

887 Grantcharov, T. P., Bardram, L., Funch-Jensen, P. and Rosenberg, J. Laparoscopic performance after one night on call in a surgical department: prospective study. *BMJ* **323**, 1222–3, doi:10.1136/bmj.323.7323.1222 (2001).

888 Eastridge, B. J. et al. Effect of sleep deprivation on the performance of simulated laparoscopic surgical skill. *Am J Surg* **186**, 169–74, doi:10.1016/s0002–9610(03)00183–1 (2003).

889 Baldwin, D. C., Jr and Daugherty, S. R. Sleep deprivation and fatigue in residency training: results of a national survey of first- and second-year residents. *Sleep* **27**, 217–23, doi:10.1093/sleep/27.2.217 (2004).

890 Fargen, K. M. and Rosen, C. L. Are duty hour regulations promoting a culture of dishonesty among resident physicians? *J Grad Med Educ* **5**, 553–5, doi:10.4300/JGME-D-13–00220.1 (2013).

891 Temple, J. Resident duty hours around the globe: where are we now? *BMC Med Educ* **14 Suppl** 1, S8, doi:10.1186/1472–6920–14-S1-S8 (2014).

892 Moonesinghe, S. R., Lowery, J., Shahi, N., Millen, A. and Beard, J. D. Impact of reduction in working hours for doctors in training on postgraduate medical education and patients' outcomes: systematic review. *BMJ* **342**, d1580, doi:10.1136/bmj.d1580 (2011).

893 O'Connor, P. et al. A mixed-methods examination of the nature and frequency of medical error among junior doctors. *Postgrad Med J* **95**, 583–9, doi:10.1136/postgradmedj-2018–135897 (2019).

894 McClelland, L., Holland, J., Lomas, J. P., Redfern, N. and Plunkett, E. A national survey of the effects of fatigue on trainees in anaesthesia in the UK. *Anaesthesia* **72**, 1069–77, doi:10.1111/anae.13965 (2017).

895 Giorgi, G. et al. Work-related stress in the banking sector: a review of incidence, correlated factors, and major consequences. *Front Psychol* **8**, 2166, doi:10.3389/fpsyg.2017.02166 (2017).

896 Blackmer, A. B. and Feinstein, J. A. Management of sleep disorders in children with neurodevelopmental disorders: a review. *Pharmacotherapy* **36**, 84–98, doi:10.1002/phar.1686 (2016).

897 Wasdell, M. B. et al. A randomized, placebo-controlled trial of controlled release melatonin treatment of delayed sleep phase syndrome and impaired sleep maintenance in children with neurodevelopmental disabilities. *J Pineal Res* **44**, 57–64, doi:10.1111/j.1600–079X.2007.00528.x (2008).

898 Owens, J. A. and Mindell, J. A. Pediatric insomnia. *Pediatr Clin North Am* **58**, 555–69, doi:10.1016/j.pcl.2011.03.011 (2011).

899 Blumer, J. L., Findling, R. L., Shih, W. J., Soubrane, C. and Reed, M. D. Controlled clinical trial of zolpidem for the treatment of insomnia associated with attention-deficit/hyperactivity disorder in children 6 to 17 years of age. *Pediatrics* **123**, e770–76, doi:10.1542/peds.2008-2945 (2009).

900 Landsend, E. C. S., Lagali, N. and Utheim, T. P. Congenital aniridia – a comprehensive review of clinical features and therapeutic approach. *Surv Ophthalmol*, doi:10.1016/j.survophthal.2021.02.011 (2021).

901 Abouzeid, H. et al. PAX6 aniridia and interhemispheric brain anomalies. *Mol Vis* **15**, 2074–83 (2009).

902 Hanish, A. E., Butman, J. A., Thomas, F., Yao, J. and Han, J. C. Pineal hypoplasia, reduced melatonin and sleep disturbance in patients with PAX6 haploinsufficiency. *J Sleep Res* **25**, 16–22, doi:10.1111/jsr.12345 (2016).

903 Auger, R. R. et al. Clinical practice guideline for the treatment of intrinsic circadian rhythm sleep–wake disorders: advanced sleep–wake phase disorder (ASWPD), delayed sleep–wake phase disorder (DSWPD), non-24-hour sleep–wake rhythm disorder (N24SWD), and irregular sleep–wake rhythm disorder (ISWRD). An update for 2015: an American Academy of Sleep Medicine clinical practice guideline. *J Clin Sleep Med* **11**, 1199–1236, doi:10.5664/jcsm.5100 (2015).

904 Emens, J. S. and Eastman, C. I. Diagnosis and treatment of non-24-h sleep–wake disorder in the blind. *Drugs* **77**, 637–50, doi:10.1007/s40265-017-0707-3 (2017).

905 Burke, T. M. et al. Combination of light and melatonin time cues for phase advancing the human circadian clock. *Sleep* **36**, 1617–24, doi:10.5665/sleep.3110 (2013).

906 Smith, M. R., Lee, C., Crowley, S. J., Fogg, L. F. and Eastman, C. I. Morning melatonin has limited benefit as a soporific for daytime sleep after night work. *Chronobiol Int* **22**, 873–88, doi:10.1080/09636410500292861 (2005).

907 Warren, W. S. and Cassone, V. M. The pineal gland: photoreception and coupling of behavioral, metabolic, and cardiovascular circadian outputs. *J Biol Rhythms* **10**, 64–79, doi:10.1177/074873049501000106 (1995).

908 Fisher, S. P. and Sugden, D. Endogenous melatonin is not obligatory for the regulation of the rat sleep–wake cycle. *Sleep* **33**, 833–40, doi:10.1093/sleep/33.6.833 (2010).

909 Quay, W. B. Precocious entrainment and associated characteristics of activity patterns following pinalectomy and reversal of photoperiod. *Physiol Behav* **5**, 1281–90, doi:10.1016/0031–9384(70)90041–7 (1970).

910 Deacon, S., English, J., Tate, J. and Arendt, J. Atenolol facilitates light-induced phase shifts in humans. *Neurosci Lett* **242**, 53–6, doi:10.1016/s0304–3940(98)00024-x (1998).

911 Cox, K. H. and Takahashi, J. S. Circadian clock genes and the transcriptional architecture of the clock mechanism. *J Mol Endocrinol* **63**, R93-R102, doi:10.1530/JME-19–0153 (2019).

912 Jagannath, A. et al. The CRTC1-SIK1 pathway regulates entrainment of the circadian clock. *Cell* **154**, 1100–1111, doi:10.1016/j.cell.2013.08.004 (2013).

913 Pushpakom, S. et al. Drug repurposing: progress, challenges and recommendations. *Nat Rev Drug Discov* **18**, 41–58, doi:10.1038/nrd.2018.168 (2019).

914 Goldstein, I., Burnett, A. L., Rosen, R. C., Park, P. W. and Stecher, V. J. The serendipitous story of Sildenafil: an unexpected oral therapy for erectile dysfunction. *Sex Med Rev* **7**, 115–28, doi:10.1016/j.sxmr.2018.06.005 (2019).

915 Burton, M. et al. The effect of handwashing with water or soap on bacterial contamination of hands. *Int J Environ Res Public Health* **8**, 97–104, doi:10.3390/ijerph8010097 (2011).

916 Jefferson, T. et al. Physical interventions to interrupt or reduce the spread of respiratory viruses: systematic review. *BMJ* **336**, 77–80, doi:10.1136/bmj.39393.510347.BE (2008).

## References

917  Zasloff, M. The antibacterial shield of the human urinary tract. *Kidney Int* **83**, 548–50, doi:10.1038/ki.2012.467 (2013).

918  McDermott, A. M. Antimicrobial compounds in tears. *Exp Eye Res* **117**, 53–61, doi:10.1016/j.exer.2013.07.014 (2013).

919  Valore, E. V., Park, C. H., Igreti, S. L. and Ganz, T. Antimicrobial components of vaginal fluid. *Am J Obstet Gynecol* **187**, 561–8, doi:10.1067/mob.2002.125280 (2002).

920  Edstrom, A. M. et al. The major bactericidal activity of human seminal plasma is zinc-dependent and derived from fragmentation of the semenogelins. *J Immunol* **181**, 3413–21, doi:10.4049/jimmunol.181.5.3413 (2008).

921  Nicholson, L. B. The immune system. *Essays Biochem* **60**, 275–301, doi:10.1042/EBC20160017 (2016).

# Acknowledgements

I would like to thank the following kind and generous individuals for their guidance, help and support in the preparation of this book: Elizabeth Foster, Victoria Foster, Renata Gomes, Aarti Jagannath, Glenn Leighton, William McMahon, Peter McWilliam, Mariya Moosajee, Andrea Nemeth, Stuart Peirson, David Ray, and Sridhar Vasudevan. For their medical expertise, considerable advice and critical eye I would also like to thank Alastair Buchan, Ben Canny, David Howells and Christopher Kennard. All errors and mistakes are mine and I apologize in advance should any have sneaked through. I would also like to thank my many friends and colleagues over the past 40 years with whom I have shared countless discussions and undertaken very many experiments. Special thanks go to my PhD supervisor, Professor Sir Brian Follett FRS, who taught me how to do science, and my mentor at the University of Virginia, the late Professor Michael Menaker, who gave me the confidence to trust my judgement. All of these countless interactions have framed my views on the science of biological time. Importantly, these discussions also generated much laughter and deep friendships, for which I am immensely grateful. I would also like to thank my editor, Tom Killingbeck at Penguin Books, for his encouragement, help, gentle guidance and tolerance. And, finally, my agent, Rebecca Carter at Janklow & Nesbit, for her forbearance and belief that I could, and indeed should, write this book.

# Index

*Page numbers in italics indicate figures and those in bold indicate tables.*

Index